INTENSIONALITY

LECTURE NOTES IN LOGIC

A Publication of

THE ASSOCIATION FOR SYMBOLIC LOGIC

LECTURE NOTES IN LOGIC 22

INTENSIONALITY

Edited by

Reinhard Kahle
Departamento de Matemática
Universidade de Coimbra, Portugal

ASL
ASSOCIATION FOR SYMBOLIC LOGIC

CRC Press
Taylor & Francis Group
Boca Raton London New York

CRC Press is an imprint of the
Taylor & Francis Group, an informa business
AN A K PETERS BOOK

CRC Press
Taylor & Francis Group
6000 Broken Sound Parkway NW, Suite 300
Boca Raton, FL 33487-2742

First issued in paperback 2020

ISBN-13: 978-1-56881-268-7 (pbk)
ISBN-13: 978-1-56881-267-0 (hbk)

Visit the Taylor & Francis Web site at
http://www.taylorandfrancis.com

and the CRC Press Web site at
http://www.crcpress.com

Addresses of the Editors of Lecture Notes in Logic and a Statement of Editorial Policy may be found at the back of this book.

Library of Congress Cataloging-in-Publication Data

Intensionality / edited by Reinhard Kahle.
p. cm. – (Lecture notes in logic ; 22)
Includes bibliographical references.
ISBN 1-56881-276-1 – ISBN 1-56881-268-X (paperback)
1. Logic, Symbolic and mathematical–Congresses. 2. Mathematics–Philosophy–Congresses. I. Kahle, Reinhard, 1967- II. Series.

QA9.A1L55 2005
511.3–dc22 2005048873

Publisher's note: This book was typeset in LaTeX, by the ASL Typesetting Office, from electronic files produced by the authors, using the ASL document class `asl.cls`. The fonts are Monotype Times Roman. The cover design is by Richard Hannus, Hannus Design Associates, Boston, Massachusetts.

PREFACE

As indicated in the title of this volume, our aim is to launch a discussion about the concept of *Intensionality*. The authors approach the discussion of intensionality from different perspectives. In addition to the philosophical issues raised by the standard possible worlds approach to intensionality, essays in this volume also address technical aspects of modal logic. The result is a volume that highlights the particular interdisciplinary nature of intensionality with articles spanning the areas of philosophy, linguistics, mathematics and computer science.

THOMAS FORSTER's contribution grows out of three essays which question the role of possible worlds in modern philosophy. In the first essay he argues that in a possible worlds approach some necessary truths — those concerning relations between the worlds — are not justified by truth in all possible worlds. They need the *modal aether*. The second essay discusses *actuality and indexicality* and the third one argues that *there is no logical proximity relation* which is needed for "the closest world".

In contrast to the *extensionalist* version of the predicate account to modalities, VOLKER HALBACH, HANNES LEITGEB, and PHILIP WELCH investigate an *intensionalist* version. In the extensionalist version, modalities are regarded as unary predicates of sentences. It turns out that this approach suffers from serious inconsistency problems. This is not the case for the intensionalist version, where necessity is considered as a predicate of intensional entities. The paper contains a philosophical discussion of a possible worlds semantics for this approach. Following the common possible worlds semantics for operators, the authors work out semantical conditions to single out the appropriate frames which admit a possible worlds model. It turns out that the crucial condition is that "any converse illfounded world must have a large depth".

In his contribution WILFRID HODGES shows that any notion of meaning for sentences can be canonically extended to a semantics for the whole language (assuming only some simple book-keeping). The resulting semantics is called the *fregean extension* and "is determined up to the question which expressions have the same meaning". Starting from Tarski's definition of truth, the fregean extension is discussed as a context principle with respect to the linguistic and logical concerns of Frege and Husserl.

MARCUS KRACHT and OLIVER KURZ present a *coherence semantics* of modal predicate logic. Coherence frames refine counterpart frames by capturing modal predicate logic with different logics of identity within the same semantical framework. The paper contains completeness results with respect to models for all modal predicate logics which are extensions of free quantified K with equality.

VAN LAMBALGEN and HAMM reconstruct *coercion* as an intensional phenomenon. The formal framework is the event calculus, a formalism that allows an encoding of different *aspectual classes*. Coercion is then described as a map from one *sense* of an expression to a different but related sense. Following Moschovakis, the sense is viewed as an algorithm which "transforms an episode into the denotation of the expression in a model".

KARL-GEORG NIEBERGALL discusses the extent to which intensionality plays a role in metamathematics, in particular, with respect to arithmetizations used for self-reflecting reasoning in Peano Arithmetic and related systems. As a result, two versions of a *Serious Intensionality Thesis* are stated for a theory T (in the language of Peano Arithmetic). It involves the complexity of the arithmetization, conditions for the corresponding proof-predicates, and, in one version, a condition for the consistency statements.

ANDREJA PRIJATELJ investigates models of *Dynamic Intensional Logic* (DIL). This logic is an extension of Intensional Logic which captures dynamic aspects. Mainly, three postulates are added about *rigidness*, *distinctness* and *update*. The contribution contains a universal construction of a DIL model. In the main theorem it is shown that every DIL model is isomorphic to one obtained by the universal construction.

KRISTER SEGERBERG gives a formal approach to *intention*, as it is understood in everyday usage, in the context of action. Based on a formalization of an agent environment with a notion of history, term operators are introduce which express *intentional* or *mental* propositions about the agent's mind. The given theory is discussed with respect to its contribution to the philosophical debate about actions.

KAI WEHMEIER discusses the question of proper names and descriptions in intensional contexts, e.g., counterfactuals. Observing the subtle difference of changing the *mood* in natural language for different purposes, he proposes a more careful formalization respecting the mood. For it, an extension of the modal logic S5 by an operator A for *actuality* is introduced and formally investigated.

The contribution of THOMAS EDE ZIMMERMANN concerns the semantics of opaque verbs. Opaque verbs allow unspecific readings of indefinite objects. One can attribute two senses to them, the *notional* one with respect to unspecific objects and the *objectual* one for their ordinary transparent use. In the traditional approach the objectual sense is derivable from the notional by a logical transformation. Refining this transformation by taking into account mental attitudes leads to the question of whether these extralogical components are part of the lexical meaning of the verb or whether they are grammatically induced. The former approach is called the *coercion* approach and the latter one the *indeterminateness* approach. The paper sheds light on this question by discussing three topics that are closely related to it: *Higher-order specificity*, *co-predications* and *mixed-coordinate objects*, and *unattested ambiguities*.

Since the contributions cover various aspects of different areas, we did not divide the volume in sections. A rough classification could run as follows:

- *General philosophical considerations.* The contributions of FORSTER, HODGES and WEHMEIER contain philosophical discussions about the proper approaches to intensional phenomena.
- *Formal theories.* HALBACH, LEITGEB, WELCH, as well as KRACHT, KURZ discuss suitable formalizations of intensional aspects in the context of possible worlds semantics. VAN LAMBALGEN, HAMM (Event calculus), PRIJATELJ (Dynamic Intensional Logic), and SEGERBERG (Logic of Actions) also present formal theories to deal with intensional phenomena. To a certain extent, NIEBERGALL's paper also belong to this group when he investigates intensionality in formalized mathematics.
- *Linguistics.* The contributions of VAN LAMBALGEN and ZIMMERMANN investigate intensional features in linguistics.
- *Mathematics.* Intensionality in mathematics is covered by NIEBERGALL.
- *Computer Science.* The paper of SEGERBERG has, to a certain extent, relations to computer science, where the formalization of agent environments is still a major task in artificial intelligence.

This volume goes back to a workshop on *Intensionality*, held in October, 2000 at the Ludwig-Maximilians University of München, Germany. A second, related workshop took place as part of the ESSLLI summer school 2002 in Trento.

The speakers of the workshop in München were Peter Aczel (Manchester), Matthias Baaz (Wien), Andrea Cantini (Firenze), Volker Halbach (Konstanz), Wilfrid Hodges (London), Gerhard Jäger (Bern), Reinhard Kahle (München), Hans Kamp (Stuttgart), Marcus Kracht (Berlin), Hannes Leitgeb (Salzburg), Per Martin-Löf (Stockholm), Yiannis Moschovakis (Los Angeles/Athens), Karl-Georg Niebergall (München), Krister Segerberg (Uppsala), Holger Sturm (Konstanz), Kai Wehmeier (Leiden), and Thomas Ede Zimmermann (Frankfurt). I am very grateful to the speakers for their contribution at the workshop as well as to all the participants, who took part in a very vivid discussion about intensionality.

The same holds for the speakers and participants of the workshop in Trento, which was co-organized by Fritz Hamm (Tübingen). The list of speakers in Trento comprises: Ralph Albrecht (Tübingen), C. Anthony Anderson (Santa Barbara), Michael Arndt (Tübingen), Jens Brage (Stockholm), Hans Kamp (Stuttgart), Uwe Mönnich (Tübingen), Yiannis Moschovakis (Los Angeles/Athens), Nikolaj Oldager (Lyngby), Orin Percus (Milano/Tübingen), Thomas Ede Zimmermann (Frankfurt), and Sandro Zucchi (Milano).

The workshop in München was sponsored by the *Graduiertenkolleg "Sprache, Information, Logik"* (Language, Information, Logic) of the Ludwig-Maximilians-Universität. Its original planning was carried out together with Peter Schroeder-Heister (Tübingen), to which I am very grateful for his extensive support. I thank Godehard Link, the head of the Graduiertenkolleg SIL, for the opportunity to organize the workshop in Munich, as well as the other members of the Graduiertenkolleg which helped in the organization, in particular, Andrea Schalley (cf. also the workshop web page `http://www.cis.uni-muenchen.de/sil/tagung/`.)

For the workshop in Trento most of the thanks is due to Fritz Hamm as the initiator and co-organizer. We both thank the ESSLLI-2002 organizers for their support of the event. More information about this workshop can found under the web page `http://www-ls.informatik.uni-tuebingen.de/kahle/trento/`.

The editor received reviewing help from Aldo Antonelli, Michael Arndt, Jeremy Avigad, Ulf Friedrichsdorf, Theo M.V. Janssen, Manfred Kupffer, Uwe Mönnich, Ulrich Pardey, Arnim von Stechow, Volodya Shavrukov, and Albert Visser which is gratefully acknowledged.

At several stages of the preparations of this volume I got helpful support of Volker Halbach and Fritz Hamm.

While preparing the volume I received the sad news that Andreja Prijatelj died on March 31, 2002, after an illness of 6 months. She submitted her paper already at time of the workshop in München. But she did not get the chance to revise the submission for the final publication. Therefore, it is included here as it was submitted. I am grateful to Andreja's PhD (co-)supervisor, Anne Troelstra, who wrote an obituary notice that I have added to her paper. We miss her.

The Editor
Reinhard Kahle
Coimbra

TABLE OF CONTENTS

THE MODAL AETHER

THOMAS FORSTER

This essay owes its appearance here to the good offices of Reinhard Kahle who, at short notice, allowed me the opportunity to turn into something publishable a rat's nest of clandestine fragments hitherto available only to the author's friends and students. I had always assumed that they were in any case so heretical that nobody would publish them even if I tidied them up and I am grateful to Kahle for giving me the chance to be burnt rather than merely ignored. The one major regret I now have about the delay caused by my timidity is that in the dozen-or-so years that have passed since the first draughts of this essay were circulated David Lewis has died. Readers familiar with the literature will immediately recognise Lewis as the most important single creator of the Augean stables I am reporting on, and as the writer most likely to wish to maintain them intact as a World Heritage Site. It is true that I thought his ideas terrible, but he defended them with personal integrity and without malice: I enjoyed his company, and am sorry that he is not around to reply.

The three essays out of which it grew were entitled "The modal æther", "Indexicality" and "The closest possible world". In the first essay I argue that if possible worlds are to be used at all to explain necessary truth, then at least some truths (those concerning relations between worlds) are necessary in virtue of something other than truth in all those worlds. In the second I argue that the idea that actuality as indexical is in need of a lot of explanation. In the third I argue that there is no logical notion of closest possible world. The essays get progressively more technical, but I preface them with an introductory essay whose general drift can be caught even by those with no formal background, and insert between the first and second essays a brief sketch of the topological ideas on which ideas of indexicality and proximity (of the two final essays) presumably ultimately rely.

§1. How did it all start? I once heard a philosophical colleague telling a lecture-threatre full of first year students that an inference was valid if "there was no possible world in which" the antecedent was true and the conclusion was false. Perhaps my readers are not as shocked by this as I am. They should

Intensionality
Edited by Reinhard Kahle
Lecture Notes in Logic, 22

be: the idea of a possible world is not needed for an explanation of the concept of valid argument, since we knew what valid arguments were before we had possible world semantics.

However the point is not that this is a shocking story, and that people teaching first-year logic ought to know better. The point is not even that this represents an unwarranted intrusion of possible world jargon into an area where it has nothing to offer. The worry is that there are nowadays many other applications of possible world imagery in more complex settings than this (counterfactual, fictions etc.) and this story raises the possibility that they all may be fully as misconceived as this one was. I fear they are: in fact over the years I have been driven reluctantly to the conclusion that all the uses known to me of possible world semantics beyond formal logic are misconceived, and that the light that possible world imagery appears to have shed on various philosophical problems in recent years will be seen in the years to come to have been entirely spurious.

I want to emphasise both that my concern is a philosophical concern not a mathematical concern. I am not a mathematician annoyed at someone else playing with my toys: I am a philosopher who is concerned that bad use of mathematics makes for bad philosophy.

Anyone with a naturalistic philosophy who wishes to say that a research programme is misconceived had better have not only evidence of the misconception, but a theory about how the mistake came to be made: the fact that people spin the theories that they do is itself a part of nature and deserves an explanation just as much as do the phenomena those theories are theories of. I shall be offering such a theory.

By the middle of the last century the soil out of which possible world analyses were to grow had been prepared and mulched. By then the concept of a mathematical structure was more-or-less settled. A structure was a set ("carrier set" or "domain") with some associated relations or functions. For example a *group* was a set of elements with a multiplication operation defined on it satisfying certain rules and so on. This picture is useful and unproblematic and is serving mathematics well, but it was never going to give any way of describing a logic internal to a structure that is different from that being used by the researcher externally to erect it.

In possible world semantics there is not merely one single carrier set, but a whole host, and the host is equipped with relation(s) which hold *between* the sets (worlds) on top of the old relations that hold *within* the worlds. The most important of these relations is a binary relation of *accessibility*. This relation of accessibility can be used to give more complicated definitions of what it is for a formula to be true in a world than had been possible before. For example we are no longer compelled to say that W believes $\neg p$ iff W does not believe p: we can say W believes $\neg p$ iff no world accessible from W believes p. This can give us a semantics for a logic in which the law of double negation fails.

Suppose there is a world W that does not believe p, but that from every world accessible from W there is a world that *does* believe p then W believes $\neg\neg p$ — even though it doesn't believe p. A particularly pleasing illustration can be given where the accessibility relation is transitive, and can be thought of loosely as succession in time. Then $\neg\neg(\exists x)\ (S(x))$ is "One day there will be a saviour"; $(\exists x)\ \neg\neg(S(x))$ is "the one who will become the saviour is already among us" and $(\exists x)\ (S(x))$ is "The saviour is here!". A transitive accessibility relation is associated with intuitionistic logic, and this illustration reminds us that we should think of intuitionistic double negation as *prophecy*.

Since the several worlds in the family of possible worlds can disagree about the truth values of particular formulæ, if we wish to ascribe a truth-value to a formula on the basis of its behaviour in such a family, it is customary to designate one of the worlds to speak up when verdicts on truth-values are required. Accordingly the world so designated is called the *designated* world.

The expression "laboratory toy" is information technology slang for a physical device that is understood from a theoretical point of view, and which will behave according to specification when handled in a laboratory by an expert, but which cannot be let loose on the public because the conditions for its safe or reliable use cannot be guaranteed. Those who use laboratory toys outside the laboratory do so not only at their own risk, but to a certain extent at the risk of others too. One is reminded not only of Goethe's *Zauberlehrling* but also of Hergé's delightful possible world *Les Bijoux de la Castafiore* wherein the jewels of the title are stolen by magpies who are entirely unaware of their significance. Possible world semantics is a laboratory toy: safe in the hands of logicians perhaps, but not in the beaks of magpies, nor in the beaks of their students.

The basic assumptions that make possible-world analysis work in the logician's laboratory are roughly as follows: (i) The language for which the semantics is being provided is agreed and defined in advance; (ii) There is a completely unproblematic transworld identity relation; (iii) The relation of identity between components worlds is unproblematic, and (iv) The web of worlds is being created by fiat in some well-controlled non-circular process, and not happening inside any of the component worlds.

All these assumptions are too basic ever to have been spelled out by the logicians who use this style of semantics. The result is that the health warnings never got written. Let us consider these assumptions in turn: what they mean and how they do not hold outside the laboratory. The discussions will overlap, since the conditions fail in related ways.

I can well imagine that many philosophers reading this will irritably exclaim that these assumptions are distorting oversimplifications which they do not make when applying possible world semantics to their concerns. However these are assumptions that have to be made if the techniques of possible world semantics are to be applied, and if these assumptions are wrong then the point is being conceded that possible world semantics are not applicable to their concerns. If this is so, then the uses of possible world discourse outside logic are rhetorical not substantial.

1.1. An agreed language. It is a matter of record that there is no agreement on what might be the formal language in use during any of these applications of possible world semantics to philosophy. Nor indeed is there any attempt to secure agreement. This looks defensible: can it really matter what the language is? Can we not use this gadgetry whatever canonical language we finally decide on?

The important point about the canonical language is that every possibility must be a consistent theory in it, and *vice versa*. Thus there is a problem about what the constant symbols of this language are to be. Is there to be one for each inhabitant of the union of all the worlds? There are numerous reasons for the answer to be "yes". If not all objects have names then those that do are specially privileged and we have thereby built in a *de re* modality from the outset. We can permute unnamed members of a carrier set for a possible world to obtain a new possible world. Or is it new? If not all objects have names then we can construct elementarily equivalent worlds which are nevertheless palpably distinct: they are non-isomorphic for example. Is this one possibility or two?[1]

But if every object in any carrier set for any possible world has a name in the canonical language it is no longer possible to invent new constants[2]. Inventing new constants is an essential technique in the standard construction of saturated models of theories[3] by means of the completeness theorem. For any finitely satisfiable set of formulæ $\{\Phi_i(x) : i \in I\}$ of formulæ we invent a new constant symbol a and an axiom $\Phi_i(a)$ for every i. We would find ourselves in a situation where although saturated models of appropriate theories remain possible, we cannot appeal to the usual devices of the completeness theorem to create them. Thus it would no longer be the case that possibilities are all reified into worlds.

Most philosophers would probably want for the canonical language something a great deal richer than the spavined first-order languages which are essentially the only languages which admit the completeness theorem, and

[1]Two structures are elementarily equivalent for a language if the language cannot express the difference between them.

[2]"The set of all constants" sounds like a denoting term from the Age of Nightmares of 100 years ago. It is true that consistent set theories with universal objects like the set of all sets and the set of all ordinals can be found, but they disallow certain manipulations of such sets which would certainly have to be allowed in a theory-of-everything. However, I shall not pursue the matter, as I believe that the incoherence of philosophical possible world talk has nothing to do with the logical or semantical paradoxes.

[3]A structure is *saturated* if whenever $\{\Phi_i(x) : i \in I\}$ is a set of formulæ with "x" free such that for any finite $J \subseteq I$ there is an object a such that $\bigwedge_{i \in J} \Phi_i(a)$ (we say $\{\Phi_i(x) : i \in I\}$ is *finitely satisfiable*) then there is an object a satisfying *all* the Φ_i simultaneously ($\{\Phi_i(x) : i \in I\}$ is *satisfiable*).

would feel that a higher-order language is called for. But the completeness theorem is what tells us that existence of a model is the same as freedom from contradiction, and if it is not available there is no longer any reason to believe that to every possibility there corresponds a world.

1.2. Unproblematic transworld identity. The mere fact that transworld identity excites debate at all shows that this condition for the application of possible world semantics is not met. But even if the philosophical community were agreed on the nature of transworld identity it could still have got it wrong. Philosophy has a way of throwing up new problems which have interesting parallels with familiar old ones[4]. There are obvious parallels between the problem of ascertaining who I am in other possible worlds and the familiar problem of who I was in a previous life in *this* possible world. It might be an idea to use the familiar problem as a point of departure. After all, if I can find out who I am in some *other* possible world, finding out who I was in a previous life in *this* possible world should be comparatively straightforward. Even this appears to be quite hard: at any given time there are plenty of people who think they were Napoleon in a previous life (I saw many when I worked in a psychiatric hospital); at any time at most one of them can be right, and there doesn't seem to be any obvious way to ascertain which[5].

If the canonical language is one of a sort for which we can prove the completeness theorem then we can indeed reify possibilities into models. But the completeness theorem does not enable us to reify *de re* possibilities into worlds. In situations where we are wondering if *some given object* might be Φ the completeness theorem has nothing to say at all. Consider theory of groups with a constant "a" denoting an involution, and theory of groups with a constant "a" denoting an object of order 4. These are both possibilities, and according to the completeness theorem there are models (which of course are groups) manifesting these possibilities. But, as every user of the completeness theorem knows, this doesn't mean that there is an object that might be an involution and also might be an element of order four. That's not to say that we cannot put these two groups into a possible world structure and identify the two designata of "a" by fiat, but it is precisely *by fiat* that identification arises, and not from the two applications of the completeness theorem. This difference between the obscurity of transworld identity in the philosophical

[4]See Cresswell, *op. cit.*

[5]And now we'll never know which. The psychiatric hospital in question has been long since closed down and all the possible-Napoleons untraceably dispersed into care-in-the-community. Thinking you might have been Napoleon in a previous life is apparently no longer grounds for being judged a menace to yourself or others and confined at public expense in a psychiatric hospital. On the other hand thinking that you are (were? might be?) Napoleon in another world is apparently still grounds for being maintained in a philosophy department. Perhaps this will change once the *Nouveau Right* realise that philosophers, too, can be rehabilitated by care-in-the-community at great savings to higher-rate taxpayers.

applications of possible world semantics and the utter straightforwardness of transworld identity by fiat in the original setting is very striking.

1.3. Noncircularity and stipulation. Consider the formula $\Diamond\exists x\Phi(x) \rightarrow \exists x\Diamond\Phi(x)$. Suppose we wish to falsify it. We want a possible world model in which it is possible that there is a thing which is Φ, but there is no thing which is possibly Φ. It is easy to arrange this: merely put in one of the worlds something which is Φ-in-the-sense-of-that-world, and ensure that no such object inhabits the designated world. This is feasible because all the worlds in a model — the designated world included — are constructs into which we can put whatever we please: the world we inhabit is a toolkit that has things just lying around and we can choose to put them into any given world W or not, as the spirit moves us. In particular, we can choose what to put into the designated world, and indeed which world to designate. However the *actual* world is actual because it just *is* actual, not because we have designated it as actual. We are most certainly *not* free to put whatever we choose into the actual world, still less are we free to choose to leave things out of it. And the situation is rather worse than that, for on any sensible account of what possibility is, we are not going to be able to stipulate what any other possible worlds contain either. We do not control possible worlds by stipulation in the way we control the components of a model. Possible worlds are not brought into existence by our stipulations, but are brute facts of life.

This difference from the original setting doesn't of course mean that possible world semantics cannot be applied elsewhere, but it does serve as a warning that the setting is very different.

The question of what happens if the possible worlds can individually describe the relations between them is complex enough to deserve an entire section to itself.

§2. The modal æther. There is an argument due to Aristotle that is designed to show that there is only one universe: if there were two, they would just be two parts of one. This does not apply straightforwardly to the modern theory of possible worlds, for these are conceived expressly in response to the compulsion we are under (given the administrative complexity of life in a world with possibilities) of submitting universes in multuplicate. In this context, Aristotle's point is the slightly more subtle one that nothing happens *outside* possible worlds — *there is no modal æther*. If a possible world is a way the world might be, then it must furnish answers to all questions about the way the world might be. In particular it must have answers to all questions about possible worlds, since there are such things.

The invention of multiple possible worlds (over the old-style models I wrote about above where there is only one carrier set) creates an enormous amount of new paperwork. There are various accessibility relations between possible worlds; relations of satisfaction between propositions and worlds; there are relations of *habitation* between objects and worlds; relations of correspondence or identity across worlds; the property of being a possible world and so on. The machinery that I am trying to point to can be recognised by its two salient features: (i) none of it was needed under the old dispensation (ii) it doesn't appear to be located inside any one possible world, and nothing seems to be gained by attempting to locate discussions of it inside possible worlds at all. Indeed it could be characterised as that part of our theory of nature that remains when all information internal to possible worlds is ignored altogether, rather in the way in which the geometry of space-time is what remains once we expunge events. Let us call it the *machinery*.

What is the metaphysical status of the machinery? Let us say that a property $\Phi(\vec{x})$ is noncontingent iff $\forall\vec{x}\ (\Phi(\vec{x}) \leftrightarrow \Box\Phi(\vec{x}))$. The list of free variables may be empty, so we can talk of noncontingent *propositions* as well as noncontingent *properties*. It's pretty clear that all the properties the machinery is concerned with are going to have to be noncontingent, and this is the received view: the property of being a possible world; the binary property of being an inhabitant of a possible world; all relations in the style "$\Psi(\vec{x})$ holds in W" where Ψ is atomic and quite possibly even when it isn't, are one and all noncontingent. If we do not make the assumption that the property of being a possible world is noncontingent, it would seem that one of the questions which will occupy us below, "does the truth value of "$\Psi(\vec{x})$ holds in W" depend for its answer on the world in which it was being asked?" could not itself even be asked, for the W we are asking about might not even be a world in the other world we are asking the question! And what is this other world anyway? Who says it is a world? It is easy to feel that failure to make the assumption that "W is a world" is noncontingent lets loose an infinite regress. It is indeed fortunate that the possible worlds tradition, by choosing the horn of the dilemma that represents truth-in-a-possible-world as neccessary, spares us the need to explore this regress.

Can the machinery be properly described inside all possible worlds, despite appearances? Or does it go on outside them, in a modal *aether*? In what follows I shall go along with the received view that facts about the machinery are noncontingent, and explore what happens if one attributes their noncontingency not to the æther but to their having the same truth-value in all possible worlds. I extract consequences that seem to me absurd, and conclude that at least some necessary truths are true not in virtue of what happens in possible worlds, but true in virtue of what happens in the æther. In other words, possible world semantics doesn't provide a uniform account of necessary truth.

2.1. Truth in a possible world. It is a simple matter to devise a paradox along the lines of Tarski's theorem on the undefinability of truth, by requiring that the actual world contain truth-definitions for the possible worlds, and *a* *fortiori* for itself. One might be tempted to claim that this demonstrates the incoherence of the enterprise. This would be unfair, for the whole point of Tarski's theorem is that *any* attempt to construct truth-definitions in this style for an entire language (with certain minimal closure conditions) will result in paradox, and this will happen even if there is no plurality of possible worlds. Mostly this problem is simply ignored. I shall follow general practice, and will try to reason in such a way as to ensure that any absurdities that do crop up do not arise merely because of Tarski's theorem.

If semantics is to go on inside possible worlds rather than the æther then we have to make sense primarily not of "ϕ is true in W_1" but rather "It is true in W_2 that ϕ is true in W_1". There is an immediate and obvious possibility of an infinite regress here: the same argument will establish that we have to make sense primarily not of "It is true in W_2 that ϕ is true in W_1", but rather "It is true in W_3 that it is true in W_2 that ϕ is true in W_1".

How vicious is this regress? The ωth stage presents us with a situation in which each original n-place predicate has become an $\omega^* + n$-place predicate[6] where all argument places except the last n are occupied by variables ranging over possible worlds. A language with predicate letters that have infinitely many argument places is an alarming prospect to non-logicians, but even more alarming is the thought that the regress might not halt there. After all, the same step of asking whether or not this (by now infinitary) atomic relation between individuals and worlds holds can be done in any possible world, and the regress resumes and will continue transfinitely.

This has to be nipped in the bud before the Burali-Forti paradox comes up over the horizon. We absolutely must have a proof that for all W_1 and W_2, and for all ϕ, ϕ is true in W_1 iff it is true in W_2 that ϕ is true in W_1. The obvious way to do this is by structural induction on formulæ.

Presumably noncontingency will tell us that, for atomic ϕ at least, W_1 thinks that W_2 thinks that ϕ iff W_2 does indeed think that ϕ. But even this comes at a price.

THEOREM 1. *All worlds and all individuals inhabit all worlds.*

PROOF. W_1-believing-$\phi(x)$ is a relation between W_1 and x. Noncontingency tells us that every W_2 must believe W_1 and x to be related in this way. But for W_2 to have any beliefs about W_1 and x they must both inhabit W_2. ⊣

So far so good. Let us now consider the induction step for quantifiers and connectives. By way of illustration consider the induction step for $\rightarrow$. Suppose W_1 thinks that W_2 thinks that $A \rightarrow B$. So W_1 thinks that every possible world accessible from W_2 that thinks A also thinks B. But by noncontingency of possible-worldhood and of accessibility (both from the machinery) something

[6] ω^* is the length of the negative integers in their natural order. Thus $\omega^* + n$ is in fact equal to n, but I write it this way to keep in mind the fact that the last n places are different from the first ω^* of them.

is a possible world accessible from W_2 iff W_1 thinks it is. And by induction hypothesis such a world will satisfy A iff W_1 thinks it does. So if for all W_1 and W_2, W_1 thinks that W_2 thinks that A iff W_2 does indeed think that A, and the same holds for B, then it holds for $A \to B$.

The other propositional connectives are straightforward too: let us just check the existential quantifier. Let us consider the inductive step for the existential quantifier. Suppose W_2 thinks that W_1 thinks that $(\exists x)\ (\phi(x))$. The semantics for the existential quantifier are that W thinks that $(\exists x)\ (\phi(x))$ iff there is an x such that W thinks that $\phi(x)$. Applying this twice we arrive at: there is an x such that W_2 thinks that W_1 thinks that $\phi(x)$ and by induction hypothesis that will be W_1 thinks that $(\exists x)\ (\phi(x))$ as desired. The universal quantifier is similar.

2.2. Interworld relations. We have just seen an argument to the effect that every world and every individual inhabits every world. This relies on the assumption that the truth-value of "W_1 believes ϕ" for ϕ atomic, doesn't depend on the world it is evaluated in. If we weaken the assumption we can obtain a weakened conclusion by a different argument.

THEOREM 2. *Some objects belong to more than one world.*

PROOF. If x (in W_1 but not W_2) is a counterpart of y (in W_2 but not W_1) then this assertion has to be made in some possible world, W_3, say, which is different from W_1 and W_2. So $W_3 \models x$ is the W_1 counterpart of the W_2 object y. But if objects belong to only one possible world, then $W_1 = W_2 = W_3$, and there is only one world. ⊣

THEOREM 3. *$K5$ is true.*

PROOF. If I assert that p is (logically) necessary I am claiming that p is true in all possible worlds. *All* possible worlds, note. This is in sharp contrast to the standard situation where I am asserting that, say, all dodos have died, where I don't actually mean *all* dodos, merely all dodos in this world. This difference is very important! The standard situation is captured by the recursive definition above: $W \models \forall x\ \Phi$ iff $\forall x \in W\ W \models \Phi$. So if we are to stick to the recursive definition and successfully produce the intended effect of the universal quantification implicit in "p is necessary" we conclude that these two accounts have to give the same result. Therefore (at least in the case where we are quantifying over worlds) it cannot make any difference whether our domain of quantification is the world we inhabit or the union of all possible worlds. That is to say

$$\forall W\ \left[((\forall W')\ (W' \models \Phi)) \leftrightarrow ((\forall W' \in W)\ W' \models \Phi)\right]$$

or equivalently

$$\forall W\ \left[((\exists W')\ (W' \models \Phi)) \leftrightarrow ((\exists W' \in W)\ W' \models \Phi)\right].$$

In other words, if there is a W in which Φ holds, then in any possible world there is such a W. Thus, if Φ is possible, every world thinks it is possible. This is the $K5$ principle $\diamond\Phi \rightarrow \Box\diamond\Phi$. ⊣

Thus we have shown that logical necessity obeys $K5$. Notice that none of this depends on Φ being *closed*: the same proof works for Φ with $\vec{x}$ free if for "all possible worlds" we read "all possible worlds inhabited by $\vec{x}$".

2.2.1. *Do all possible worlds inhabit themselves?* A possible world is merely a world that could be actual, so for every W there is W' that thinks W is actual. This W' is presumably W itself, since no world can simultaneously believe that two distinct worlds are both actual, and the idea that we can have two worlds W_1 and W_2 each of which thinks the other is the actual world is unpalatable. All this does is tell us that all questions about actuality can be easily solved: $\forall W\ \forall W'\ (W \models \text{Actual}(W') \leftrightarrow W = W')$.

What can we say about relations between objects that belong to different possible worlds? Do objects ever belong only to different worlds, or for any x and y is there some possible world they both inhabit? The problem is that, in any possible world W in which the question is asked, the answer is trivial. For any two objects there is a world they both inhabit, namely W, since inside W we cannot quantify over anything outside W. *And that is the only answer we are going to get.* This is a totally unsatisfactory state of affairs. The only way of getting any other answer would be to admit that there is a modal æther.

There is a strong temptation to believe that these questions have answers, that is to say that the answers do not depend on which possible world we ask them in. That is to say, the answers are logically necessary. But it does not seem to be *logically* necessary that any two objects inhabit a common possible world, nor that there should be two objects that don't. We seem forced to the conclusion that these questions do not have answers.

2.3. An amuse-gueule and a conclusion. I close with a simple argument discovered only in the process of final revision of this essay.

Let T be the theory consisting of all facts about the machinery. Suppose T had an axiomatisation with an axiom ψ and remaining axioms T' not entailing ψ. Then $T' \cup \{\neg\psi\}$ would be a consistent theory, and therefore true in one of the possible worlds[7]. So there would be a possible world that thought that the truth about relations between possible worlds was represented by $T' \cup \{\neg\psi\}$ rather than T. But then ψ would be a truth about relations between possible worlds that was not neccessary. But ψ was necessary. So there is no axiomatisation of T with ψ as an independent axiom. But it is standard that for any theory T and any theorem ψ of T the set $\{\psi\} \cup \{\psi \rightarrow \sigma : \sigma \in T\}$ is an axiomatisation wherein ψ is independent. We will need this last fact again,

[7]I hope nobody will object to this in the grounds that $T' \cup \{\neg\psi\}$ can't be consistent because ψ is neccessary. Such an objector would have us believe that there are no independent axiomatisations of any theory consisting of necessary truths.

so we may as well have a proof now. Suppose *per impossibile* that ψ followed from $\{\psi \to \sigma : \sigma \in T\}$. Then it would follow from finitely many $\psi \to \sigma_i$ and we would have $(\bigwedge_{i \in I}(\psi \to \sigma_i)) \to \psi$ for some finite set I. But this last is equivalent to $(\psi \to \bigwedge_{i \in I} \sigma_i) \to \psi$. Now $(((\psi \to \bigwedge_{i \in I} \sigma_i) \to \psi) \to \psi)$ is a truth-table tautology (Peirce's law) so we can deduce ψ outright.

(I suspect that a reply to this last argument can be given along the following lines. By all means $T' \cup \{\neg\psi\}$ would be a consistent theory, and therefore true in one of the possible worlds: the point is that no problem arises because in those worlds in which it is true it is not a theory of the machinery, but a pointless theory describing only some artifice wished on us by the completeness theorem.)

To summarise: if we want possible world semantics we are stuck with the machinery; we have to give an account of truths about the machinery; there seems to be no sensible alternative to the received view that makes propositions about the machinery noncontingent; there doesn't seem to be any way that this necessity can arise from truth in all possible worlds. It must come from the æther.

This is messy rather than catastrophic, but accepting the reality of the modal æther commits one to a philosophy in which the mediæval distinction between knowledge from reason (of the innards of the possible worlds) and knowledge from faith (about the æther) play a rather larger rôle than one might think desirable. Surely we have made more progress since the renaissance?

§3. Topology and possible worlds. The two remaining essays treat issues in possible world theory that are normally approached topologically when encountered elsewhere. In spacetime consideration of indexicality and proximity involves topology. What can we say about topologies on possible worlds? A brief sketch of some background topology may be in order.

A topology on a set X is a family O of "open" subsets of X. Any set that is a union of open sets is open; the empty set is open, and any intersection of finitely many open sets is open. A function from X to X is *continuous* iff $\{f^{-1}(x) : x \in X'\}$ is open whenever $X' \subseteq X$ is open. In euclidean space an open set is either a *ball* (for every point p and real number α, the set of all points within α of p form a ball) or obtained from such balls by (arbitrary) union and (finite) intersection as above. The balls are said to be a *basis*. A continuous map from the space onto itself, whose inverse map is also continuous is said to be an *autohomeomorphism*.

If, for any two points x and y in the space, there is a way of splitting the space into two open subsets, one of which contains x and the other y, then the space is said to be *totally disconnected*.

These ideas evolved first in connection with attempts to describe transformations of space, but there is a natural topology on spaces of models too.

An *ultrafilter* in a boolean algebra is family F of elements satisfying the three conditions (i) if a and b are both in F, so is $a \wedge b$. (ii) if $a \geq b$ and b is in F, so is a. Finally (iii) either a or $\neg a$ is in F. The stone space of a boolean algebra B is the space whose points are the ultrafilters of B, and where a basic open set (corresponding to the balls in euclidean space) arise from elements of B. For each b in B, the set of ultrafilters containing b form a basic open set.

If one is looking to topology to provide a weapon for the analysis of relations between possible worlds the natural thing to reach for is the stone space of all complete theories extending the canonical theory T containing all necessary truths.

This extension of topological ideas from physical space to Logic is due to Tarski, though curiously spaces like this are usually known as "Stone Spaces", after M. H. Stone.

Stone spaces tend to be totally disconnected: the set of extensions of T that prove ϕ and the set of extensions of T that prove $\neg\phi$ are complementary and both open. Euclidean space is not totally disconnected: indeed it is not possible to split euclidean space into two disjoint open sets at all: the complement of an open set is never open.

Lindenbaum algebras — at least of first-order theories — tend to be *homogenous*: for any two points of the algebra other than 0 and 1 there is an automorphism sending one of them to the other. This will ensure that the Stone space of complete theories extending T is homeogeneous in an analogous sense: given any two points of the space, there is an autohomeomorphism sending one of them to the other. Euclidean space, too, is homogeneous in this sense. The question of the homogeneity of the Stone topology on possible worlds is one that hasn't been addressed, and I suspect the answer will depend sensitively on the canonical language. Believing, as I do, that there is very little sense to be made of the idea of a space of all possible worlds in the sense envisaged by some philosophers, any attempt on my part to examine the question of the homogeneity of its space would be an exercise in *mauvaise foi*.

In the two following essays I examine two ideas for applications of possible worlds that turn out to pull in opposite directions. To argue that actuality is indexical one wants the stone space on possible worlds to be homogeneous. On the other hand if it is homogeneous then there is no nontrivial logical structure on it and one cannot look to topology to provide a notion of closeness of possible worlds.

§4. **Actuality and indexicality.** The thesis that actuality is indexical is the thesis that the difference between a possible world and the actual world is just like the difference between here and there, or between now and then. The thoughtful reader will immediately suspect there might be a parallel with McTaggart's arguments about the unreality of time. Such a reader should cast an eye over Cresswell [1990].

Part of the difference between there and here (or rather the relation between there and here) is that one can *get from* there to here. Or from then to now. These ideas are developed in point-set topology and have given rise over the years to a well-understood notion of a *connected space*. In layman's terms a space is connected iff any two points in it can be connected by a continuous line lying entirely within the space. If one believes that the relation between possible worlds is indexical, it is natural to seek a topology on the family of possible worlds which will give us a similar analysis. There is a natural topology on the family of possible worlds, but unfortunately it is not connected. However it turns out that connectedness is only part of the story anyway.

The picture below shows a standard illustration from point-set topology of a connected space. Although this topology is connected (one can get from any point in it to any other point in it by following a line lying within the space) the difference between pairs of points is not always what we would consider indexical. The difference between any two points in the disk is indexical: not only is there a path between them, but there is an autohomeomorphism of the space that sends one of them to the other. Likewise the difference between any two points on the excrescence: there is a path between them and there is an autohomeomorphism of the space sending one to the other. The relation between a point on the excrescence and a point in the disk is not indexical: there is a path between them all right, but no autohomeomorphism that moves one to the other. The point where the excrescence meets the disk is not related indexically to any other point. There are four bundles of points: (i) the points in the disk, (ii) the points on the excrescence, and (iii) the single point at the junction, and (iv) the remaining points on the boundary of the disk. The relations between points within each bundle is indexical: the relation between points in different bundles is not.

This reminds us that another important ingredient of indexicality is the idea of *indistinguishability* between indexically-related points so beguilingly alluded to by David Lewis who would always happily tell all comers that "Possible worlds are *just like* this one, only they're possible not actual" (my italics). The idea is that there is no way of distinguishing between addresses in spacetime (or between worlds) on the basis solely of information that refrains from mentioning events at those addresses.

This idea of indistinguishability can be captured in various ways: for example by the idea of an automorphisms, where we say that two things are indistinguishable if there is an automorphism sending one to the other. One could also approach it logically by saying that two things are indistinguishable if in the appropriate language there is no formula that tells one from the other. If there is no connectedness analysis available to us (remember that the usual "Stone" space on possible worlds is not connected) we will have to look to a logical analysis like this to explain what indexicality of worlds is.

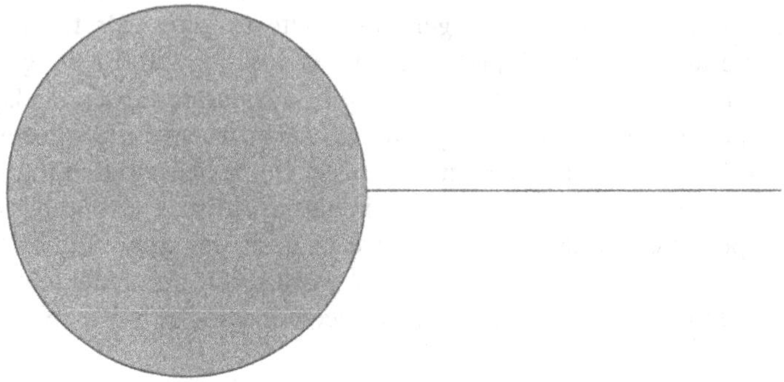

FIGURE 1. A connected space.

On this account, claiming that the relation between any two worlds is indexical is to say that the machinery has no way of telling two worlds apart: any one-place predicate from the machinery holds of all worlds or of none. But then we can ask: can the machinery distinguish *tuples* of worlds? There is some point to this question, as the following example will make clear. Consider a universe consisting of two causally disconnected spacetimes, as in the illustration below:

The relation between the two members of a pair both of which are in one of the two halves is indexical (like the pair $\{a, b\}$) and the relation between the two members of a pair which belong to different components is not ($\{a, c\}$). When (as is the case here) we have a connectedness analysis available, we can say that the relation between the components of the first pair is indexical (we can travel from a to b) but the second is not. However, even if we do not have a connectedness analysis available, we can detect the difference between pairs whose two components come from the same half of the space and those whose two components come from different halves becuase there is no automorphism sending a pair of the first flavour to a pair of the second flavour. So if we have no connectedness analysis available, we will want to explain the indexical relation between all points not by saying merely that any two *points* are indistinguishable, but that any two *pairs of points* are indistinguishable.

There is a temptation at this point to say that the machinery must be logically trivial. If the machinery cannot distinguish pairs of worlds then if even one world can see one other world then every world can see every other world! This would indeed make for a trivial machinery. However this would be a bit hasty: it is probably only a part of the machinery that has to be logically

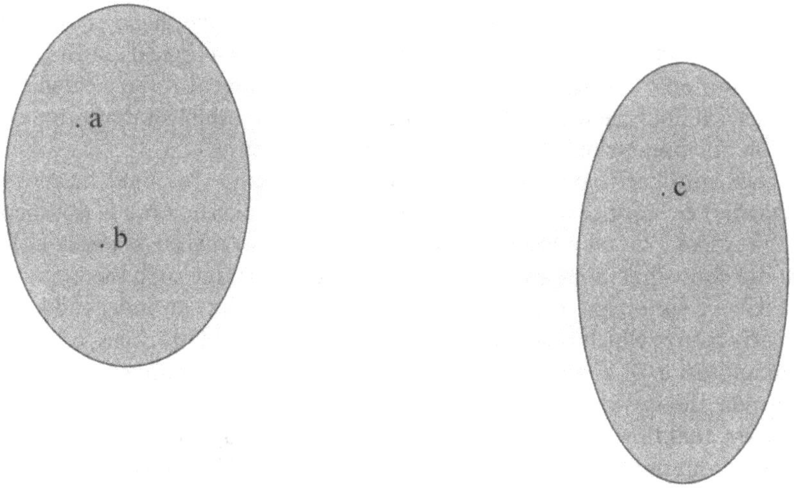

FIGURE 2. Two causally disconnected spacetimes.

degenerate in this way. After all, relations between dates is indexical even though there is an order relation on them: when thinking about indexicality one discards time and surveys matters *sub specie æternitatis*. Defenders of the idea that relations between possible worlds are indexical must be given the chance to offer a reasoned explanation of which details can be discarded and why: after all, the process of judicious discarding that took us from cartesian geometry to point-set topology took several hundred years and a great deal of ingenuity. We cannot expect the analogous task for possible world theory to be done overnight. A start would be nice.

§5. **There is no logical proximity relation.** The most baroque of the extravagant claims made by possible world theorists was the claim that possible world theory had something useful to say about counterfactuals. The plan was that a counterfactual would be true at a world W if the consequent was true at the closest world to W in which the antecedent held. Sadly this project had no logical underpinning.

For these purposes we may take possible worlds to be complete theories. Let T be an arbitrary theory and ψ an arbitrary theorem of T. A "closest world to T in which ψ is false" will be a theory which is like T in all possible respects except that ψ is false in it instead of true. We will explore the possibility of a logical basis for this notion.

There are many conceptions of logic, but we need distinguish here only between a *narrow* one (logic as deduction) and a *broad* one (logic as the centre of the web). The question of whether or not there is a logical (in the broad sense) notion of closeness is too extensive and technical to be treated here. However it does seem to be fairly straightforward to establish that there is no notion of closeness which is logical in the narrow sense.

If we construe logic narrowly, then the only thing we could mean by a (logically) closest possible (ersatz) world to T in which $\neg\psi$ holds is a theory which is like T in all possible respects except that it proves $\neg\psi$ instead of ψ. We can show that in no interesting cases is there a unique such theory.

First we show that any theory T which proves ψ has an independent axiomatisation in which ψ is an axiom, and this we have already done. In such an axiomatisation of T we can replace ψ by $\neg\psi$. The result is an axiomatisation of a new theory which is a desired closest possible (ersatz) world. Finally we show that this can be done in (infinitely) many ways, with no (narrowly) logical grounds for preferring any one to the others. The construction is very elementary, uses only manipulations of the propositional calculus and accordingly is unaffected by enrichment through quantifiers, higher-order variables, predicate modifiers etc.

There will remain the question of whether or not there is a notion of proximity which behaves in the way we want and is logical in some more broadly construed way. Of course if Logic is merely the stuff at the centre of the web we can choose where to draw the line between the core and the periphery, and simply rule that the answer is yes, by assigning the necessary proximity relations to Logic. A more interesting question is whether there is some natural pre-existing notion of Broad Logic for which the answer is yes.

It is a standard result that every theory has an axiomatisation in which none of the axioms can be derived from the remaining axioms. Sadly there is no space here to provide a proof. Now let $\{B_i : i \in A\}$ be an independent axiomatisation of T, let ψ be an arbitrary theorem of T and let $\{B_i : i \in I\}$ be a maximal subset which does not prove ψ. (We are not making any assumptions about the size of A). Let us call these *maximal subsets*. For the moment we will duck the question of how many maximal subsets there are, since this depends in complicated but fundamentally uninteresting ways on the language in which T is expressed.

Now let I be such a maximal set, and consider the new axiomatisation of T devised as follows: let ψ be an axiom, B_i is an axiom if $i \in I$, and $\psi \to B_i$ is an axiom for $i \notin I$.

It is clear that this is an axiomatisation of T. We wish ψ to be an independent axiom in this presentation. Suppose ψ is *not* independent.

Then there is a finite J disjoint from I and a finite $I' \subset I$ so that

$$\{B_i : i \in I'\} \cup \{\psi \to B_j : j \in J\} \vdash \psi$$

which is to say

$$\{B_i : i \in I'\}, \left(\psi \to \bigwedge_{j \in J} B_j\right) \vdash \psi$$

$$\bigwedge_{i \in I'} B_i \vdash \left(\psi \to \bigwedge_{j \in J} B_j\right) \longrightarrow \psi.$$

But "$((\psi \to \bigwedge_{j\in J} B_j) \to \psi) \to \psi$" is an instance of Peirce's Law so we can infer

$$\bigwedge_{i \in I'} B_i \vdash \psi$$

contradicting the assumption that $\{B_i : i \in I\}$ is a subset which does not prove ψ, so ψ is independent as desired. Accordingly we can form an axiomatisation of a new theory $T*$ by replacing ψ by $\neg\psi$ in this axiomatisation.

Now fix an independent axiomatisation of T, and for each I a maximal subset as above let T_I be the theory $T*$ with axiomatisation derived as above from the given axiomatisation of T. Evidently for $I \neq J$ we have $I \not\subseteq J \not\subseteq I$. We can also prove $T_I \not\subseteq T_J \not\subseteq T_I$ as follows.

Since $I \not\subseteq J$ there is some axiom B in I which is not in J. We shall show that $T_J \nvdash B$. Suppose $T_J \vdash B$, then

$$\langle B_i : i \in J\rangle, \neg\psi, \langle \psi \to B_i : i \notin J\rangle \vdash B.$$

Now $\neg\psi \to (\psi \to B_i)$ so we don't need $\langle \psi \to B_i : i \notin J\rangle$ anyway. This simplifies the last assertion to

$$\langle B_i : i \in J\rangle, \neg\psi \vdash B.$$

Now J is *maximal* not proving ψ so $\langle B_i : i \in J\rangle \cup \{B\} \vdash \psi$. Therefore

$$\langle B_i : i \in J\rangle, \neg\psi \vdash \psi$$

whence

$$\langle B_i : i \in I\rangle \vdash \psi$$

contradicting assumption on J.

Therefore any two distinct maximal subsets of the axiomatisation will give rise to distinct theories.

There is of course no guarantee that T_I is a complete theory when I is a maximal set. If it isn't we will not even need the fact that $I \neq J \to T_J \neq T_J$ to show that there are many "closest" theories to T in which $\neg\psi$, for any incomplete theory (of this sort, at least) has many complete extensions. For theories with countably many constants there will be uncountably many such completions.

Model theory and logic have more to tell us than that there is no sensible logical notion of proximity. A *logical* notion pertaining to a structure is one preserved by all automorphisms of that structure. Thus a logical relation is one which — considered as a set of n-tuples — is fixed setwise by all automorphisms of the structure. This is of course the Erlanger programm view of Logic, and since I wrote the earlier versions of this draught I have learned of some interesting literature on the subject: Tarski [1986], Vann McGee [1996] and Keenan [2001]. Frobenius in 1897 gave the definition of a *characteristic* subgroup of a group G as one fixed setwise by all automorphisms of G. There is also a discussion in Rogers [1967] chapter 4.

What does this show? Well, only that there is no *logical* basis (given a world W) for designating any particular world in which $\neg\psi$ to be that world which is closest (for example) to W. We need something else. This something else is presumably a metric arising from a relative proximity relation, or perhaps merely a relative proximity relation itself. Each way of Gödel numbering the canonical language gives rise to a metric on the stone space: let the distance between two theories be the sum of all 2^{-n} where n is the Gödel number of a sentence in the symmetric difference. All these metrics give rise to the same Stone topology, but the metrics themselves can be quite different and give rise to different truth-values for counterfactuals.

The point is the simple one that possible worlds do not come naturally equipped with the relations necessary to explain the phenomena for whose supposed explanation it was that possible worlds were invoked. We have to supply the metrics and proximity relations ourselves, since there is no *logical* reason for preferring one to the other. Batteries not included. This will not surprise thoughtful users of this machinery, but it should serve as a warning to others that possible world semantics is not a candidate for an *explanation* of (for example) counterfactuals, but is merely a procedural device for presenting the theoretical entities (the metrics) that are. We could equally well try providing a description of the problem in iambic pentameters. In short, it is question-begging. There is nothing wrong with this as long as we retain a sense of the difference between *expressing* something and *explaining* it. Over the years far too many people have been seduced by the beguiling imagery of possible world semantics into thinking that once a problem has been expressed in that formalism then progress has automatically been made. The virtue of this simple illustration is that it reminds us that this is not so.

§6. Coda. I acknowledged at the outset of this essay a responsibility as a naturalistic philosopher to provide an explanation of how the errors I allege are being made get made. If possible world semantics outside logic is such a huge mistake, why do so many clever people make it? There's no mystery about that. It was obvious from the outset that possible world semantics was going to be very useful in formal logic, and it didn't take long for philosophers with an interest in logic to recognise an affinity between its ideas and a strand of thought in philosophy going back to Leibniz and beyond and jump to the conclusion that it would be useful to them[8]. As well as looking like a useful technique, possible world semantics came with a lot of appealing imagery — worlds *seeing* one another, vicariously enjoying the wicked adventures of our counterparts in less humdrum worlds and so on; by reifying possibilities into worlds it collaborates insidiously with our natural inclination to realism, and finally in its allusions to formal logic it attracts not only those who value logic and have high expectations of it, but also those who — while having no exalted view of the rôle Logic can play in Philosophy — are happy to appear inclusive. Finally, it's fun without guilt: it allows us to combine the delight of reading Dunsany or Tolkien while pretending that we are doing Philosophy. An unbeatable combination!

REFERENCES

[1] M.J. Cresswell, *Modality and Mellor's McTaggart*, ***Studia Logica***, vol. 49 (1990), pp. 163–170.

[2] E. Keenan, *Logical objects*, ***Logic, meaning and computation, essays in memory of Alonzo Church*** (Anderson and Zelëny, editors), Kluwer, 2001, pp. 149–180.

[3] V. McGee, *Logical operations*, ***Journal of Philosophical Logic***, vol. 25 (1996), pp. 567–580.

[4] H. Rogers, ***Theory of recursive functions and effective computation***, McGraw-Hill, 1967.

[5] A. Tarski, *What are logical notions?*, ***History and Philosophy of Logic***, vol. 7 (1986), pp. 143–154.

DEPARTMENT OF PURE MATHEMATICS AND MATHEMATICAL STATISTICS
CENTRE FOR MATHEMATICAL SCIENCES
UNIVERSITY OF CAMBRIDGE
WILBERFORCE ROAD
CAMBRIDGE, CB3 0WB, UK
E-mail: t.forster@dpmms.cam.ac.uk

[8]This is a very human thing to do: fault-tolerant pattern-matching has long been useful for spotting lions in undergrowth or edible fruit in thick canopy, but we pay for the speed by sacrificing reliability. Normally this is not a problem, as one does not lose one's life by jumpily mistaking something for a lion, and a hastily misidentified fruit can be spat out.

POSSIBLE WORLDS SEMANTICS FOR PREDICATES

VOLKER HALBACH, HANNES LEITGEB, AND PHILIP WELCH

Abstract. We develop possible worlds semantics for $\Box$ as a predicate rather than as an operator of sentences. The unary predicate symbol $\Box$ is added to the language of arithmetic (or an extension thereof); this yields the language $\mathcal{L}_\Box$. Every world in our possible worlds semantics is the standard model of arithmetic plus an interpretation of $\Box$. We investigate possible–worlds models where $\Box\ulcorner A \urcorner$ is true at a world w if and only if A is true in all worlds seen by w. The paradoxes exclude certain frames from being frames for $\Box$ as a predicate. We provide some sufficient and also some necessary conditions on frames that are allowed to act as frames for the predicate approach. Completeness results for certain infinitary systems corresponding to well known modal operator systems are established. We draw some conclusions concerning the current state of the predicate approach to modalities.

§1. **Modalities as predicates.** Modalities like necessity and possibility, may be analysed logically in essentially two ways: either as predicates, or as operators. In the first case they are applied to singular terms, whereas in the second case they are applied to formulæ, but in both cases the application gives us new formulæ. Thus the distinction between the operator and the predicate conception of necessity is made on the syntactical level at first. Both conceptions are tied to certain semantics respectively. If "necessary" and "possible" are regarded as predicates, they are interpreted as properties of objects and a decision has to be made concerning what precisely they should be predicates of: syntactical entities like sentences, or contents of syntactical entities like propositions (let us ignore further options like utterances or mental objects). In either case, necessity and possibility are properties of such entities, or, perhaps, relations between such entities and further objects. If "necessary" and "possible" are regarded as operators, they do not express properties or relations like predicate and relation expressions; necessity does not apply to anything — much like the logical connectives or the quantifiers. In this sense

---

The closure ordinal γ for the operator Φ was suggested to Volker Halbach independently by Peter Aczel, and by Philip Welch who later proved Theorem 4.7, and agreed to be a co-author of the paper. The work of Volker Halbach was partially sponsored by the research group *Logic in Philosophy* at the Universities of Tübingen and Konstanz.

Intensionality
Edited by Reinhard Kahle
Lecture Notes in Logic, 22

the operator conception of necessity is radically deflationary. Similar considerations apply not only to necessity but also to the notions of knowledge, belief, future and past truth, obligation and so on, which have been treated in analogous fashions as necessity.

There are various arguments in favour of, or against, each of these options of logical representation, but we shall not dwell on the pros and cons in more detail because this would require a careful analysis (see Bealer [4] and Halbach [16]). Instead we are going to concentrate on a particular problem affecting modalities if regarded as unary predicates of sentences, i.e., according to the extensionalist version of the predicate view from above. This position has been held and defended most notably by Quine [24, 23]. We think that there would be very good reasons for preferring this account if it was not threatened by inconsistency: Montague [21] was the first to notice that the predicate version of the acknowledged modal system T is inconsistent if it is joined with arithmetic. This is due to the existence of self-referential expressions, which is a consequence of Gödel's diagonalisation lemma, which is in turn derivable from even weak systems of arithmetic. From this Montague concluded:

> Thus if necessity is to be treated syntactically, that is, as a predicate of sentences, as Carnap and Quine have urged, then virtually all of modal logic [...] must be sacrificed.

Note that the intensionalist version of the predicate view on modality, i.e., where necessity is considered a predicate of intensional entities, is not necessarily affected by analogous problems. Propositions might e.g., be much more different from their syntactical counterparts than sometimes thought of, and therefore there might not be any self-referential propositions; perhaps self-referential sentences do not express propositions at all (for a recent account following that line of thought see Slater [29]). On the other hand, if propositions have a syntactical structure similar to sentences, and if every well-formed sentence expresses a proposition, then inconsistency strikes again (as has been pointed out by Asher and Kamp [2]).

According to the operator account of modality, there are simply no self-referential sentences rising from the use of modal expressions, since in order to be self-referential the sentences with operators would have to contain themselves as a proper part, which they certainly do not in the standard operator languages (below we shall make a remark on operator languages with additional devices like propositional quantification). Thus self-reference is abandoned from the start by suppressing referential clauses and by that move also the inconsistencies are avoided. This is one of the reasons why the majority of philosophers and logicians have opted to reject the predicate view altogether.

For the security from inconsistency, however, one has to pay a high price: quantification is not expressible in operator languages unless one adds further devices like propositional quantifiers. For instance, there is no way of expressing "All tautologies are necessary" in a single sentence of the operator language. If auxiliary devices like propositional quantification are employed, we expect that these additional devices reintroduce the difficulties of the predicate approach (Grim [13] argues for this view).

We doubt that the inconsistencies justify the general preference of the operator view over the predicate conception. We shall sketch some arguments for this claim but leave a thorough discussion of these matters to another paper.

First of all, by far not every system of modal logic has an inconsistent predicate counterpart, e.g., the predicate version of the minimal normal operator system K is indeed consistent. Secondly, des Rivières & Levesque [10] and Belnap & Gupta [5, p. 240ff] have shown that sentences of modal operator systems can be translated into sentences of corresponding and consistent predicate systems such that provability is preserved. Thus, roughly speaking, everything that can be said with operators can (consistently) be expressed with a predicate. Only if the provability of axiom *schemes* is to be preserved, troubles are emerging: we can have $\Box\ulcorner A \urcorner \rightarrow A$ for any translation of a modal operator formula into the language with $\Box$ as a predicate, but we cannot have it for all sentences of the predicate language. The translations of operator sentences are always grounded in the set of sentences which are modality-free, and therefore self-referential instantiations of the T-scheme are excluded.

Along the main stream of the operator conception of the modalities there has been some work on the predicate approach. One "modality", however, has even been treated as a predicate by most authors, namely truth. The motivation for preferring the predicate approach in this case is obvious: the trivial modal system with the "T-sentences" $\Box p \leftrightarrow p$ as axioms for $\Box$ will hardly qualify as a satisfying analysis of truth. Again we leave a discussion of the involved relations holding between the truth predicate, modal operators, and substitutional quantification to another paper. For our present purposes, however, it is important that several logicians have provided various models for axiom systems with predicates of sentences (or their codes). Friedman and Sheard [12], for instance, have studied a variety of systems built over an arithmetical base theory plus axioms for an additional unary (truth) predicate. Although the additional predicate is usually labelled as a truth predicate, some systems rather resemble systems for necessity or other typical modalities. We think that this applies, e.g., to one important derivative of the Friedman & Sheard systems, namely Cantini's [7] system VF. Therefore we surmise that some work on axiomatic theories of truth is also relevant for the predicate conception of other notions.

Truth theorists have investigated a variety of axiom systems (see, e.g., Cantini [8], Halbach [15] and Sheard [28] for overviews). Consistency proofs for these systems have been provided by model constructions. One aim of the present paper is a generalisation of these model constructions: we do no longer consider a single system and construct a model for this system. Rather we investigate a natural family of systems and provide a general pattern for generating models. And vice versa we aim at a generalisation of the inconsistency results. By showing that there is no model of a certain kind, we shall be able to conclude that a (semiformal — in the sense of Schütte [27], Sect.28) system is inconsistent or that a certain formal system does not have an ω-model.

Of course, we cannot hope to capture all the sophisticated model constructions found in the literature in a single approach, simply because they are too varied. Here we take some inspiration from the operator approach. Modal logicians do not investigate arbitrary modal systems (i.e. logic plus arbitrary axioms and rules for the modal operator $\Box$) but they restrict their focus on certain well behaved theories. The most common class of such well behaved theories is known as the set of normal modal systems (see, e.g., Boolos [6]). We follow exactly the same route for predicates. In this way we shall obtain results on the semantics of a comprehensive and important class of systems with a modality predicate.

Opting for possible world semantics in a more or less standard form implies certain restrictions. For instance, necessitation holds in all normal modal systems. Our possible world semantics for predicates will verify an analogous rule. Thus axiom systems which are not compatible with necessitation (like the Kripke-Feferman system Ref(PA) of Feferman [11]), which is also known as KF, fall outside the scope of our present paper. By no means do we wish to claim that these systems are less interesting or natural than "normal" systems, but it makes sense to restrict attention to a class of well behaved system, that is, "normal" systems first. Just as in the case of modal logic one finally can go on and explore non-normal systems and their semantics in a general framework.

Our approach does not only include a restriction to normal modal systems or, more exactly, to the predicate analogues of normal modal systems (see Lemma 2.3), but also a restriction to systems consistent in ω-logic. This restriction is brought about by our choice of the standard model of arithmetic as the basis for our model constructions. First of all this restriction serves the purpose of avoiding the intricacies of nonstandard models of arithmetic, but furthermore systems consistent in ω-logic are certainly quite natural. (However, there are natural and philosophically significant ω-inconsistent systems like the system FS mentioned below.)

In our approach we want to stick as closely as possible to the common possible worlds semantics for operators, i.e. we use frames of worlds such that every world is a classical model. Since we shall apply the predicate $\Box$ to numerical codes of sentences, we assume that every world is actually a standard model of arithmetic which is only extended by assigning some extension to the Box — thus we restrict ourselves at first to just the arithmetical vocabulary plus a unary modal predicate. Later we are going to extend this language by non-arithmetical and non-logical vocabulary. Our treatment of necessity by means of a predicate extends to other modalities as well, but for the sake of simplicity we shall only speak of necessity and not enumerate all other modalities to which our approach may be applied.

On the semantical level, and as is common in modal logic, necessity is considered as truth in all accessible worlds, i.e., $\Box\ulcorner A\urcorner$ is true in w if and only if A is true in all worlds accessible from w, where accessibility is a binary relation (defined by the frame) on the class of worlds. At this point the fundamental difference between possible worlds semantics for modal operator and modal predicate languages shows up: according to the *operator* conception, the satisfaction clause for $\Box\ulcorner A\urcorner$ can be turned into a proper *definition*, but not so in the predicate case. If a frame is given and a function that assigns a model for the $\Box$-free language to any world, then truth or falsity for all sentences without $\Box$ extends to truth for all sentences with $\Box$ as an operator in the usual recursive way. Thus one can always build a model on any given frame if the operator conception is presupposed. But if $\Box$ is treated as a predicate, the recursive definition of truth at a world cannot be carried out anymore, because a sentence $\Box\ulcorner A\urcorner$ is atomic, while on the operator account $\Box A$ has the complexity of A plus one. There is also no way to provide a definition of the (possibly transfinite) complexity of a formula that would allow for a recursive definition of truth at a world. The satisfaction clause for modal sentences now works just as a condition on models, such that every model that meets the conditions counts as a proper instance of a possible worlds semantics for predicates. For some frames a recursive definition of truth at a world is not only impossible, but there is not even a model based on the frame at all due to self-referentiality. Montague's [21] Theorem, for instance, shows that one cannot build possible worlds models on reflexive frames (see Example 3.1 below) if necessity is expressed by a predicate. This means that there is no way to assign truth and falsity to all sentences in such a way that $\Box\ulcorner A\urcorner$ comes out as true if and only if A holds at any world (and if some conditions are met which will be explained below). Another example is provided by McGee's [20] theorem on ω-inconsistent systems. It is a direct consequence of this theorem that a frame consisting of the natural numbers ordered by the predecessor relation (so that $0R1R2R3R\dots$) does not support a possible worlds model (see Example 3.2 below).

Our question from above may finally be stated in the following form: which frames support a possible worlds model, if $\Box$ is a predicate, and if the object language allows for self-referentiality? We call the problem of characterising the frames which support possible worlds models for predicates the *Characterisation Problem*.

A satisfying answer to the characterisation problem would exhibit a graph-theoretical description of the frames supporting possible worlds models. The restrictions imposed by theorems like those due to Montague and McGee should be consequences of such a characterisation result. A system with a modal predicate might thus be checked for consistency (or rather consistency in omega logic) by investigating what its corresponding possible worlds semantics would look like and whether the latter satisfies our characterisation criteria. In this way paradoxes and inconsistency results for modalities as predicates can be modelled in a systematic and informative way.

Our development of possible worlds semantics for predicates also yields some further insights: it sheds new light on hierarchies of metalanguages; some results of provability logic and modal logic show up in a different fashion; we can state the sound and complete systems for necessity as a predicate corresponding to certain types of frames; our treatment of transitive converse wellfounded frames can be seen as a variant of the revision theory of truth.

We regard this paper as a preliminary view on the issues sketched above. We also omit the more involved proofs, which we give in [17], but we outline some of the essential ideas.

§2. **Formal development.** In this section we develop formally the account set out in the previous section.

$\mathcal{L}_{PA}$ is the language of arithmetic; $\mathcal{L}_{\Box}$ is $\mathcal{L}_{PA}$ augmented by the one-place predicate symbol $\Box$. We shall identify $\mathcal{L}_{PA}$ with the set of all arithmetical formulæ and $\mathcal{L}_{\Box}$ with set of all formulæ that may contain $\Box$. We do not distinguish between expressions (like formulæ and terms) and their arithmetical codes. We use uppercase roman letters $A, B, \dots$ for sentences of $\mathcal{L}_{\Box}$. If a formula may have a free variable we always indicate this by writing $A(v)$ etc.

We shall assume that the language $\mathcal{L}_{PA}$ features certain function symbols. This will render the notation somewhat more perspicuous. Of course, these function symbols can be eliminated in the usual way and thus are not really required.

On our account all arithmetical truths come out necessary. Thus one might wish to include further vocabulary that allows for more contingent sentences, but we shall only do so later, in order to keep the exposition of the main definitions and results as simple as possible. Once they have been established they can be carried over easily to languages with additional vocabulary.

We shall consider *standard* models only, i.e., worlds with the set of natural numbers in the domain and the standard interpretation of the arithmetical vocabulary. Hence models for $\mathcal{L}_{\Box}$ have the form $(\mathbb{N}, X)$ where $\mathbb{N} = \langle \mathbb{N}, +, \times, 0, ', \dots \rangle$ is the standard model of arithmetic and $X \subseteq \omega$ the extension for $\Box$. Since models differ only in the extension of $\Box$, we may identify such a model with an extension for $\Box$. Thus we write $X \models A$ for $(\mathbb{N}, X) \models A$.

$\Diamond x$ is defined as $\neg \Box \dot{\neg} x$ where $\dot{\neg}$ is a function symbol for the function sending the code of a sentence of $\mathcal{L}_{\Box}$ to the code of its negation.

The definition of frames and possible worlds models (PW-models, for short) parallels the usual definitions for the operator approach as given, e.g., in Hughes and Cresswell [18, p. 37f.] Boolos [6].

DEFINITION 2.1. A *frame* is an ordered pair $\langle W, R\rangle$ where W is nonempty and R is a binary relation on W.

The elements of W are called *worlds*, R is the *accessibility* relation. w *sees* v if and only if wRv; alternatively we say that w is below v.

DEFINITION 2.2. A *PW-model* is a triple $\langle W, R, V\rangle$ such that $\langle W, R\rangle$ is a frame and V assigns to every $w \in W$ a subset of $\mathcal{L}_\Box$ such that the following condition holds:

$$V(w) = \{A \in \mathcal{L}_\Box | \forall u\, (wRu \Rightarrow V(u) \models A)\}.$$

If $\langle W, R, V\rangle$ is a model, we say that the frame $\langle W, R\rangle$ *supports* the PW-model $\langle W, R, V\rangle$ or that $\langle W, R, V\rangle$ is based on $\langle W, R\rangle$. A frame $\langle W, R\rangle$ *admits* a valuation if there is a valuation V such that $\langle W, R, V\rangle$ is a PW-model.

V assigns to every world a set of sentences, the extension of $\Box$ at that world. The condition on the function V in the definition says that a sentence is in the extension of $\Box$ at a world w if and only if it is true in all worlds seen by w. Thus if $\langle W, R, V\rangle$ is a PW-model, the following holds:

$$V(w) \models \Box\ulcorner A\urcorner \text{ iff } \forall v \in W\big(wRv \Rightarrow V(v) \models A\big).$$

If we had to include further "contingent" vocabulary, the valuation V had not only to interpret $\Box$ but also this additional vocabulary. Since we are only concerned with the language $\mathcal{L}_\Box$, V needs to interpret only $\Box$ at every world and therefore it needs to assign only a set of sentences to every world, i.e., the extension of $\Box$ at the world.

As pointed out above, the definition of truth at a world cannot be carried out recursively as in the operator case. On the operator account one can build models on every frame. Thus the question whether a frame supports a model does not arise for $\Box$ as an operator. For this problem is trivial: every frame admits a model for $\Box$ as an operator. However, if $\Box$ is a predicate, there are restrictions and the following question is sensible:

CHARACTERISATION PROBLEM. Which frames support PW-models?

An answer to this problem exhibits the restrictions imposed by the "paradoxes" (like Montague's) on possible worlds semantics for predicates. We are only able to present partial solutions to the Characterisation Problem in this paper.

We shall now show that possible worlds semantics for predicates is in many respect similar to the usual possible worlds semantics for operators.

Analogous definitions of frames and models for operators lead to the so called *normal* systems of modal logic with the minimal system K (see, e.g., Boolos [6]). These systems are closed under necessitation and the necessity operator $\Box$ distributes over material implication. For the predicate account something similar can be shown.

LEMMA 2.3 (Normality). *Suppose $\langle W, R, V\rangle$ is a PW-model, $w \in W$ and $A, B \in \mathcal{L}_\Box$. Then the following holds:*

(i) *If $V(u) \models A$ for all $u \in W$, then $V(w) \models \Box\ulcorner A\urcorner$.*
(ii) $V(w) \models \Box\ulcorner A \rightarrow B\urcorner \rightarrow (\Box\ulcorner A\urcorner \rightarrow \Box\ulcorner B\urcorner)$

Since we are dealing with standard models only, we do not only obtain schemata like (ii) but also their universal closure, that is

$$\text{(1)} \qquad \forall x \forall y(\text{Sent}_{\mathcal{L}_\Box}(x) \wedge \text{Sent}_{\mathcal{L}_\Box}(y) \rightarrow \Box(x \dot{\rightarrow} y) \rightarrow (\Box x \rightarrow \Box y)).$$

$\dot{\rightarrow}$ represents the function that yields, when applied to two codes of sentences, the code of their material implication. $\text{Sent}_{\mathcal{L}_\Box}(x)$ represents the set of sentences of $\mathcal{L}_\Box$. In order to avoid confusing notation, we shall state schemata like (ii) instead of the universally quantified sentences like (1). This does not make a difference because of the use of standard models.

§3. **Some limitative results.** In order to illustrate the limitations imposed by the main condition in Definition 2.2, we present some examples of frames that do not support PW-models.

As has been noted in the introduction, several restrictions on the class of frames supporting a PW-model are known. These restrictions correspond to several well known inconsistency results.

EXAMPLE 3.1 (Montague). If $\langle W, R\rangle$ admits a valuation, then $\langle W, R\rangle$ is not reflexive.

The trivial proof resembles Montague's proof showing that the predicate version of the modal system T is inconsistent.

Montague's theorem is a generalisation of Tarski's theorem which says that the scheme $\Box\ulcorner A\urcorner \leftrightarrow A$ for all sentences $A \in \mathcal{L}_\Box$ is inconsistent with a theory proving diagonalisation. The possible worlds analogue of the scheme $\Box\ulcorner A\urcorner \leftrightarrow A$ is a frame with only one world that sees itself. Of course, such a frame does not admit a valuation, because the frame is reflexive, and thus Example 3.1 implies the possible worlds analogue of Tarski's theorem which states that a frame with one single reflexive world seeing itself does not support a PW-model.

The arguments showing that a frame does not admit a valuation proceed by diagonal arguments (in our last example by a Liar sentence).

McGee's main theorem in [20] on the ω-inconsistency of a certain theory of truth imposes a restriction on the existence of PW-models based on a non-transitive frame. Pred is the predecessor relation: for any natural number n it holds only between n and $n + 1$.

EXAMPLE 3.2 (McGee, Visser). $\langle \omega, \text{Pred} \rangle$ does not support a PW-model.

In order to prove this, one can adapt McGee's proof in [20], which uses a certain self-referential sentence with quantification. Instead of McGee's theorem Visser's theorem [31] on illfounded hierarchies of languages can be used for deriving the above result. The relations between the different ω-inconsistency results and their implications is studied by Leitgeb [19].

Also Gödel's Second Incompleteness Theorem can be rephrased in terms of possible worlds semantics. One may prove an analogue of Löb's Theorem, which will be very useful in later sections.

Transitivity yields the predicate analogue of the 4 axiom:

LEMMA 3.3. *Let $\langle W, R, V \rangle$ be PW-model based on a transitive frame. Then*

$$V(w) \models \Box \ulcorner A \urcorner \longrightarrow \Box \ulcorner \Box \ulcorner A \urcorner \urcorner$$

obtains for all $w \in W$ and all sentences $A \in \mathcal{L}_\Box$.

LEMMA 3.4 (Löb's Theorem). *For every world w in a PW-model based on a transitive frame and every sentence $A \in \mathcal{L}_\Box$ the following holds:*

$$V(w) \models \Box \ulcorner \Box \ulcorner A \urcorner \longrightarrow A \urcorner \longrightarrow \Box \ulcorner A \urcorner.$$

The proof is just the usual proof of Löb's Theorem with the provability predicate replaced by the primitive symbol $\Box$.

Löb's Theorem is a kind of an induction principle (see Boolos [6]). Since we cannot assign arbitrary truth values to the sentences of $\mathcal{L}_\Box$ in any given world, we cannot conclude that R is converse wellfounded in general.

We shall also use the following version of Löb's Theorem:

COROLLARY 3.5. *Assume $\langle W, R \rangle$ is transitive and $A \in \mathcal{L}_\Box$. If $V(w) \models \Box \ulcorner A \urcorner \rightarrow A$ for all $w \in W$, then also $V(w) \models A$ for all worlds w.*

From (the modal version of) Löb's Theorem 3.4 (the modal version of) Gödel's Second Incompleteness Theorem can be derived by setting $A = \bot$ for a fixed contradiction $\bot$. A dead end is a world that does not see any world. $V(w) \models \Box \ulcorner \bot \urcorner$ holds if and only if w is a dead end. It follows:

EXAMPLE 3.6 (Gödel's Second Incompleteness Theorem). In a transitive frame admitting a valuation every world is either a dead end or it can see a dead end.

PROOF. Since the frame is transitive the predicate analogue of the 4 axiom scheme holds. This suffices for a proof of the formalized incompleteness theorem and we obtain $V(w) \models \Box\ulcorner\bot\urcorner \vee \Diamond\ulcorner\Box\ulcorner\bot\urcorner\urcorner$. ⊣

In the sequel we shall provide a theorem (Theorem 6.1) that implies all the above results and does away with the need for all the different diagonal sentences.

§4. Inductive definitions of valuations.

A world w is converse wellfounded in a frame $\langle W, R\rangle$ if and only if every subset of $\{v \mid wR^*v\}$ has an R-maximal element, where R^* is the transitive closure of R. A set of worlds is converse wellfounded if and only if all its elements are converse wellfounded. A set $W' \subseteq W$ of worlds is converse wellfounded if and only if $\{u \in W \mid \exists v \in W'(u = v \vee vRu)\}$ is wellfounded with respect to R^{-1}.

So far we did not show that there exist any PW-models. As Belnap and Gupta [5, Theorem 6E.5] have shown, the extension of $\Box$ can easily be defined by recursion on R^{-1} in converse wellfounded frames. It applies also to frames that are not transitive.

THEOREM 4.1. *If $\langle W, R\rangle$ is converse wellfounded, then $\langle W, R\rangle$ admits exactly one valuation V.*

PROOF. The extension for $\Box$ is defined by transfinite recursion on R^{-1}.

Assume $V(v)$ has been defined for all v such that wRv. Then $V(w)$ is defined in the obvious way:

$$V(w) := \{A \in \mathcal{L}_\Box \mid \forall v\ (wRv \Rightarrow V(v) \models A)\}.$$

Clearly there is no choice in the definition of $V(w)$, which is thus uniquely determined. ⊣

As we shall show later, there are frames that admit more than one valuation. Of course, these frames are converse illfounded (i.e., not converse wellfounded).

We turn to the transitive frames and investigate the following special case of Characterisation Problem: which transitive frames support PW-models? Throughout the rest of the section we assume that $\langle W, R\rangle$ is transitive.

In order to fit the present definition in the general framework of inductive definitions, it is convenient to define the extension of $\Diamond$ rather than the extension of $\Box$. The choice of $\Diamond$ as the basic predicate does not make a difference, because $\Box$ and $\Diamond$ are interdefinable in the following way:

$$\Box x :\Longleftrightarrow \neg\Diamond\dot\neg x,$$

$$\Diamond x :\Longleftrightarrow \neg\Box\dot\neg x.$$

We employ the following operation $^{\diamond}$ turning an extension of $\Box$ into the corresponding extension of $\Diamond$ and vice versa:

$$X^{\diamond} = \{A \in \mathcal{L}_{\Box} | \neg A \notin X\}.$$

We have $X^{\diamond\diamond} = X$ for deductively closed X.

We define a satisfaction relation $\models_{\Diamond}$:

$$(\mathbb{N}, X) \models_{\Diamond} A :\iff (\mathbb{N}, X^{\diamond}) \models A.$$

Thus $(\mathbb{N}, X^{\diamond}) \models_{\Diamond} A$ holds if and only if $(\mathbb{N}, X) \models A$. Usually we shall write $Y \models_{\Diamond} A$ for $(\mathbb{N}, Y) \models_{\Diamond} A$.

DEFINITION 4.2. We define an operator Φ whose domain is all sets of Gödel numbers of sentences of $\mathcal{L}_{\Box}$: $X \mapsto \Phi(X)$.

$$\Phi(X) := X \cup \{A \in \mathcal{L}_{\Box} | (\mathbb{N}, X) \models_{\Diamond} A\}.$$

The operator Φ is not monotone, that is, $X \subseteq Y$ does not generally imply $\Phi(X) \subseteq \Phi(Y)$. For instance $\neg\Diamond\ulcorner 0 = 0\urcorner$ is an element of $\Phi(\varnothing)$, but it fails to be an element of $\Phi(\{0 = 0\})$. In the following, $\Phi^{\alpha}(X)$ designates the α-fold application of Φ to X; at limit ordinals λ, $\Phi^{\lambda}(X)$ is defined as the union $\bigcup_{\alpha<\lambda} \Phi^{\alpha}(X)$ of all previous stages. Clearly then: for any X $\alpha < \beta \rightarrow \Phi^{\alpha}(X) \subseteq \Phi^{\beta}(X)$; there is a least κ such that $\Phi^{\kappa}(\varnothing) = \Phi^{\alpha}(\varnothing)$ for all $\alpha \geq \kappa$.

LEMMA 4.3. *Assume $\langle W, R, V\rangle$ is a PW-model based on a transitive converse wellfounded frame. If w has R^{-1}-rank α, then $V(w)^{\diamond} = \Phi^{\alpha}(\varnothing)$.*

A set $X \subseteq \mathcal{L}_{\Box}$ is a fixed point (of Φ) if and only if $\Phi(X) = X$. Since Φ is not monotone, $\Phi^{\kappa}(\varnothing)$ is not necessarily the smallest fixed point.

REMARK 4.4. $X \subseteq \mathcal{L}_{\Box}$ is a fixed point if and only if $X \models_{\Diamond} \Box\ulcorner A\urcorner \rightarrow A$ (or, equivalently, $X \models_{\Diamond} A \rightarrow \Diamond\ulcorner A\urcorner$) for all $A \in \mathcal{L}_{\Box}$.

The existence of fixed points implies that there are frames that are converse illfounded but which admit a valuation nevertheless, as we shall show in the following theorem.

In order to state the theorem, we need some definitions. The *converse wellfounded part* of a set of the form $\{v | wRv\}$ is the largest R-upwards closed set X (i.e., the largest X such that $\forall y \in X(yRx \Rightarrow x \in X)$) on which R^{-1} is wellfounded. The *depth* of a world w is the rank of R^{-1} restricted to the converse wellfounded part of $\{v | wRv\}$. (Thus a world does not have to be converse wellfounded to have a depth.) We shall apply these notions only to transitive frames.

THEOREM 4.5. *Assume $\langle W, R\rangle$ is transitive and every converse illfounded world in $\langle W, R\rangle$ has depth at least κ. Then $\langle W, R\rangle$ admits a valuation.*

PROOF. The definition of V for converse wellfounded w is forced by Lemma 4.3. If $w \in W$ has depth at least κ we put $V(w) = \Phi^{\kappa}(\varnothing)$. This holds in particular for all converse illfounded worlds.

As κ is a fixed point, the condition

$$V(w) = \{A \in \mathcal{L}_\Box \mid \forall u\,(wRu \Rightarrow V(u) \models A)\}$$

is satisfied by all worlds $w \in W$. ⊣

This and Löb's Theorem imply the following corollary:

COROLLARY 4.6. *If $\langle W, R, V\rangle$ is a transitive converse illfounded model, then there is no formula A that holds exactly in the converse wellfounded worlds of $\langle W, R\rangle$.*

Using an ordinal notation system for the ordinals below ω_1^{CK}, the first non-recursive ordinal (see Rogers [26]), it is possible to construct for each $\alpha < \omega_1^{\mathrm{CK}}$ sentences that hold at a world w of a transitive frame if and only if the world is converse wellfounded and w has depth α. However, in general one cannot express converse wellfoundedness in $\mathcal{L}_\Box$ according to Corollary 4.6.

In the following we discuss the closure ordinal κ of Φ and thereby show that there is a gap between ω_1^{CK} and κ.

We abbreviate by setting $X_\alpha = \Phi^\alpha(\emptyset)$. The function $\delta \mapsto X_\delta$ can be defined by a Σ_1-recursion over any sufficiently closed model $\mathcal{M}$ containing the standard model $\mathbb{N}$ of arithmetic. In particular this is true over any model $\mathcal{M}$ of Kripke-Platek (KP) set theory containing the integers as a standard set. Transitive models of KP are called admissible sets. An ordinal α is admissible if L_α is a model of KP. ADM is the set of all admissible ordinals greater than ω. ADM* is the class ADM together with its limit points.

The idea of the proof of the following theorem is that for infinite ordinals δ below the closure ordinal κ of this operation, the sets $X_{\delta+1}$ can be construed essentially as truth sets for the δ'th level of the L-hierarchy. The point where this identification breaks down is precisely where the L-hierarchy requires parameters in the first order definitions that are required to make up the elements of the next level.

Let γ be least so that L_γ has a transitive Σ_1-end extension (that is, there is a transitive M with L_γ as an element, and $\langle L_\gamma, \in\rangle \prec_{\Sigma_1} \langle M, \in\rangle$). This ordinal is in some senses a large one: it is recursively Mahlo, (indeed a fixed point in a recursively "hypermahlo" hierarchy). It is easy to see that $L_\gamma \models \forall x(|x| = \omega)$ and from (i) below, that γ is less than the least ordinal ν so that $L_\nu \models$ KP $+\Sigma_1$-separation. Further one may show the following facts:

(i) that $\langle L_\gamma, \in\rangle \prec_{\Sigma_1} \langle L_{\gamma+1}, \in\rangle$ (and is the least such γ with this property), and

(ii) that L_γ is an admissible set that is the Skolem Hull of Σ_1 parameter free terms inside itself. Thus any element x of L_γ is named by a parameter free Σ_1 Skolem term, and the $\in$-diagram of L_γ is essentially given by the Σ_1-truth set for L_γ. This fails first at $L_{\gamma+1}$. (For background on the theory KP and on admissible sets, see Barwise [3].) These two facts are used in the proof of 4.7 below.

It is also well known that γ is least so that L_γ is *first-order reflecting*, that is, for all n, if φ is any Π_n formula (with parameters from L_γ allowed) in the language of set theory, then the following holds (see Aczel [1]):

$$L_\gamma \models \varphi \implies \exists \alpha < \gamma \; L_\alpha \models \varphi.$$

(As indicated above, L_γ is a model of a strong extension of *KP* — stronger than the theory *KPi* which is proof theoretically equivalent to ACA + Bar Induction + Δ^1_2-Comprehension Scheme; however it is known, by consideration of (i) above, that it is weaker than the latter strengthened with the Π^1_2-Comprehension Scheme — see Rathjen [25].)

THEOREM 4.7. $\gamma = \kappa$.

In particular, this shows that $\kappa > \omega_1^{\mathrm{CK}}$ (ω_1^{CK} is Π_2-reflecting but not Π_3-reflecting).

For $\mathcal{A}$ an admissible set we let $\mathcal{WFP}(\mathcal{A})$ denote the *wellfounded part* of $\mathcal{A}$, that is, the largest structure $\mathcal{B} \subseteq \mathcal{A}$ such that $\in\restriction |\mathcal{B}| \times |\mathcal{B}|$ is wellfounded. If α is the ordinal rank of the wellfounded part $\mathcal{WFP}(\mathcal{A})$, then it is known that $L_\alpha \models$ KP. If $\mathcal{A}$ is a model of KP, $On(\mathcal{A})$ denotes the class of ordinals in the sense of $\mathcal{A}$. We identify the standard part of $On(\mathcal{A})$ with the standard ordinals. Hence $On(\mathcal{WFP}(\mathcal{A})) \supseteq \omega + 1$ holds if and only if the integers of $\mathcal{A}$ are standard. We shall only have to consider in what follows admissible sets in which the integers are standard.

We can justify the following existence theorem for valuations, which yields valuations for converse illfounded frames if it is applied to nonstandard models $\mathcal{A}$ of KP.

THEOREM 4.8. *If $\langle W, R\rangle$ is transitive, $\mathcal{A}$ is a model of* KP, $\langle W, R\rangle \in \mathcal{A}$ *and either*

(i) $\exists a$ *($a \in On(\mathcal{A})$ with $\mathcal{A} \models$ "$a = \mathrm{rank}(R^{-1})$") or*
(ii) $\mathcal{A} \models \forall w$*($w$ is converse illfounded $\Rightarrow$* $\mathrm{depth}(w) > \kappa^*$*) where κ^* is the closure ordinal of the operator Φ"; then $\langle W, R\rangle$ admits a valuation.*

Note: there is no requirement that the model $\mathcal{A}$ be wellfounded in the above. The theorem is the non-standard counterpart to Theorem 4.5 (which it generalises). The point is simply the observation that the operation $w \rightarrowtail X_w$, for $w \in W$, is sufficiently absolute between $\mathcal{A}$ and the universe of all sets, that the valuation defined by a Σ_1-recursion inside $\mathcal{A}$ really does serve as a true valuation.

§5. A necessary condition for the existence of a valuation. As we have explained above, a frame need not be converse wellfounded for it to admit a valuation. However, Löb's Theorem can be employed for showing the following:

THEOREM 5.1. *Let $\langle W, R, V\rangle$ be a transitive model. Then the following must hold for the depth α of every converse illfounded world in $\langle W, R\rangle$: either $\alpha \in$ ADM* or $\alpha \geq \kappa$.*

The partial characterisation of the class of transitive frames admitting valuations yielded by Theorem 5.1 and Theorem 4.5 is incomplete because it does not say whether worlds of depth α with $\omega_1^{CK} \leq \alpha \leq \kappa$ have to be converse wellfounded. Proposition 5.2 shows that for any admissible ordinal $\alpha > \omega$, there exists frames with converse illfounded worlds of depth α.

To summarise: For the following classes of transitive frames $\langle W, R\rangle$:

(i) If every world $w \in W$ has a depth $\alpha < \omega_1^{CK}$ then R is converse wellfounded iff $\langle W, R\rangle$ admits a valuation;
(ii) if every converse illfounded world has depth $\alpha \geq \kappa$ (without any restriction being placed on the ordering R for the part containing those worlds of depth $\geq \kappa$, i.e., arbitrary loops and infinitely descending chains are possible) then the frame admits a valuation. Thus for frames where all depths are smaller than ω_1^{CK} or larger than κ the Characterisation Problem for transitive frames is settled. Between these two ordinals the depth of every converse illfounded world of a transitive frame admitting a model must be admissible or a limit of admissibles according to Theorem 5.1. In the next proposition we shall see that there actually are frames admitting models with converse illfounded worlds of depth ω_1^{CK} — the least possible value by 5.1.

In the following proposition $\mathcal{WFP}(R^{-1})$ denotes the largest wellfounded initial segment of R^{-1}.

PROPOSITION 5.2. *For any $\alpha \in$ ADM there is a model $\langle W, R, V\rangle$ with R^{-1} illfounded, with $\mathcal{WFP}(R^{-1}) \cong \alpha$ and $V(w)$ is a fixed point for some $w \in W$.*

This follows from Barwise's [3] Compactness Theorem and a standard omitting types argument.

We conclude this section with a note on the uniqueness of valuations.

Are there two distinct PW-models based on the same frame? That is, are there frames that admit at least two distinct valuations? Converse wellfounded frames have exactly one valuation by Theorem 4.1; thus the valuation is always unique. Therefore two valuations for the same frame can differ only on converse illfounded worlds. One can show the following again by employing an illfounded model of KP.

THEOREM 5.3. *There is a linearly ordered frame $\langle W, R\rangle$ which admits two different valuations V, V'.*

§6. The general characterisation problem. So far we have focused on transitive frames. In this section we generalise some results to other frames.

Theorem 5.1 yields information also on frames that are not transitive. For the transitive closure of R can be defined within the language $\mathcal{L}_\Box$.

THEOREM 6.1. *If $\langle W, R\rangle$ supports a PW-model, then in the transitive closure of $\langle W, R\rangle$ every converse illfounded world in W has depth α with $\alpha \in \mathrm{ADM}^*$ or $\alpha \geq \kappa$.*

Theorem 6.1 can be used for demonstrating that the frame $\langle \omega, \mathrm{Pred}\rangle$ in Example 3.2 does not admit a valuation. Thus Theorem 6.1 implies all negative results in § 3. Therefore, in a sense, Löb's Theorem implies the possible worlds analogues of Tarski's, Montague's, McGee's and Visser's theorems. In § 7 we shall show how to get from these analogues back to ω-inconsistency results.

By Theorem 4.5 a transitive frame admits a valuation if every converse illfounded world $w \in W$ has at least depth κ. One might hope to generalise Theorem 4.5 by showing that a frame admits a valuation if all converse illfounded worlds are of depth at least κ in the transitive closure of the frame. However, this generalisation fails. Thus there is no characterisation of the class of frames that admit valuations in terms of the transitive closure of the frames.

THEOREM 6.2. *There are frames that do not support a PW-model, although their transitive closure does.*

Obviously, if $\langle W, R, V\rangle$ is a PW-model and wRw, then $V(w)$ is a fixed point of Φ. So far we have used only the fixed point $\Phi^\kappa(\emptyset)$. However, other fixed points of Φ are required for valuations of certain frames that are not transitive. There is a model with two reflexive worlds v and w such that $V(v) \neq V(w)$. Again we omit the proofs.

§7. Completeness. In modal logic completeness theorems play a central role. For instance, the sentences valid in all frames at all words under any valuation are exactly the theorems of the modal system K. Similarly, the system K4 is associated with the class of transitive frames, the Gödel system G with the class of converse wellfounded frames (see Boolos [6]). Can we obtain analogous results in our setting?

Since we are dealing with standard models only, the set of all sentences valid in all PW-models $\langle W, R, V\rangle$ contains all arithmetical truths and is closed under the ω-rule. Of course this set is not recursively enumerable.

DEFINITION 7.1. A set *Th* of sentences of $\mathcal{L}_\Box$ is *$\Box$-closed* if and only if it satisfies the following conditions:

(i) *Th* contains PA and is closed under logic, necessitation and the ω-rule:

$$\frac{A(\overline{0}), A(\overline{1}), A(\overline{2}), \ldots}{\forall x A(x)}$$

(ii) $\Box\ulcorner A \to B\urcorner \to (\Box\ulcorner A\urcorner \to \Box\ulcorner B\urcorner)$ is in Th for all A and B in $\mathcal{L}_\Box$.
(iii) All instances $\forall x\Box\ulcorner A(\dot{x})\urcorner \to \Box\ulcorner\forall x A(x)\urcorner$ of the Barcan formula (i.e., the formalised ω-rule) are in Th for all formulæ $A(v)$ of $\mathcal{L}_\Box$.

For a set Th of sentences of $\mathcal{L}_\Box$ we write $(\mathbb{N}, S) \models Th$ if for all sentences $A \in \mathcal{L}_\Box$ $(\mathbb{N}, S) \models A$ holds.

THEOREM 7.2 (Completeness). *If Th is $\Box$-closed and consistent, then there is a PW-model $\langle W, R, V\rangle$ such that $(\mathbb{N}, V(w)) \models Th$ holds for all $w \in W$.*

Our completeness proof resembles the usual proofs for operator modal logic (see, e.g., Chagrov and Zakharyaschev [9, chapter 5]). However, since we are dealing with standard models only, we have to employ ω-logic rather than pure first-order logic.

From the Completeness Theorem we can also derive several inconsistency results. For instances, we can prove that the truth theory FS in Halbach [14], which is equivalent to a system introduced by Friedman and Sheard [12], is inconsistent in ω-logic. Here we present still another example.

$\mathcal{D}$ is the smallest $\Box$-closed set containing all instances of $\Box\ulcorner A\urcorner \to \Diamond\ulcorner A\urcorner$ for all $A \in \mathcal{L}_\Box$. Obviously $\mathcal{D}$ is the predicate analogue plus the ω-rule of the system D of deontic logic. Note that, according to the definition of $\mathcal{D}$, the ω-rule and the rule of necessitation may be applied to consequences of $\Box\ulcorner A\urcorner \to \Diamond\ulcorner A\urcorner$ for all $A \in \mathcal{L}_\Box$.

EXAMPLE 7.3. $\mathcal{D}$ is inconsistent.

This follows from the Completeness Theorem and the fact that every frame admitting a PW-model contains a dead end, which contradicts the axiom scheme $\Box\ulcorner A\urcorner \to \Diamond\ulcorner A\urcorner$ of $\mathcal{D}$.

If the ω-rule is dropped from $\mathcal{D}$ the system becomes consistent (though ω-inconsistent because of McGee's [20] result) because it is a subtheory of the consistent system FS of Halbach [14].

We define the predicate counterparts of the operator modal systems K and K4.

DEFINITION 7.4. $\mathcal{K}$ is the smallest $\Box$-closed set. $\mathcal{K}_4$ is the smallest $\Box$-closed set that contains all instances of $\Box\ulcorner A\urcorner \to \Box\ulcorner\Box\ulcorner A\urcorner\urcorner$.

By applying the theorem to $Th = \mathcal{K}$ and $Th = \mathcal{K}_4$ we get the following two corollaries:

COROLLARY 7.5. *The following are equivalent for all $A \in \mathcal{L}_\Box$:*

(i) *A is valid at all worlds in all PW-models, i.e., $V(w) \models A$ for all PW-models $\langle W, R, V\rangle$ and $w \in W$.*
(ii) *A is in $\mathcal{K}$.*

PROOF. In order to prove that (i) implies (ii), we apply our Completeness Theorem. If $A \notin \mathcal{K}$ there is a world w in the PW-model $\langle W, R, V\rangle$ constructed in the proof of the completeness result such that $A \notin w$, i.e., $V(w) \not\models A$.

The Normality Lemma 2.3 ensures that (ii) implies (i). ⊣

We obtain also an analogous result for $\mathcal{K}4$:

COROLLARY 7.6. *The following are equivalent for all $A \in \mathcal{L}_\Box$:*

(i) *A is valid at all worlds in all transitive PW-models, i.e., $V(w) \models A$ for all PW-models $\langle W, R, V\rangle$ and $w \in W$.*

(ii) *A is in $\mathcal{K}4$.*

Not all completeness theorems known from operator modal logic carry over to their predicate counterparts. For instance, the logic G obtained from K_4 by adding the operator version $\Box(\Box p \rightarrow p) \rightarrow \Box p$ is complete with respect to the class of all transitive converse wellfounded frames. The predicate system $\mathcal{K}4$ has Löb's Theorem already as a theorem. However, it does not prove all sentences valid in all transitive converse wellfounded frames, as we shall show now by recursion-theoretic considerations, more precisely by comparing the complexities of $\mathcal{K}4$ and the fixed point $\Phi^\kappa(\emptyset)$.

By Corollary 7.6 $\mathcal{K}4$ is the set of sentences valid in all transitive PW-models; $\Phi^\kappa(\emptyset)^\diamond$ is the set of all sentences valid in a all transitive *converse wellfounded* PW-models. Therefore $\Phi^\kappa(\emptyset)$ is a subset of $\mathcal{K}4$. We shall show that the converse does not hold: there are sentences that are valid in all transitive converse wellfounded PW-models but that fail at a world of some transitive converse illfounded PW-model. In particular, Corollary 7.6 does not hold if (i) is restricted to converse wellfounded frames.

$\mathcal{K}4$ is obviously Π^1_1. The following Theorem shows that $\Phi^\kappa(\emptyset)$ has a much higher complexity than $\mathcal{K}4$. Thus converse illfounded PW-models really do matter.

THEOREM 7.7. *The set $\Phi^\kappa(\emptyset)$ is Δ^1_2 but neither Σ^1_1 nor Π^1_1.*

§8. Extensions to other languages and ground models. In this section we generalise our approach and the methods developed so far to languages extending $\mathcal{L}_\Box$.

Our results carry over to acceptable models in a straightforward way. Thus we can use other well behaved models $\mathcal{M}$ (with the appropriate language $\mathcal{L}_\mathcal{M}$) instead of the standard model of arithmetic. In this case $\Box$ has to be conceived as a relation obtaining between formulas and variable assignments. Thereby we introduce genuine de re-modality.

DEFINITION 8.1. Let $\mathcal{M} = \langle M, R, \ldots\rangle$ be any structure. Let $\kappa_\mathcal{M}$ be the least ordinal so that $L_{\kappa_\mathcal{M}}(\langle M, R, \ldots\rangle)$ is the first level of the relativised Gödel L-hierarchy built over M, using elements of M as *urelemente*, which is first order reflecting.

The methods of § 4 can be used to show the following. Let $\mathcal{M}$ be a countable acceptable structure (in the sense of [22], that is essentially $\mathcal{M}$ has a definable coding scheme). Consider the class of frames where now $\mathbb{N}$ has been replaced by $\mathcal{M}$ at each world of the frame. One may show:

(i) The fixed point $\Phi^{\kappa_{\mathcal{M}}}(\emptyset)$ occurs in sufficiently long converse wellfounded frames $\langle W, R\rangle$ at the ordinal $\kappa_{\mathcal{M}}$.

(ii) We may look at admissible sets $\mathcal{A}$ over the structure $\mathcal{M}$ (much as Barwise does in [3]), and take orderings $R \in \mathcal{A}$ with $\mathcal{A} \models$ "R^{-1} is wellfounded" and construct valuations V for the worlds of W.

(iii) The Characterisation Theorem 6.1 of the wellfounded parts of frames admitting valuations holds with κ replaced by $\kappa_{\mathcal{M}}$ and the class ADM replaced by ADM($\mathcal{M}$) — the class of ordinals admissible with respect to the structure $\mathcal{M}$.

In general the results of § 4 and § 5 go through *mutatis mutandis*.

The formulæ of the language $\mathcal{L}_{\mathcal{M}}$ are interpreted in every world in the same way. In particular, in the special case $\mathcal{L}_{\mathcal{M}} = \mathcal{L}_{\mathrm{PA}}$ we have considered in the previous sections, the arithmetical vocabulary does not allow for any variation in its evaluation, at least not on our account: it is interpreted in every world by the standard model.

Moreover, we could allow the universe to vary from world to world, as long as the codes of sentences and codes of sequences of all objects existing in the respective world are in the universe. However, we do not go that far afield from our original approach and we illustrate the effects of contingent vocabulary by considering a slight modification of our approach with $\mathcal{L}_{\mathrm{PA}}$. For our simple examples we need only an extension of $\mathcal{L}_{\mathrm{PA}}$ by an additional propositional parameter.

Let $\mathcal{L}^p_{\mathrm{PA}}$ be the language $\mathcal{L}_{\mathrm{PA}}$ of arithmetic extended by an additional propositional variable p. By adding $\Box$ we obtain the language $\mathcal{L}^p_{\Box}$. A PW-model is defined as in Definition 2.2 except that V has also to tell whether p is true at that world. More precisely $V(w)$ is a pair $\langle S, k\rangle$ where $S \subseteq \mathcal{L}^p_{\Box}$ is the extension of $\Box$ and k is the truth value 0 or 1 of p at w. If $V(w) = \langle S, k\rangle$, S has to satisfy the following condition:

$$S = \{A \in \mathcal{L}^p_{\Box} \mid \forall u\,(wRu \Rightarrow V(u) \models A)\}.$$

Of course $V(u) \models A$ for $A \in \mathcal{L}^p_{\Box}$ is defined in the obvious way.

In the following theorem we show that we are not completely free to assign truth values to p if the frame is converse illfounded.

PROPOSITION 8.2. *Assume $\langle W, R\rangle$ supports a PW-model. Then the transitive closure of a frame $\langle W, R\rangle$ is converse wellfounded if and only if it allows for arbitrary interpretations of p.*

By the latter we mean that given a frame and given a function $g : W \to \{0, 1\}$ we have a model $\langle W, R, V\rangle$ based on that frame such that for all $w \in W$ and some $S \subseteq \mathcal{L}^p_{\Box}$ the equation $V(w) = \langle S, g(w)\rangle$ holds.

Proof. Assume that the frame is transitive.

Löbs Theorem 3.4 applies also to p, that is, the following holds for any valuation V and world w:

$$V(w) \models \Box\ulcorner\Box\ulcorner p\urcorner \longrightarrow p\urcorner \longrightarrow \Box\ulcorner p\urcorner.$$

By stipulating $V(w) \models p$ for converse wellfounded w and $V(w) \models \neg p$ for converse illfounded w, one can exclude converse illfounded worlds for transitive frames (see Boolos [6, p. 75f] and Corollary 4.6).

Since the transitive closure of R can be defined in $\mathcal{L}_\Box$, the argument can be generalised to non-transitive frames. ⊣

Although we are not completely free to evaluate p in any world of a converse illfounded model in any arbitrary way, we can use p in order to express certain features of the frame (for instance, we can force a frame to branch). Here we do not explore what can be expressed with additional vocabulary, but it should be obvious that interesting conditions are expressible with more propositional parameters and by quantifying over these parameters.

If we allow also for additional objects in every world beyond the natural numbers and we change our setup accordingly, we shall encounter the usual problems of quantified modal logic. In particular, the Barcan formula will fail if we allow for increasing domains etc.

Each of these extensions is worth to be studied and philosophically relevant — indeed these extensions are more relevant than the setting that we have followed: it is certainly not satisfying, from a philosophical point of view, to concentrate on worlds which are standard models of arithmetic with an additional interpretation of the necessity predicate. But independent of what language is chosen and what the set of worlds looks like semantically, if a modality is to be expressed by a predicate, then every world has to have an interpreted theory of syntax in the background, and that means that every world has to comprise arithmetic (modulo coding). Thus, we have actually concentrated on the minimal prerequisites of a possible worlds semantics for predicates, and that is why our results also apply to the philosophically more relevant cases.

§9. Some preliminary conclusions. In this final section we try to draw some conclusions from our results.

We have shown how all limitative results of § 3 may finally be reduced to a single limitative result. The possible worlds analogues of Tarski's, Montague's, McGee's, Visser's theorems, which have been mentioned in § 3, as well as the analogues of some results not mentioned here like those by Thomason [30] and partially those by Friedman and Sheard [12] may be reduced to the single condition on frames in Theorem 6.1. Löb's Theorem is the origin of all the inconsistency and ω-inconsistency results. On the semantical side this is mirrored by the condition that any converse illfounded world must have a large depth; this condition excludes all simple loops, reflexive frames etc. from being frames that admit a PW-model. Thus we conclude that the mentioned paradoxes flow from a common source.

Our semantical investigations underline the central role of converse wellfoundedness of frames. Provability logic (see, e.g., Boolos [6]) thus appears in a new light, at least to us. The main results of provability logic might be understood as suggesting that there is a profound relation between provability and converse wellfoundedness. In contrast, our results show that, if one sets up possible worlds semantics for predicates, converse wellfoundedness always plays an important role. All frames supporting a PW-model have to be converse wellfounded — except for worlds of great depth. Thus all predicates with reasonable possible worlds semantics are much like the provability predicate. This holds even in the absence of the transitivity axiom as Theorem 6.1 shows.

Surprisingly the condition of Theorem 6.1 does not rule out all converse illfounded models. As Theorem 7.7 shows, our completeness result Corollary 7.5 for $\mathcal{K}$ would fail, if we restricted our attention to converse wellfounded models only. That is, $\mathcal{K}$ is incomplete with respect to the class $\Phi^{\kappa}(\emptyset)^{\diamond}$ of all converse wellfounded models. Consequently there are sentences that hold in all converse wellfounded models but fail at some converse illfounded model and are therefore not theorems of $\mathcal{K}$. This shows that the common semantical source of the paradoxes is not precisely converse illfoundedness of the frames but rather the failure of the more complex condition stated in Theorem 6.1.

The condition of Theorem 6.1 implies that many frames familiar from operator modal logic do not support PW-models. In particular, the frames for T, S_4 and S_5, i.e., all reflexive frames do not admit a valuation. It seems that we have to sacrifice the core systems of modal logic for the predicate approach. We do not conclude from this, however, that the predicate approach is mistaken altogether. We recommend to approach necessity and other notions much like truth in order to ban inconsistency from these notions. For truth a wide array of semantical and axiomatic approaches has been offered in the literature. For necessity there should be even more.

REFERENCES

[1] Peter Aczel and Wayne Richter, *Inductive definitions and reflecting properties of admissible ordinals*, ***Generalized recursion theory*** (Jens E. Fenstad and Peter Hinman, editors), North-Holland, 1973, pp. 301–381.

[2] Nicholas Asher and Hans Kamp, *Self-reference, attitudes, and paradox*, ***Properties, types and meaning*** (Gennaro Chierchia, Barbara H. Partee, and Raymond Turner, editors), vol. 1, Kluwer, Dordrecht, 1989, pp. 85–158.

[3] Jon Barwise, ***Admissible sets and structures***, Perspectives in Mathematical Logic, Springer-Verlag, Berlin, 1975.

[4] GEORGE BEALER, ***Quality and concept***, Clarendon Press, Oxford, 1982.

[5] NUEL BELNAP and ANIL GUPTA, ***The revision theory of truth***, MIT Press, Cambridge, 1993.

[6] GEORGE BOOLOS, ***The logic of provability***, Cambridge University Press, Cambridge, 1993.

[7] ANDREA CANTINI, *A theory of formal truth arithmetically equivalent to* ID_1, ***The Journal of Symbolic Logic***, vol. 55 (1990), pp. 244–259.

[8] ———, ***Logical frameworks for truth and abstraction. an axiomatic study***, Studies in Logic and the Foundations of Mathematics, vol. 135, Elsevier, Amsterdam, 1996.

[9] ALEXANDER CHAGROV and MICHAEL ZAKHARYASCHEV, ***Modal logic***, Oxford Logic Guides, Oxford University Press, Oxford, 1997.

[10] JIM DES RIVIÈRES and HECTOR J. LEVESQUE, *The consistency of syntactical treatments of knowledge*, ***Theoretical aspects of reasoning about knowledge: Proceedings of the 1986 conference*** (Joseph Y. Halpern, editor), Morgan Kaufmann, Los Altos, 1986, pp. 115–130.

[11] SOLOMON FEFERMAN, *Reflecting on incompleteness*, ***The Journal of Symbolic Logic***, vol. 56 (1991), pp. 1–49.

[12] HARVEY FRIEDMAN and MICHAEL SHEARD, *An axiomatic approach to self-referential truth*, ***Annals of Pure and Applied Logic***, vol. 33 (1987), pp. 1–21.

[13] PATRICK GRIM, ***The incomplete universe: Totality, knowledge, and truth***, MIT Press, Cambridge, Mass., 1991.

[14] VOLKER HALBACH, *A system of complete and consistent truth*, ***Notre Dame Journal of Formal Logic***, vol. 35 (1994), pp. 311–327.

[15] ———, *Truth and reduction*, ***Erkenntnis***, vol. 53 (2000), pp. 97–126.

[16] ———, ***Semantics and deflationism***, Habilitationsschrift, Universität Konstanz, 2001.

[17] VOLKER HALBACH, HANNES LEITGEB, and PHILIP WELCH, *Possible worlds semantics for modal notions conceived as predicates*, ***Journal of Philosophical Logic***, vol. 32 (2003), pp. 179–223.

[18] G. E. HUGHES and M. J. CRESSWELL, ***A new introduction to modal logic***, Routledge, London and New York, 1996.

[19] HANNES LEITGEB, *Theories of truth which have no standard models*, ***Studia Logica***, vol. 21 (2001), pp. 69–87.

[20] VANN MCGEE, *How truthlike can a predicate be?*, ***Journal of Philosophical Logic***, vol. 14 (1985), pp. 399–410.

[21] RICHARD MONTAGUE, *Syntactical treatments of modality, with corollaries on reflexion principles and finite axiomatizability*, ***Acta Philosophica Fennica***, vol. 16 (1963), pp. 153–167.

[22] YIANNIS MOSCHOVAKIS, ***Elementary induction on abstract structures***, North-Holland, Amsterdam, 1974.

[23] WILLARD VAN ORMAN QUINE, *Three grades of modal involvement*, ***The ways of paradox***, Harvard University Press, Cambridge, Mass., revised and enlarged ed., 1976, pp. 158–176.

[24] ———, *Intensions revisited*, ***Contemporary perspectives in the philosophy of language*** (French, Uehling, and Wettstein, editors), University of Minnesota Press, Minneapolis, 1977, pp. 5–11.

[25] MICHAEL RATHJEN, *Proof theory of reflection*, ***Annals of Pure and Applied Logic***, vol. 68 (1994), pp. 181–224.

[26] HARTLEY ROGERS, ***Theory of recursive functions and effective computability***, McGraw-Hill Book Company, New York, 1967.

[27] KURT SCHÜTTE, ***Proof theory***, Springer, Berlin, 1970.

[28] MICHAEL SHEARD, *A guide to truth predicates in the modern era*, ***The Journal of Symbolic Logic***, vol. 59 (1994), pp. 1032–1054.

[29] BARRY HARTLEY SLATER, *Natural language's semantic closure*, forthcoming.

[30] RICHMOND H. THOMASON, *A note on syntactical treatments of modality*, ***Synthese***, vol. 44 (1980), pp. 391–396.

[31] ALBERT VISSER, *Semantics and the liar paradox*, ***Handbook of philosophical logic***, vol. IV, Reidel, Dordrecht, 1989, pp. 617–706.

NEW COLLEGE
OXFORD
OX1 3BN, UNITED KINGDOM
E-mail: volker.halbach@philosophy.oxford.ac.uk

UNIVERSITÄT SALZBURG
INSTITUT FÜR PHILOSOPHIE
FRANZISKANERGASSE 1
5020 SALZBURG, AUSTRIA
E-mail: Hannes.Leitgeb@sbg.ac.at

DEPARTMENT OF MATHEMATICS
UNIVERSITY OF BRISTOL
BRISTOL BS8 1TW, UNITED KINGDOM
and
GRADUATE SCHOOL OF SCIENCE & TECHNOLOGY
KOBE UNIVERSITY
ROKKO-DAI, NADA-KU
KOBE 657, JAPAN
Current address: Institut für formale Logik, Währinger Str. 25, 1090 Wien, Austria
E-mail: welch@kobe-u.ac.jp

A CONTEXT PRINCIPLE

WILFRID HODGES

Abstract. Starting from any language provided with sentence meanings and a grammar, and using the principle that the meaning of a phrase is what it contributes to the meanings of sentences containing it, we derive a semantics for the whole language. The semantics is necessarily compositional and carries a structure of semantic categories. With a further assumption on the grammatical heads of phrases, we can assume that the meanings of head words are functions from meanings to meanings, in the Frege style. The paper sketches these results and considers how far the semantics of Husserl, Frege and Tarski can be seen as examples of this pattern.

When Alfred Tarski wrote his famous definition of truth [20] (1933) for a formal language, he had several stated aims. His chief aim was to define truth of sentences. Giving correct meanings of other expressions of the language was nowhere in his list of aims at all; it was a happy accident that a general semantics fell out of his truth definition.

So the following facts, all very easily proved, came to me as a surprise. Given any notion of meaning for sentences (for example, a specification of when they are true and when not), and assuming some simple book-keeping conditions, there is a canonical way of extending this notion to a semantics for the whole language. I call it the *fregean extension*; it is determined up to the question which pairs of expressions have the same meaning. Tarski's semantics for first-order logic is the fregean extension of the truth conditions for sentences. A few more book-keeping conditions guarantee that the fregean extension can be chosen to have good functional properties of the kind often associated with Frege and with type-theoretic semantics.

Tarski himself was certainly interested in the question how far his solution of his problem was canonical, and we can learn useful things from his discussion of the issue. But the main results below on fregean extensions come closer to the linguistic and logical concerns of Frege and Husserl, a generation earlier than Tarski. Husserl has been unjustly neglected by logicians, and Frege's innovations in linguistics deserve to be better known.

The formal content of the first section below comes from my paper [13], where I also spell out the grammatical assumptions more precisely than I do here. The commentary is different here, and it mostly records and expands

Intensionality
Edited by Reinhard Kahle
Lecture Notes in Logic, 22

my talk to the conference on Intensionality at Munich in October 2000. My warmest thanks to Reinhard Kahle and his fellow organisers, and to the other participants for their stimulus and wise comments. I am grateful to the referee for some thoughtful suggestions.

§1. Sentence meanings and word meanings. In §60 of his *Grundlagen der Arithmetik*, Gottlob Frege [4] makes the following pronouncement, which I shall refer to as *Frege's Dictum*:

(1) Es genügt, wenn der Satz als Ganzes einen Sinn hat; dadurch erhalten auch seine Theile ihren Inhalt.
(It is enough if the sentence as a whole has a meaning; it is this that confers on its parts also their content.)

I shall take Frege's statement at face value as a recipe for finding expression meanings from sentence meanings (up to synonymy). The resulting expression meanings are what I shall call the *fregean cover* of the sentence meanings.

I shall assume that we are discussing a *language* L, and that it is determinate what the (grammatical) *expressions* of L are, and when an expression occurs as a *constituent* of another expression. (Frege's Dictum refers to constituents as "parts".) Every expression is a constituent of itself, and every word in an expression is a constituent of the expression. Every expression is a finite string of words. For an easy life we can suppose that when an expression e occurs as a constituent of another expression f, e consists of one or more consecutive words from f. But strictly this is not necessary. Thus for greater realism we can allow that in French the consecutive constituents "a" and "il" yield "a-t-il", and in English the consecutive constituents "you" and "did go?" yield "did you go?". But in what follows I generally ignore such complications, on the assumption that they can be dealt with separately.

Following Frege I shall assume that a particular class of expressions is picked out as fundamental; Frege says "sentences", but for greater generality I shall call the selected expressions the *target expressions*. Still following Frege, I shall assume that each target expression e has a *meaning* $\mu(e)$; here μ is a function whose domain is the set of target expressions.

Frege's Dictum assumes also that every expression e has a *content* $\nu(e)$ (where ν is a function whose domain is the set of all expressions). At the time of the *Grundlagen* Frege had not yet distinguished between Sinn, Inhalt and Bedeutung, and it is probably safe to read Sinn and Inhalt (meaning and content) as the same thing. In that case Frege is also assuming that $\mu(e) = \nu(e)$ for every target expression, in other words that ν is an extension of μ to the set of all expressions. But again for greater generality I won't assume that until the end of this section.

Frege's Dictum says that for every expression e, $\nu(e)$ is determined by the restriction of μ to the set of target expressions containing e. Taken in isolation, the Dictum says nothing at all about how the one determines the other. But even in isolation it does surely imply the following:

(2) If expressions e and f have different contents, then there must be a target expression containing one of them, say e, which does not go over into a target expression of the same meaning when we replace some occurrence of e in it by f.

Strictly we can't deduce the converse. It's consistent with the letter of Frege's Dictum that, for example, $\nu(e) = \mu(e)$ for all target expressions e and $\nu(e)$ is the Mandelbrot set for all other e. But to make Frege's Dictum interesting, we need a better idea than this. If we are not to import irrelevances, the natural gloss is to ask for ν to make as many distinctions as possible, subject to condition (2) above.

So our task is to define the relation $\nu(e) = \nu(f)$ in a way which expresses that e and f make the same contribution to the meanings of target expressions containing them. We note that this relation must be an equivalence relation on the set of all expressions. I shall write $\equiv_\nu$ for the equivalence relation of "having the same ν-value"; and likewise with μ.

In fact there seems to be just one reasonable way of doing this. Namely we define:

(3) For any two expressions e and f, $\nu(e) = \nu(f)$ if and only if:
for every target expression $s(e)$ containing e as a constituent as shown, there is a target expression $s(f)$ got by replacing this constituent by f, and $\mu(s(e)) = \mu(s(f))$;
and likewise with e, f transposed.

I leave it to the reader to check that this defines an equivalence relation $\equiv_\nu$ on the class of expressions.

When ν is as in (3), I shall say that ν is a *fregean cover* of μ. Note that there always is a fregean cover of μ; for example take the equivalence classes of the equivalence relation defined in (3). Also it is unique in the sense that any two fregean covers agree about whether any given pair of expressions are synonymous (i.e. have the same content). In view of this uniqueness I shall refer to ν as *the* fregean cover of μ.

REMARK 1. Much of what follows would become pointless if there was some other reasonable way of defining ν. I can't see one. For example you might try defining $\nu(e) = \nu(f)$ to be true if and only if whenever $s(e)$ and $s(f)$ are target expressions that agree except for the substitution of f for e or vice versa, $\mu(s(e)) = \mu(s(f))$. This works very badly. Suppose for example that e and f are expressions which can never be substituted for each other in expressions of L. Then the proposed definition says that e and f have the same content! For this and similar reasons, this proposed definition does not in general give an equivalence relation.

Or you might try restricting the definition to $s(e)$ and $s(f)$ where $s(f)$ comes from $s(e)$ by replacing e everywhere by f, and vice versa. This gives an equivalence relation, but it throws away any information to be gained from target expressions that contain both e and f, which is unreasonable.

No doubt there are other variants. I should be interested to hear of any serious alternative to (3) that does not invoke other kinds of structure on the language L.

REMARK 2. We have surreptitiously added one item to the assumptions about L, by assuming we know how to substitute one expression for another inside a third expression. This could just be substitution of strings, in which case the assumption is trivial. But nothing changes if we allow more savvy kinds of substitution. For example we might allow ourselves to know that when you put "I" for "He" in

He loves her.

the result is

I love her.

Here I note a remark of Peter Ramus (quoted on p. 335 of [17]):

(4) It's incredible but true ... that there have been lecturers in [the University of Paris] who clenched their teeth and insisted that *I loves* is just as correct speech as *I love* ...

The theory that Ramus attacks here was probably that "love" and "loves" mean the same thing and the difference between them is purely syntactic.

In any case we shall assume that an allowed substitution never affects the structure of constituents that don't overlap with the substituted expression, even though it may alter the forms of individual constituents. Thus if $r(e_1, \ldots, e_n)$ is an expression and f_1 can be substituted for e_1 in it, then the expression resulting from the substitution is $r(f_1, e_2, \ldots, e_n)$.

Now that we have the fregean cover, what is it good for? The following three propositions state properties of the fregean cover ν of a semantics μ on target expressions.

PROPOSITION 1. *If e, f are target expressions and $e \equiv_\nu f$ then $e \equiv_\mu f$.*

PROOF. Immediate, replacing e by f. ⊣

The next proposition needs two book-keeping assumptions:

(5) [Cofinality] Every expression is a constituent of some target expression.

(6) [Closure under subterms] If $e(f(g))$ is an expression of L which has a constituent $f(g)$ with a constituent g, then g is a constituent of $e(f(g))$, and if $e(f(g'))$ is also an expression of L with g' as constituent then so is $f(g')$.

I assume henceforth that L satisfies both Cofinality and Closure under Subterms.

PROPOSITION 2 (Compositionality). *If $r(e_1, \dots, e_n)$ is an expression and for each i, $e_i \equiv_\nu f_i$, then $r(f_1, \dots, f_n)$ is a expression and*

$$r(e_1, \dots, e_n) \equiv_\nu r(f_1, \dots, f_n).$$

PROOF. By making one substitution at a time we reduce to the case $n = 1$. We must show that if $r(e)$ is a expression then so is $r(f)$ and $\nu(r(e)) = \nu(r(f))$. By Cofinality there is a target expression $t(r(e))$, which has e as a constituent by Closure under Subterms. Then by considering $t'(e) = t(r(e))$, we deduce from $\nu(e) = \nu(f)$ that $t'(f)$ is a target expression, and so again by Closure under Subterms $r(f)$ is a expression. Then a similar argument shows that $\nu(r(e)) = \nu(r(f))$. ⊣

We say that ν is an *extension* of μ if for all target expressions e, f,

$$e \equiv_\mu f \iff e \equiv_\nu f.$$

Right to left holds always, by Proposition 1. The fregean cover of μ is called the *fregean extension* of μ if it's an extension of μ.

PROPOSITION 3. *μ has a fregean extension* (*viz.* ν) *if and only if the following holds*:

> *For all target expressions p, q and $t(p)$, if $p \equiv_\mu q$ then $t(q)$ is also a target expression and $t(p) \equiv_\mu t(q)$.*

PROOF. This is just unpacking the definition of ν. ⊣

EXAMPLE 1. Let μ be the standard Tarski semantics for first-order logic, but restricted to sentences. (Then $\mu(\phi) = \mu(\psi)$ if and only if ϕ and ψ are logically equivalent.) The full Tarski first-order semantics (which assigns to each formula and structure the class of all assignments satisfying the formula in the structure) is the fregean extension of μ if we arrange that the meaning of each formula includes the information what variables are free in the formula.

EXAMPLE 2. Jaakko Hintikka devised a linear notation (called IF for "independence friendly") for branching quantifiers. In [10, p. 371] Hintikka and Sandu claim that there can't be a compositional semantics for IF that is as "fine grained" (i.e. roughly, makes as many distinctions) as the standard game semantics.

The problem here is to find a semantics that extends to all formulas of IF the game semantics μ on sentences. Since the condition of Proposition 3 holds for IF, μ has a fregean extension, which is automatically compositional by Proposition 2. In [12] I described such an extension explicitly.

EXAMPLE 3. W. Zadrozny [23] published a proof that every semantics is "equivalent to" a compositional semantics. People objected that he had a strange and undefined notion of "equivalent to".

A recent unpublished reply of Lappin and Zadrozny [15] says, in our terminology, that every semantics defined on all expressions has a fregean cover. One can read this off from (3) above by taking all expressions to be target expressions. (Zadrozny shows more for the languages he considers, using non-well-founded sets.)

Dag Westerståhl [22] has some information about analogues of the fregean extension that exist when the conditions on L and μ are relaxed.

§2. The husserlian level. We consider the following equivalence relation $\sim$ on expressions:

> $e \sim f$ if and only if for every target expression $s(e)$, $s(f)$ is also a target expression, and vice versa.

A semantics v defined on all expressions of L is said to be *husserlian* (for reasons that will follow shortly) if

> whenever e and f are expressions with $e \equiv_v f$, then $e \sim f$.

These notions are important for us because of the following, which one can extract at once from the definition of fregean covers:

COROLLARY 4 (The husserlian property). *If v is a fregean cover, then v is husserlian.*

So Frege's Dictum forces us to consider husserlian semantics. (Recall Remark 2 above. The Parisian scholars that Ramus attacked seem to have had a non-husserlian semantics for Latin, unless they were also sophisticated about substitutions.)

For future reference I state here another property of $\sim$, which is implicit in the proof of Proposition 2 above.

PROPOSITION 5. *If $r(e_1, \ldots, e_n)$ is an expression and for each i, $e_i \sim f_i$, then $r(f_1, \ldots, f_n)$ is a expression and*

$$r(e_1, \ldots, e_n) \sim r(f_1, \ldots, f_n).$$

It seems that Husserl was the first person to call attention to the equivalence relation $\sim$, in his Fourth Logical Investigation [14]. In §10 of this Investigation he offers the name *Semantic Category* (*Bedeutungskategorie*) for the equivalence classes of the relation $\sim$. He gives "Dieser Baum" and "Dieses Gold" as examples of two expressions in the same category, and he calls this category "nominale Materie". He is well aware that in order to get the machinery of categories to work for a natural language, we may have to allow substitutions to make mechanical changes in the context, for example modifying case endings (his §13, see my Remark 2 again).

For his definition of semantic categories, Husserl's target expressions are those whose meaning is "selbständig", i.e. constitutes "die volle und ganze Bedeutung eines konkreten Bedeutungsaktes" (the full and complete meaning of a concrete act of meaning, §7). Readers of Frege will assume that Husserl means sentences; but he does not. For example in the next section he describes "Röte" (redness) as a word whose meaning is selbständig. We note that the notion of a "full and complete meaning" is useless for explanatory purposes.

Husserl argues (§13 for example) that there are "Gesetze der Bedeutungsverknüpfung" (laws of combination of meanings) which determine what categories of phrases can be combined. In fact this is a mathematical consequence of our definitions and assumptions. Each compound expression e_0 is a concatenation of other expressions, say

$$e_0 = e_1 \dots e_n$$

(for example, taking $e_1, \dots, e_n$ to be the words of e_0 in order, or alternatively the immediate constituents of e_0 in order). Write $\mathrm{cat}(f)$ for the category of f. Then we claim that any concatenation of expressions of categories

$$\mathrm{cat}(e_1), \dots, \mathrm{cat}(e_n)$$

respectively is an expression of category $\mathrm{cat}(e_0)$. This follows at once from Proposition 5. So L satisfies a context-free grammar rule

$$C_0 \implies C_1 \ \dots \ C_n$$

where each C_i $(i \geq 0)$ is $\mathrm{cat}(e_i)$. The set of these rules, together with lexical rules

$$C \implies e$$

wherever e is a word and C is $\mathrm{cat}(e)$, determines the language L. (Here we are ignoring surface adjustments, as in Section 1 above.)

Husserl illustrates his "laws of combination of meaning" with the example (§13):

noun-meaning + adjective-meaning gives noun-meaning,

as for example "rotes" and "Haus" gives "rotes Haus". Disappointingly he gets the two terms in the wrong order. This warns us that for Husserl the category structure all lies at the level of meaning, and he is not too interested in the accidental ways it may show up in any particular language.

The fact that L has a context-free grammar is less significant than it might seem, because we placed no restriction on the number or length of the context-free rules. Husserl (§13) sees it as one of the jobs of a "science of meanings" to reduce the rules to as small a number as possible. He seems to assume the number is finite. In fact he lays great emphasis on the infinity of the set of meaningful expressions, and it is hard to see why he does this unless he wants to stress that his theory explains how we can use finite means to answer infinitely many questions.

Let me develop this for a moment. A central theme of Husserl's Fourth Investigation is that the laws of combination of meanings allow us to distinguish an infinite number of meaningful expressions from the meaningless expressions that could be formed. Thus for example, on schemas of sentences built up with "and" (§13 p. 332):

> Man versteht ohne weiteres, dass die Komplikationen in kombinatorisch überschaubarer Weise *in infinitum* [his italics] fortschreiten, dass jede neue Form an dieselbe Bedeutungskategorie, als Sphäre der Variabilität für ihre Termini, gebunden bleibt, und dass solange diese Sphäre eingehalten wird, alle darnach zu bildenden Bedeutungsverbindungen notwendig **existieren**, d.i. einen einheitlichen Sinn darstellen müssen.
>
> (We see at once that the compoundings go on *in infinitum*, that we can survey the possible combinations, that for each new schema the class from which its variables can take their values must always be the same semantic category, and that as long as they stay in this class, all combinations of meanings that result from this construction necessarily **exist**, i.e. must represent a unified sense.)

The central fact here, that finite means allow us to generate and parse arbitrarily many sentences, and that this tells us something useful about the form of a possible grammar, is familiar enough from writings of Chomsky. It seems that Husserl was the first person to say it with any precision. (And Frege's first remarks in this direction, in a letter to Jourdain, are fourteen years later than Husserl. Frege's version is often cited, Husserl's never.)

By the middle of the twentieth century, Husserl's category idea — no doubt rediscovered by several other people — lay close to the heart of the movement called *structural linguistics*. It was widely agreed that the idea does not work as it stands. Thus Zellig Harris [9], who refers to the contexts in which a word is allowed as the "co-occurrents" of the word:

> On the other hand, to describe a language in terms of the co-occurrences of the individual morphemes is virtually impossible: almost each morpheme has a unique set of co-occurrents; the set varies with individual speakers and with time ... ; it is in general impossible to obtain a complete list of co-occurrents for any morpheme; and in many cases a speaker is uncertain whether or not he would include some given morpheme as a co-occurrent of some other one.

So the structuralists made various adjustments that need not concern us here. For us it is more important to note that we needed the relation $\sim$ in order to cash in Frege's Dictum. Hence Harris's arguments against the usability of $\sim$ are indirectly arguments against the usability of Frege's Dictum too.

§3. Tarski's view. Harris's arguments stand as a warning against trying to use category analysis naively on natural languages. But none of his arguments apply to the formal languages of logic.

Alfred Tarski, in §4 "The concept of true sentence in languages of finite order" of his paper on the definability of truth [20] (1933), set out to describe the kinds of language for which his methods would yield a truth definition. The theory of semantic categories played an essential role in his discussion. For the English translation of 1935 Tarski added a Postscript which began as follows:

> In writing the present article I had in mind only formalized languages possessing a structure which is in harmony with the theory of semantical categories and especially with its basic principles.... Today I can no longer defend decisively the view I then took of this question. In connexion with this it now seems to me interesting and important to inquire what the consequence would be for the basic problems of the present work if we included in the field under consideration formalized languages for which the fundamental principles of the theory of semantical categories no longer hold.

Unfortunately Tarski's remarks had the opposite effect to what he intended. Instead of meditating on the "interesting and important" question of how one might relax those "fundamental principles of the theory of semantical categories", most readers of Tarski's paper have simply labelled the relevant §4 as obsolete and skipped it. (At this point in my talk it was heart-warming to hear Per Martin-Löf muttering "It's the best section".)

One of Tarski's first moves in this §4 is to describe a typical language L. He is rather explicit about its construction. From our point of view it will be convenient to state what he does more abstractly, and split it into two conditions. The first is that the language L must have unique parsing:

> [Unique Parsing Principle] If two constituents of an expression overlap, then one is a constituent of the other.

The constituents of an expression are always partially ordered by inclusion. The Unique Parsing Principle says that if two constituents have a common lower bound in this partial ordering, then they are comparable; in other words the partial ordering is a tree branching downwards. This tree gives the unique complete parsing.

One consequence is that the language L has a canonical context-free grammar. Its rules for compound expressions e have at the left the category of e and on the right the categories of the immediate constitutents of e, i.e. the expressions immediately below e in the tree.

Tarski assigns all formulas to the same semantic category. In our terminology, the reason is that he takes his target expressions to be formulas — or in his terminology, sentential functions — rather than sentences. (A sentence is a formula with no free variables.) But as mentioned earlier, we want to take sentences as the target expressions in order to study how far the assignment of truth values to sentences determines an assignment of meanings to other expressions. If two formulas ϕ and ψ have different free variables, then replacing one by the other in a suitable sentence yields a formula that is not a sentence. So for us there are infinitely many different categories of formula, one for each finite set of free variables. In what follows I use this more complicated context-free grammar, which is a refinement of the simpler one that Tarski's text implies.

The second condition needs some preliminary definitions. We say that a context-free grammar rule is *branching* if the list of categories on the righthand side of its arrow has length at least two. We say that the language L has a *head system* if the categories of L are partitioned into two sets, *heads* and *non-heads*, so that

1. In each branching rule, exactly one of the categories on the righthand side is a head category;
2. if a head category occurs on the righthand side of a rule, then that rule is branching.

We shall say that an expression is a *head expression* if its category is a head category. Our construction of the context-free grammar from the categories implies clause 2, so clause 1 is the significant one.

If H_1 is a head category which occurs on the righthand side of a context-free rule which has a head category H_2 on the lefthand side, we record this by writing $H_1 \succ H_2$. We say that the head system is *well-founded* if there is no infinite descending sequence

$$H_0 \succ H_1 \succ H_2 \ldots,$$

and *flat* if the relation $\succ$ is empty. The head system is flat if and only if no head expression is compound. If the head system is flat then clearly it is well-founded.

Tarski's construction of L ensures that

> [Head Principle] The language L has a head system in which no head expression is compound.

The head expressions are symbols for what Tarski calls "the *fundamental operations on expressions* by means of which composite expressions are formed from simpler ones" (his italics). Strictly I am assuming here (as Tarski seems to) that quantifiers are indecomposable grammatical units; if we could separate off the quantifier symbol from the variable, the head system would still be well-founded though it wouldn't be flat.

The existence of a well-founded head system has a dramatic impact on fregean covers. Suppose ν is a semantics defined on all the expressions of the language L, and L has a head system. For any category C we write $\nu[C]$ for the set of all $\nu(e)$ as e ranges through C. We shall say that ν *respects* the head system if the following holds: For each head category H and each expression

h of category H, $\nu(h)$ is the following set of ordered pairs $\langle R, F\rangle$: the items R are the context-free rules with H on the right, and for each such rule R, say

$$C \Longrightarrow H\ D_1\ \ldots\ D_n,$$

F is a function with domain $\nu[D_1] \times \cdots \times \nu[D_n]$ and values satisfying

$$F(\nu(d_1), \ldots, \nu(d_n)) = \nu(c)$$

whenever the d_i are any expressions of category D_i and c is the expression $hd_1 \ldots d_n$. (By Proposition 2 applied to the expression c, there is a unique such function F.)

In Tarski's case, since formulas with different free variables are in different categories, the head expression $\wedge$ (conjunction) stands on the right of infinitely many different context-free rules. This illustrates why we had to take $\nu(h)$ as a set of pairs. But the next proposition is not limited to Tarski's languages.

PROPOSITION 6. *Suppose the language L has a well-founded head system and μ is a meaning function defined on the target expressions of L. Then μ has a fregean cover which respects the head system.*

For the proof, we begin with the case where the head system is flat. First construct a fregean cover ν such that no value $\nu(e)$ is an ordered pair. Then for each head expression h replace $\nu(h)$ by the set of pairs $\langle R, F\rangle$ defined above. (Using Proposition 2 again, this set depends only on $\nu(h)$.) Since the expressions $d_1, \ldots, d_n$ and c in the definition of $\langle R, F\rangle$ are non-head expressions, their ν-values remain as they were before, so the replacement never interferes with itself. Call the resulting semantics ν'.

If the head system is not flat, then this construction needs a repair since c above may be a head expression. The repair is to define the functions $\nu'(h)$ by induction on the field of the relation $\succ$, starting with those h whose category is not $\succ$ any category. At later stages in the induction, if $\text{cat}(h) \succ \text{cat}(c)$ in the definition of F above, then by induction hypothesis we can assume $\nu'(c)$ has already been defined, and we use

$$F(\nu(d_1), \ldots, \nu(d_n)) = \nu'(c)$$

for the definition of F.

We have to show that ν and ν' make exactly the same distinctions. All distinctions involving non-head expressions remain as they were before. If g, h are of different categories, then $\nu'(g) \neq \nu'(h)$, but also $g \not\sim h$ and so $\nu(g) \neq \nu(h)$ since ν is husserlian. Finally suppose g and h are in the same head category H, and suppose for simplicity that H has the one context-free rule

$$C \Longrightarrow H\ D_1\ \ldots\ D_n.$$

Then by two applications of Proposition 2, $\nu(g) = \nu(h)$ if and only if for all expressions d_i of category D_i $(i \geq 1)$, $\nu(gd_1 \ldots d_n) = \nu(hd_1 \ldots d_n)$. (From right to left, use the definition of the fregean cover and the fact that g and h are occur only in expressions of this form.) But this is equivalent to $\nu'(g) = \nu'(h)$.

After defining his object languages, Tarski turns to semantic categories. One of his guiding ideas is that if two expressions are of the same category, then their contents should have a similar category structure. The formulation I have just given is only slightly less precise than Tarski's own ([20] p. 219). But for example, if e and f are of the same category and the content of e is a set of ordered pairs of contents of expressions of category C, then the content of f should also be a set of ordered pairs of contents of expressions of category C.

This is a sound heuristic principle. But it easily suggests the following reasoning, which is fallacious. Suppose we consider a typical fully interpreted language L (for example let L be a logical language interpreted in a structure). Then the meaning of an expression $R(v_0, \dots, v_{n-1})$, where R is an n-ary relation symbol, is a set of ordered n-tuples of individuals. A formula $\phi(v_0, \dots, v_{n-1})$, whose free variables are exactly those shown, has the same category as the formula $R(v_0, \dots, v_{n-1})$, hence its meaning should likewise be a set of ordered n-tuples of individuals. The reasoning is fallacious because logics with quantifiers of imperfect information (e.g., Hodges [11]) are counterexamples. Inspecting these cases, we see that adding the quantifiers of imperfect information bumps up the set-theoretic type of the meanings of formulas $\phi(v_0, \dots, v_{n-1})$ in general. So if we want to follow Tarski's heuristic, we have to encode the meaning of $R(v_0, \dots, v_{n-1})$ as an object of higher type too.

Tarski states a "first principle of the theory of semantical categories" that seems to translate to our setting as follows:

> [Tarski's First Principle, [20] p. 216] If there is an expression $s(e)$ which is not equal to e, and $s(f)$ is also an expression, with $s(e) \sim s(f)$, then $e \sim f$.

Unfortunately in [11] I stated this without the clause "with $s(e) \sim s(f)$"; I thank Urszula Wybraniec-Skardowska and Marcus Kracht for picking this up. My comments in [11], in particular the counterexamples in English, all refer to the corrected version above. It is not clear to me why Tarski considers the First Principle natural "from the standpoint of the ordinary usage of language" ([20] p. 216), or why he needs it for defining the concept of a sentential function ([20] p. 217).

Tarski also states a "general law", [20] p. 217. The appropriate translation of his general law into our setting is not entirely clear. But it seems to imply that each head expression lies on the righthand side of exactly one context-free rule (at least for those rules which have a category of formulas on the left).

Tarski also discusses several problems that arise if one tries to weaken the category regime. It certainly does look as if one should be able to relax his category restrictions a good deal, but I have nothing systematic to say about this.

§4. Frege's contribution. What should we credit to Frege in all this?

Frege's main writings on the foundations of mathematics were complete before Husserl's Fourth Investigation appeared. So Frege has none of Husserl's category analysis. Nevertheless some of Frege's most telling points get a sharper edge through being put in category terms.

In the first place, it was Frege in his *On Sense and Reference* [6] who pointed out the difficulties that it causes for the semantics of a language if we introduce "believe" and other words that express states of mind. One consequence that Frege himself could not have drawn is the following. If ϕ and ψ are sentences with the same meaning, then Proposition 2 implies that the two sentences

Walter Mitty believes that ϕ.
Walter Mitty believes that ψ.

have the same content in any fregean cover. So Frege's Dictum is incompatible with possible world semantics on sentences (and more generally with any semantics that gives the same value to all necessary truths), since there are people who believe some necessary truths but not others. At the meeting in Munich there was some discussion of the fact that Frege's mature notion of Sinn can't be cashed in terms of truth in possible worlds. I'm happy to report, on the basis of the observation above, that the same is already true of Frege's notion of Sinn in the *Grundlagen*.

One way of looking at Frege's examples in *On Sense and Reference* is to observe that two expressions which have the same meaning in a satisfactory semantics for L may have to be given distinct meanings in a satisfactory semantics for a language L' extending L — specifically if L' adds to L the verb "believe" and tenses on verbs. The same seems to be true for categories, and the quotation from Harris above hints at this: two expressions that have the same category in L may have different categories in L'. For a husserlian semantics, this forces the two expressions to have different meanings in L' too. For example a language might contain the two synonymous verbs "gave" and "donated". Add a few standard words to the language and we have the pair

We gave the school some books.
⋆ We donated the school some books.

In the expanded language, a husserlian semantics must give different meanings to "gave" and "donated". This does not seem to be a case that allows easy remedies like those in Remark 2 above. The moral is that fregean covers can be very unstable under extension of the language.

Second, Frege in his book *Begriffsschrift* [7] (1879) introduced a language, also called *Begriffsschrift*, which has a context-free grammar, unique parsing and a flat head system (where again we count quantifiers as unanalysable). This statement does not quite make sense, because the formulas of *Begriffsschrift* are trees rather than strings. Frege's problem was that he invented too much at once, and by inventing *Begriffsschrift* he also invented phrase markers. Let me develop this a little.

In Germany in the nineteenth century, language teachers sometimes used tree diagrams to explain the parsing of a sentence. These diagrams were not exactly precision instruments, as one can see from the following example of Billroth [3] (1832, quoted by Baum [2] p. 38):

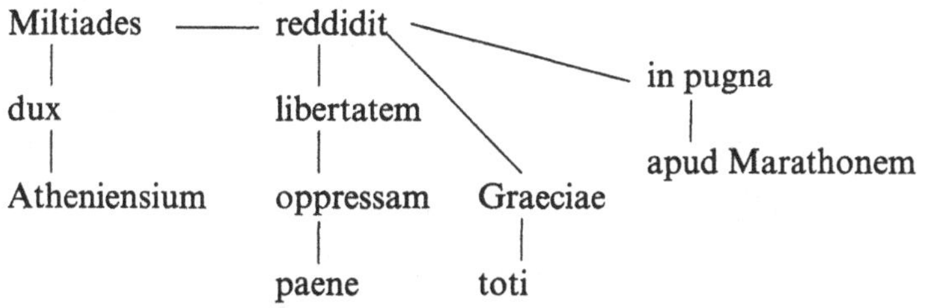

And here we recall that Frege's father Karl Alexander Frege was a language teacher — though there is no evidence that he used diagrams of this type.

In the language *Begriffsschrift*, Frege's head expressions are material implication, negation, relation symbols (including identity) and universal quantifier. In general he forms compound expressions by writing their head at the top and then lines to the right or downwards pointing to the non-head constituents. (Relation symbols are the exception; his syntax for these is the conventional one.) This would cause problems if head expressions were compound, but in *Begriffsschrift* they never are.

The more familiar Chomsky-style phrase markers are essentially the trees of inclusion of constituents. They have no distinguished head symbols, and the symbol at a nonterminal node is a category symbol rather than an expression of the language. Frege's phrase markers, with the head expressions at nonterminal nodes, are closer to Lucien Tesnière's "stemmata" ([21] p. 14f). Also Frege's phrase markers branch out from top left rather than top centre, but this is a triviality.

Frege does not have a name for "head expression", and it is only in the *Grundgesetze* [5] of 1893 that he brings his various head expressions together under one roof. There he explains that each head expression expresses a function; he has to develop the notion of a function in order to explain (§21) how this is true for the universal quantifier.

In a more usual syntax, Frege would have written his head symbols to the left or the right of their argument expressions, or maybe between them. We can imagine them so written, and then Frege's language becomes one that we can apply the notions of this paper to. Frege holds that each compound expression f is got by combining one of his head symbols, say h, with a string $e_1 \dots e_n$ of non-head expressions. Here h names a function, and f names the result of applying this function to whatever the expressions $e_1, \ldots, e_n$ are names for (*Grundgesetze* §1).

Of course this anticipates Proposition 6 above. It seems to be a highly original view. Looking back through the history of western linguistics, the only comparable earlier theory of syntactic composition seems to be that of the thirteenth and fourteenth century Modists. According to them, each compound expression is a binary combination of a "dependens" expression (the head?) and a "terminans" expression. For example Martin of Dacia parses "homo albus currit bene" along the lines

currit(bene) (albus(homo))

where $A(B)$ means A is the dependens and B the terminans. (Maierù [16] p. 298.) Martin's choice of heads does not match the intuitions of a modern linguist. As Pinborg comments [19] p. 259: "Unfortunately, the notion of dependency seems to have been intuitional for the Modistae: no purely linguistic rules are formulated in modistic grammar which help in deciding which of two semantic features is dependent on the other."

I said "western linguistics" advisedly. On the surface at least, some of the remarks of the eleventh century Arab grammarian Abd ul-Qahir al-Jurjani seem remarkably close to Frege. For example "You derive a single meaning from the sentence, not a series of meanings based on the individual words". Earlier in the same work, Jurjani has explained that words contribute to a sentence through their argument structure within the sentence. (Owens [18] pp. 35, 249.)

Finally, what did Frege himself mean by his Dictum? Clearly not everything that we have been deducing from it.

We have to look at the context in which Frege said it. The *Grundlagen* [4] is about how to define "number". The Dictum comes at the end of a section of the *Grundlagen* in which Frege establishes that a definition of "number" needs to explain the meaning of the whole expression

The number of Xs is Y.

So, in the notation of *Begriffsschrift*, what is needed is a definition of an expression of the form $N(X, Y)$ where X and Y are variables of appropriate type. The meaning of "number" is what is common to the meanings of all the sentences of the form $N(a, b)$. In our terminology, "number" is an expression of head category which combines with other expressions to form a sentence. It seems to me not unreasonable to sum up this achievement by saying that it is enough to look for the meanings of sentences of the form $N(a, b)$, because this will yield the meaning of "number".

Granted, this is not exactly what the Dictum says. In particular the Dictum seems to be about all the constituents of any sentence, not just words of head category. In the particular place where Frege utters the Dictum, he is defending his thesis against the possible objection that it supplies no mental picture of a number. His point is that you can define any word with a suitable sentence containing the word; you don't have to give a picture. In the *Grundgesetze* §33 Frege takes this point much deeper and separates it completely from psychological irrelevances. What he does — and I believe he is the first to do it — is to describe canonical forms for the sentences used to define symbols of various categories.

§5. **Appendix.** The referee kindly suggested that I should say something about the relationship of this paper to "the relevant part of formal language theory" and in particular to the important work of Haim Gaifman [1], [8]. I will be brief because the connections that I know of are all rather superficial. But the referee's request is very reasonable.

The work of Gaifman is about three types of grammar. The first is phrase-structure systems, which are what I have been calling context-free grammars. Gaifman imposes (and uses) the usual requirement that a grammar of this type has only finitely many rules; in order to deal with first-order logic I had to drop this requirement.

In [8] Gaifman proves some equivalences between phrase-structure systems and dependency systems. These latter are grammars that define trees which are essentially Tesnière's stemmata [21]. A dependency system picks out within each expression of its language a unique head word whose category forms a head category for the expression. The discussion in Section 4 above shows that Frege's *Begriffsschrift* is (except for the atomic formulas) a dependency system in which the formulas are the trees. My heads in Section 3 above are more general, because they need not be single words.

Gaifman shows in [8] that for any context-free grammar satisfying a certain condition, there is a "corresponding" dependency system (which defines the same language with almost the same constituent structures). I have no corresponding results. In Section 3 I show that any context-free grammar with a certain extra feature (a well-founded head system) allows a certain kind of semantics. My argument has to consider the case where heads are compound, a case that never arises for Gaifman. Gaifman's main effort goes into constructing a dependency system whose heads are single words, so that in general it will not correspond to the head system which I assume.

The paper [1] of Bar-Hillel, Gaifman and Shamir reports the result, originally due to Gaifman, that for every context-free grammar there is a categorial grammar which defines the same expressions (though not necessarily with the same constituent structures). Categorial grammars are a formalism proposed by Kazimierz Ajdukiewicz as an exposition of thoughts of Stanisław Leśniewski on the relationship between Husserl's semantic categories and Russell's simple theory of types. The paper [1] shows that for each context-free grammar we can construct a corresponding categorial grammar whose rules all have the form

$$\Phi \Longrightarrow [\Phi/\Psi]\,\Psi$$

and satisfy certain further conditions. Here $[\Phi/\Psi]$ is in my terminology the head category of the rule. The further conditions allow Ψ to be a head category of another rule, and they forbid Φ to be a head category. So again the kind of head system that the paper [1] constructs is not at all the same as the kind of head system that I presuppose in Section 3. I should add that Bar-Hillel et al. allow the same word to belong to more than one category, so that their notion of category is not mine either.

In short I doubt if there is any mileage in comparing these two results of Gaifman with my observations in Section 3.

REFERENCES

[1] Yehoshua Bar-Hillel, *On categorial and phrase structure grammars*, ***Language and information***, Addison-Wesley, Reading Mass., 1964, pp. 99–115.

[2] Richard Baum, ***Dependenzgrammatik***, Max Niemeyer Verlag, Tübingen, 1976.

[3] Johann Gustav Friedrich Billroth, ***Lateinische Schulgrammatik für die obern Klassen gelehrter Schulen***, Leipzig, 1832.

[4] Gottlob Frege, ***Die Grundlagen der Arithmetik***, Koebner, Breslau, 1884.

[5] ———, ***Grundgesetze der Arithmetik I***, Hermann Pohle, Jena, 1893.

[6] ———, *Über Sinn und Bedeutung*, ***Gottlob Frege, Kleine Schriften*** (I. Angelelli, editor), Georg Olms, Hildesheim, 1967, pp. 143–162.

[7] ———, ***Begriffsschrift***, Georg Olms Verlag, Hildesheim, 1998, (original 1879).

[8] Haim Gaifman, *Dependency systems and phrase-structure systems*, ***Information and Control***, vol. 8 (1965), pp. 304–337.

[9] Zellig S. Harris, *Co-occurrence and transformation in linguistic structure*, ***Language***, vol. 33 (1957), pp. 283–340.

[10] Jaakko Hintikka and Gabriel Sandu, *Game-theoretical semantics*, ***Handbook of logic and language*** (Johan van Benthem and Alice ter Meulen, editors), Elsevier, Amsterdam, 1996, pp. 361–410.

[11] Wilfrid Hodges, *Compositional semantics for a language of imperfect information*, ***Logic Journal of the IGPL***, vol. 5 (1997), pp. 539–563.

[12] ———, *Some strange quantifiers*, ***Structures in logic and computer science*** (J. Mycielski et al., editors), Lecture Notes in Computer Science 1261, Springer-Verlag, Berlin, 1997, pp. 51–65.

[13] ———, *Formal features of compositionality*, ***Journal of Logic, Language and Information***, vol. 10 (2001), pp. 7–28.

[14] Edmund Husserl, ***Logische Untersuchungen II/1***, Max Niemeyer Verlag, Tübingen, 1993, (originally published 1900); trans. by J. N. Findlay as *Logical Investigations Vol. II*, Humanity Books, Amherst, New York, 2000.

[15] Shalom Lappin and W. Zadrozny, *Compositionality, synonymy, and the systematic representation of meaning*, Preprint, 2000.

[16] Alfonso Maierù, *The grammar of the* Modistae, ***History of linguistics: II, classical and medieval linguistics*** (Giulio Lepschy, editor), Longman, London, 1994, pp. 288–302.

[17] Walter J. Ong, ***Ramus: Method, and the decay of dialogue***, Harvard University Press, Cambridge, Mass., 1958.

[18] Jonathan Owens, ***The foundations of grammar: an introduction to medieval Arabic grammatical theory***, John Benjamins, Amsterdam, 1988.

[19] Jan Pinborg, *Speculative grammar*, ***The Cambridge history of later medieval philosophy*** (Norman Kretzmann et al., editors), Cambridge University Press, Cambridge, 1982, pp. 254–269.

[20] Alfred Tarski, *The concept of truth in formalized languages*, ***Logic, semantics, metamathematics*** (J. Corcoran, editor), Hackett Publishing Co., Indianapolis, 1983, pp. 152–278.

[21] Lucien Tesnière, ***Éléments de syntaxe structurale***, Klincksieck, Paris, 1959.

[22] Dag Westerståhl, *On the compositional extension problem*, Journal of Philosophical Logic (to appear).

[23] W. Zadrozny, *From compositional to systematic semantics*, ***Linguistics and Philosophy***, vol. 17 (1994), pp. 329–342.

SCHOOL OF MATHEMATICAL SCIENCES
QUEEN MARY, UNIVERSITY OF LONDON
MILE END ROAD
LONDON E1 4NS, UK
E-mail: w.hodges@qmul.ac.uk

THE SEMANTICS OF MODAL PREDICATE LOGIC II. MODAL INDIVIDUALS REVISITED

MARCUS KRACHT AND OLIVER KUTZ

Abstract. We continue the investigations begun in [10]. We shall define a semantics that is built on a new kind of frames, called *coherence frames*. In these frames, objects are transcendental (world-independent), as in the standard constant-domain semantics. This will remove the asymmetry between constants and variables of the counterpart semantics of [10]. We demonstrate the completeness of (general) coherence frames with respect to first- and certain weak second-order logics and we shall compare this notion of a frame to counterpart frames as introduced in [10] and the metaframe semantics of [13].

§1. Introduction. In [10] we have developed a semantics that is complete with respect to first- and weak second-order modal predicate logics. This semantics was in addition quite elementary, which was already a great step forward from the previous semantics by Ghilardi [6] and by Skvortsov and Shehtman [13]. Still, from a philosophical point of view this semantics left much to be desired. The introduction of counterpart relations — although in line with at least some philosophical ideas, notably by Lewis — is not always very satisfactory since it makes the notion of an object a derived one. The things we see become strictly world bound: there is no sense in which we can talk of, say, the town hall of Berlin, rather than the town hall of Berlin in a particular world, at a particular point of time. The traditional semantics for modal predicate logic held the complete opposite view. There, objects are transcendental entities. They are not world bound, since they do not belong to the worlds. The difference between these views becomes clear when we look at the way in which the formula $\Diamond\varphi(\vec{x})$ is evaluated. In the standard semantics, we simply go to some accessible world and see whether $\varphi(\vec{x})$ holds. In counterpart semantics, we not only have to choose another world but also some counterparts for the things that we have chosen as values for the variables in this world. In the traditional semantics the question of counterparts does not arise because of the transcendental status of objects. We may view this as a limiting case of counterpart semantics, in which the counterpart relation always is the identity.

Intensionality
Edited by Reinhard Kahle
Lecture Notes in Logic, 22

Note that the addition of *constant symbols* to the language introduces further complications. In counterpart semantics, it is far from straightforward to interpret constant symbols, because we need to give an interpretation of these symbols across possible worlds that respects the counterpart relations in some appropriate sense. Variables on the other hand simply denote "objects" in the domain of a given world. In the case of traditional semantics this asymmetry appears in a similar fashion if one allows constant symbols to be non-rigid, as has been done e.g., in [5]. Then, variables denote transcendental entities, whereas constants denote something like *individual concepts*, i.e. functions from possible worlds to a domain. Facing this dilemma, one solution is to completely move to a higher-order setting, where constants and variables can be of various higher types, e.g., type-0 constants denote objects, type-1 constants individual concepts etc. (cf. [3]). In this paper, we will follow a different approach, treating constants and variables in the same way, but assuming a more sophisticated notion of a modal individual and identity-at-a-world.

It remains unsatisfactory having to choose between these competing semantics. Moreover, it would be nice if the difference between these semantics was better understood. Certainly, much research has been done into standard semantics and it is known to be highly incomplete if one aims for frame-completeness results.

However, it is known that completeness with respect to models is as easy to show as in predicate logic but that if the language contains equality, different semantics have to be chosen for different theories/logics of identity (cf. e.g., [7]). The present paper developed from the insight that if the proper semantics is introduced, modal predicate logics with different logics of identity can be treated within the same semantical framework. We call this semantics *coherence semantics*. Completeness with respect to models is then uniformly shown for all modal predicate logics that are extensions of free quantified K together with the predicate logical axioms of equality. We continue by investigating the relationships between coherence frames, counterpart frames and metaframes, discuss the treatment of identity in each of the semantics as well as the interpretation of constant symbols and finally derive a completeness result for so-called cubic generalized metaframes.

§2. Preliminaries. The language has the following symbols. Following Scott [12] we shall work with nonobjectual (possibilist) quantifiers plus an existence predicate. This allows to eliminate the objectual (actualist) quantifiers (they are now definable), and straightens the theory considerably. The existence predicate is a unary predicate whose interpretation — unlike the identity symbol — is completely standard, i.e. does not have to meet extra conditions. Hence it can actually be suppressed in the notation, making proofs even more simple.

DEFINITION 1 (Symbols and languages). The languages of modal predicate logic contain the following symbols.

1. A denumerable set $V := \{x_i : i \in \omega\}$ of *object variables.*
2. A denumerable set $C := \{c_i : i \in \omega\}$ of *constants.*
3. A set Π of *predicate constants* containing the unary *existence predicate* E.
4. The *boolean functors* $\bot$, $\wedge$, $\neg$.
5. The possibilist *quantifiers* $\bigvee$, $\bigwedge$.
6. A set $M := \{\Box_\lambda : \lambda < \kappa\}$ of *modal operators.*

Furthermore, each symbol from Π has an *arity*, denoted by $\Omega(P)$. In particular, $\Omega(E) = 1$.

The variables are called x_i, $i \in \omega$. We therefore use x (without subscript!), y, y_j or z, z_k, as metavariables. We assume throughout that we have no function symbols of arity greater than 0. However, this is only a technical simplification. Notice that in [10] we even had no constants. This was so because the treatment of constants in the counterpart semantics is a very delicate affair, which we will discuss below. Moreover, for simplicity we assume that there is only one modal operator, denoted by $\Box$ rather than $\Box_0$. Nothing depends on this choice. The standard quantifiers $\forall$ and $\exists$ are treated as abbreviations.

$$(\forall y)\varphi := \bigwedge y.E(y) \to \varphi,$$
$$(\exists y)\varphi := \bigvee y.E(y) \wedge \varphi.$$

Moreover, $\Diamond\varphi$ abbreviates $\neg\Box\neg\varphi$. The sets of *formulae* and *terms* in this language are built in the usual way. Unless otherwise stated, equality ($\doteq$) is *not* a symbol of the language.

DEFINITION 2 (First-order MPLs). A *first-order modal predicate logic* is a set L of formulae satisfying the following conditions.

1. L contains all instances of axioms of first-order logic.
2. L is closed under all rules of first-order logic.
3. L contains all instances of axioms of the modal logic K.
4. L is closed under the rule $\varphi/\Box\varphi$.
5. $\Diamond\bigvee y.\varphi \leftrightarrow \bigvee y.\Diamond\varphi \in L$.

Notice that the last of the postulates ensures that in a Hilbert-style proof all instances of the rule (MN) can be assumed to be at the beginning of the proof. ((MN) "commutes" with (MP), as is easily seen. However, it commutes with (UG) only in presence of the postulate (5).) To eliminate some uncertainties we shall note that the notions of free and bound occurrences of a variable are exactly the same as in ordinary first-order logic. A variable x occurs *bound* if this occurrence is in the scope of a quantifier $\bigwedge x$ or $\bigvee x$. We denote the simultaneous replacement of the terms s_i for x_i $(i < n)$ in χ by $[s_0/x_0, \ldots, s_{n-1}/x_{n-1}]\chi$. Or, writing $\vec{s} = \langle s_i : i < n\rangle$ and $\vec{x} = \langle x_i : i < n\rangle$, we abbreviate this further to $[\vec{s}/\vec{x}]\chi$.

If the language contains equality, the following is required of L.

Eq1. $\bigwedge x.x \doteq x \in L$.
Eq2. $\bigwedge x. \bigwedge y.x \doteq y \to y \doteq x \in L$.
Eq3. $\bigwedge x. \bigwedge y. \bigwedge z.x \doteq y \wedge y \doteq z \to x \doteq z \in L$.
Eq4. $\bigwedge y_0. \bigwedge y_1. \cdots \bigwedge y_n.y_i \doteq y_n \to \{P(y_0, \ldots, y_{n-1}) \leftrightarrow [y_n/y_i]P(y_0, \ldots, y_{n-1})\} \in L$ if $P \in \Pi$, $n = \Omega(P)$.

The axioms Eq1–Eq3 ensure that equality is interpreted by an equivalence relation. Note that the axiom Eq4 is weaker than the usual Leibniz' Law, because it only allows for the substitutability of identicals in *atomic* predicates. We start with a very basic semantics, standard constant-domain semantics. Recall that we assume no equality.

DEFINITION 3 (Frames, structures and models). A triple $\mathfrak{W} = \langle W, \lhd, U\rangle$ is called a *predicate Kripke-frame*, if W is a set (the set of *worlds*), $\lhd \subseteq W \times W$ a binary relation on W (the *accessibility relation*), and U a set (the *universe*). A *modal first-order structure* is a pair $\langle \mathfrak{W}, \mathfrak{I}\rangle$, where $\mathfrak{W}$ is a predicate Kripke-frame and $\mathfrak{I}$ a function mapping a predicate P to a function assigning to each world w an $\Omega(P)$-relation on U and a constant symbol c to a member of U. $\mathfrak{I}$ is called an *interpretation*. Further, $\mathfrak{I}_w$ is the relativized interpretation function at w, which assigns to each $P \in \Pi$ the value $\mathfrak{I}(P)(w)$ and to each constant symbol c the value $\mathfrak{I}(c)$. A *valuation* is a function $\beta : V \to U$. A *model* is a triple $\langle \mathfrak{F}, \beta, w\rangle$ such that $\mathfrak{F}$ is a modal first-order structure, β a valuation into it and w a world of $\mathfrak{F}$.

As usual, $\gamma \sim_x \beta$ means that $\gamma(y) = \beta(y)$ for all $y \in V$ different from x. If P is a predicate symbol and $\langle t_0, \ldots, t_{\Omega(P)-1}\rangle$ an $\Omega(P)$-tuple of terms, let ε be the function that assigns the tuple $\langle \varepsilon_0(t_0), \ldots, \varepsilon_{\Omega(P)-1}(t_{\Omega(P)-1})\rangle$, where $\varepsilon_i = \beta$ if $t_i \in V$ and $\varepsilon = \mathfrak{I}$ if $t_i \in C$.

DEFINITION 4 (Truth in a model). Given some modal first-order structure $\mathfrak{F} = \langle \mathfrak{W}, \mathfrak{I}\rangle$, a model $\langle \mathfrak{F}, \beta, w\rangle$, and a formula φ, we define $\langle \mathfrak{F}, \beta, w\rangle \vDash \varphi$ as follows.

$$\langle \mathfrak{F}, \beta, w\rangle \vDash P(\vec{t}) :\Longleftrightarrow \varepsilon(\vec{t}) \in \mathfrak{I}_w(P),$$
$$\langle \mathfrak{F}, \beta, w\rangle \vDash \varphi \wedge \chi :\Longleftrightarrow \langle \mathfrak{F}, \beta, w\rangle \vDash \chi; \varphi,$$
$$\langle \mathfrak{F}, \beta, w\rangle \vDash \neg\varphi :\Longleftrightarrow \langle \mathfrak{F}, \beta, w\rangle \nvDash \varphi,$$
$$\langle \mathfrak{F}, \beta, w\rangle \vDash \bigvee x.\varphi :\Longleftrightarrow \text{for some } \gamma \text{ with } \gamma \sim_x \beta : \langle \mathfrak{F}, \gamma, w\rangle \vDash \varphi,$$
$$\langle \mathfrak{F}, \beta, w\rangle \vDash \Diamond\varphi :\Longleftrightarrow \text{exists } w' \text{ such that } w \lhd w' \text{ and } \langle \mathfrak{F}, \beta, w'\rangle \vDash \varphi.$$

§3. Completeness. In this section we sketch a proof of the well-known result that every MPL as defined above can be characterized by a canonical structure. The proof presented below is a variation of proofs that can be found in the literature, see e.g., [7], and is based on the use of special, maximal-consistent sets of formulae, namely *Henkin-complete maximal consistent sets* in an extended language. This kind of completeness proof in modal predicate logic goes back to [14]. We present only the basic steps here. But note that the proof depends on the presence of the Barcan formulae for the possibilist quantifiers.

DEFINITION 5. A set T of formulae of some language $\mathfrak{L}$ of MPL is called *Henkin-complete*, if for all $\bigvee y.\chi$ in $\mathfrak{L}$ there exists a constant c such that $\bigvee y.\chi \leftrightarrow [c/y]\chi \in T$. Let $\mathfrak{L}^*$ result from a language $\mathfrak{L}$ by adding infinitely many new constant symbols; call $\mathfrak{L}^*$ a *Henkin-language* for $\mathfrak{L}$.

In what is to follow we will assume that given a language $\mathfrak{L}$, a Henkin-language $\mathfrak{L}^*$ is fixed once and for all.

DEFINITION 6. Let L be an MPL in $\mathfrak{L}$. A *Henkin-world* is a maximal L-consistent, Henkin-complete set of formulae in the language $\mathfrak{L}^*$.

LEMMA 7. *Every L-consistent set of formulae in language $\mathfrak{L}$ is contained in some Henkin-world.*

PROOF. Let Δ be L-consistent in language $\mathfrak{L}$ and let $\psi_1, \psi_2, \ldots$ be an enumeration of the formulae of type $\bigvee x.\varphi_i(x)$ in the language $\mathfrak{L}^*$. For ψ_i define $\delta_i := \bigvee y.\varphi_i \leftrightarrow [c_i/y]\varphi_i$, where c_i is a new constant not appearing in any δ_j, $j < i$. This is possible because we have an infinite supply of new variables. Define $\Delta^* := \bigcup_{i\in\omega} \Delta_i$, where $\Delta_i := \Delta \cup \{\delta_j : j < i\}$. Then $\Delta_0 = \Delta$. Δ^* is clearly Henkin-complete. By compactness, it is also L-consistent if all Δ_k are. Now, suppose there is a k such that Δ_k is inconsistent. Choose k minimal with this property. By assumption, $k > 0$. There is a finite set $\Delta' \subseteq \Delta$ such that $L \vdash \bigwedge_{\varphi\in\Delta'} \varphi \wedge \bigwedge_{i<k} \delta_i \to \neg\delta_k$. But then $L \vdash \bigwedge_{\varphi\in\Delta'} \varphi \wedge \bigwedge_{i<k} \delta_i \to (\bigvee y.\varphi_k \wedge \neg[c_k/y]\varphi_k)$ where the constant c_k does not appear in any δ_i, $i < k$. Hence, by first-order logic, we have $L \vdash \bigwedge_{\varphi\in\Delta'} \varphi \wedge \bigwedge_{i<k} \delta_i \to (\bigvee y.\varphi_k \wedge \bigwedge y.\neg\varphi_k)$, whence Δ_{k-1} is L-inconsistent, contrary to the choice of k.

Next, by a standard argument, we can turn Δ^* into a maximal L-consistent set in language $\mathfrak{L}^*$ without loosing the property of Henkin-completeness. ⊣

Notice that this method of Henkin-closure does not work for the counterpart semantics of [10]. The reason is the asymmetry between variables and constants. Instead, a slightly different definition was used, where instead of constants variables were used as witnesses.

Let C_{L^*} be the set of constant terms of $\mathcal{L}^*$. Now define W_{L^*} to be the set of all Henkin-worlds. If $\Delta \in W_{L^*}$, the following interpretation is defined.

$$\mathfrak{I}_\Delta(P) := \{\langle c_i : i < \Omega(P)\rangle : P(\vec{c}) \in \Delta\} \text{ and } \mathfrak{I}(c) = c.$$

This defines a first-order model on the world Δ. Finally, we put $\Delta \lhd \Sigma$ if for all $\Box\delta \in \Delta$ we have $\delta \in \Sigma$. $\mathfrak{I}_{L^*}$ is defined by piecing the $\mathfrak{I}_\Delta$ together; it assigns to each world Δ the function $\mathfrak{I}_\Delta$. Then we put

$$\mathfrak{Can}_{L^*} := \langle\langle W_{L^*}, \lhd, C_{L^*}\rangle, \mathfrak{I}_{L^*}\rangle.$$

This is a modal first-order structure, called the *canonical structure* for L. The following is immediate from the definitions.

LEMMA 8. *Let Δ be a Henkin-world. Then if $\bigvee y.\chi \in \Delta$, there is a constant d such that $[d/y]\chi \in \Delta$.*

Before we can prove the main result of this section, we need one more lemma whose proof does indeed depend on the presence of the Barcan formulae for the possibilist quantifiers. To state the lemma, let $\Delta^\Box := \{\varphi : \Box\varphi \in \Delta\}$.

LEMMA 9. *For every Henkin-world Δ with $\Diamond\chi \in \Delta$, there is a Henkin-world Γ such that $\Delta^\Box \cup \{\chi\} \subset \Gamma$.*

PROOF. By a standard argument from propositional modal logic, the set $\Delta^* := \Delta^\Box \cup \{\chi\}$ is L-consistent. We have to show that Δ^* can be extended to a Henkin-world. Note first that this set already contains all the constants from the Henkin-language $\mathcal{L}^*$. Let $\psi_1, \psi_2, \ldots$ be an enumeration of the formulae of type $\bigvee x.\varphi_i(x)$ in the language $\mathcal{L}^*$. For ψ_i define $\delta_i := \bigvee y.\varphi_i \leftrightarrow [c_i/y]\varphi_i$, where c_i is the first new constant such that $\Delta_i := \Delta^* \cup \{\delta_j : j < i\}$ is L-consistent. If such a constant always exists we can define Γ as the completion of $\bigcup_{i\in\omega} \Delta_i$ which is a Henkin-world. So suppose that Δ_i is L-consistent but there is no constant c_{i+1} such that Δ_{i+1} is. Then there are, for every constant c of $\mathcal{L}^*$, formulae $\alpha_0, \ldots, \alpha_{n-1} \in \Delta^\Box$ such that

$$L \vdash \bigwedge_{k\le n} \alpha_k \to \left(\left(\chi \wedge \bigwedge_{j<i} \delta_i\right) \to \left(\bigvee y.\varphi_{i+1} \wedge \neg[c/y]\varphi_{i+1}\right)\right).$$

Since $\Box\alpha_i \in \Delta$ for all i, it follows that

$$\Box\left(\left(\chi \wedge \bigwedge_{j<i} \delta_i\right) \to \left(\bigvee y.\varphi_{i+1} \wedge \neg[c/y]\varphi_i\right)\right) \in \Delta$$

for every constant c. Since Δ is Henkin-complete we can "quantify away" the constant with a variable not appearing in the formula (by using the appropriate Henkin-axiom) and apply the Barcan formula and thus obtain:

$$\Box\bigwedge z.\left(\left(\chi \wedge \bigwedge_{j<i} \delta_i\right) \to \left(\bigvee y.\varphi_i \wedge \neg[z/y]\varphi_i\right)\right) \in \Delta.$$

Distributing the quantifier, applying modus tollens and then Box-distribution we arrive at:

$$\Box\neg\Big(\bigvee y.\varphi_i \wedge \bigwedge z.\neg[z/y]\varphi_i\Big) \to \Box\neg\Big(\chi \wedge \bigwedge_{j<i} \delta_i\Big) \in \Delta.$$

Now, since $\Box(\bigwedge y.\neg\varphi_i \vee \bigvee z.[z/y]\varphi_i)$ belongs to every MPL, we thus obtain $\Box\neg(\chi \wedge \bigwedge_{j<i}\delta_i) \in \Delta$, which makes Δ_i inconsistent, contradiction. ⊣

LEMMA 10. *Let φ be a sentence in the language $\mathfrak{L}^*$ and Δ a Henkin-world. Then*

$$\langle \mathfrak{Can}_{L^*}, \Delta\rangle \vDash \varphi \iff \varphi \in \Delta.$$

PROOF. The base case, $\varphi = P(\vec{c})$, follows trivially from the definition of $\mathfrak{I}_\Delta$. The induction steps for $\bot$, $\neg$ and $\wedge$ are routine as well. Now, let $\varphi = \bigvee y.\chi$. Suppose that $\varphi \in \Delta$. Then, since Δ is Henkin-complete, there exists a constant d such that $[d/x]\chi \in \Delta$. This is a sentence, and by induction hypothesis $\langle \mathfrak{Can}_{L^*}, \Delta\rangle \vDash [d/x]\chi$. Hence, by definition, $\langle \mathfrak{Can}_{L^*}, \Delta\rangle \vDash \bigvee y.\chi$. This argument is reversible. Finally, let $\varphi = \Diamond\chi$. Assume that $\langle \mathfrak{Can}_{L^*}, \Delta\rangle \vDash \Diamond\chi$. Then there exists a Σ such that $\Delta \lhd \Sigma$ and $\langle \mathfrak{Can}_{L^*}, \Sigma\rangle \vDash \chi$. By induction hypothesis, $\chi \in \Sigma$. By definition of $\lhd$, $\Diamond\chi \in \Delta$.

Now assume $\Diamond\chi \in \Delta$. By Lemma 9 there is a Henkin-world Σ such that $\Delta^\Box \cup \{\chi\} \subset \Sigma$. Hence $\Delta \lhd \Sigma$ by definition. By induction hypothesis, $\langle \mathfrak{Can}_{L^*}, \Sigma\rangle \vDash \chi$. So, $\langle \mathfrak{Can}_{L^*}, \Delta\rangle \vDash \Diamond\chi$, as had to be shown. ⊣

Now we have given all the ingredients for a proof of the main result.

THEOREM 11. *Every modal predicate logic without equality is complete with respect to modal first-order structures, in particular*

$$\mathfrak{Can}_{L^*} \vDash \varphi \iff \varphi \in L.$$

§4. Coherence structures. Let us now see what happens if equality is introduced into the language. Evidently, if equality is just a member of Π instead of being a logical symbol, the previous proofs go through. Then the interpretation of equality is an equivalence relation in each world. But generally one requires that equality must be interpreted as identity. Nonetheless, we must ask: identity of what? Think about the example of Hesperus and Phosphorus. As for the real world they are identical, but there are some people for whom they are not. Let George be such a person. Then there is a belief world of George's in which Hesperus and Phosphorus are not identical. Many have argued that George's beliefs are inconsistent. This is what comes out if we assume standard semantics. But we could turn this around in the following way. We say that equality does not denote identity of *objects* but of something else, which we shall call the *object trace*. We say that Hesperus and Phosphorus are different objects, which happen to have the same trace in this world, but nonidentical traces in each of George's belief worlds. To make this distinction between object and object trace more acceptable we shall give a different example. Suppose someone owns a bicycle b and he has it repaired. The next day he picks it up; but then it has a different front wheel. Surely, he would consent to the statement that the bicycle he now has is that bicycle that he gave to the repair shop yesterday. But its front wheel isn't. Let's assume for simplicity that atoms are permanent, they will never cease to exist nor come into existence. Next, let us assume (again simplifying things considerably) that the trace of an object is just the collection of atoms of which it consists. Then, while the object b continued to exist, its trace has changed from one day to the other. In order not to get confused with the problem of transworld identity let us stress that we think of the objects as transcendental. b is neither a citizen of this world today nor of yesterday's world, nor of any other world. But its trace in this world does belong to this world. We may or may not assume that object traces are shared across worlds. Technically matters are simpler if they are not, but nothing hinges on that. So, in addition to the bicycle b we have two wheels w and w', and the trace of b contained the object trace of w yesterday, and it contains the trace of w' today. In the light of these examples it seems sensible to distinguish an object from its trace. Of course, we are not committed to any particular view of traces and certainly do not want to assume that object traces are simply conglomerates of atoms.

Now, in the classical semantics, identity across worlds was a trivial matter. Objects were transcendental, and in using the same letter we always refer to the same object across worlds. However, identity is not relative to worlds. If Hesperus is the same object as Phosphorus in one world, it is the same in all worlds. The distinction between object and trace gets us around this problem as follows. Denote the objects by h and p; further, let this world be w_0 and let w_1 be one of George's belief worlds. Then the traces of h and p are the same in this world, but different in w_1. This solves the apparent problem. In our words, equality does not denote identity of two objects, but only identity of their traces in a particular world.

DEFINITION 12 (Coherence frames and structures). By a *coherence frame* we understand a quintuple $\langle W, \lhd, U, T, \tau\rangle$ where $\langle W, \lhd, U\rangle$ is a predicate Kripke-frame, T a set, the set of *things*, and $\tau : U \times W \rightarrow T$ a surjective function. We call τ the *trace function* and $\tau(o, w)$ the *trace of o in w*. An *interpretation* is a function $\mathfrak{I}$ mapping each $P \in \Pi$ to a function from W to $U^{\Omega(P)}$ and each constant symbol c to a member of U. $\mathfrak{I}$ is called *equivalential* if for all $\vec{a}, \vec{b} \in U^{\Omega(P)}$ and $w \in W$, if $\tau(a_i, w) = \tau(b_i, w)$ for all $i < \Omega(P)$ then $\vec{a} \in \mathfrak{I}(P)(w)$ iff $\vec{b} \in \mathfrak{I}(P)(w)$. A *coherence structure* is a pair $\langle \mathfrak{W}, \mathfrak{I}\rangle$ where $\mathfrak{W}$ is a coherence frame and $\mathfrak{I}$ an equivalential interpretation.

Note that since trace functions are assumed to be surjective, every trace has to be the trace of some object. This is a natural condition, because objects are considered to be the primary entity, and traces a derived notion. The notion of equivalence is perhaps a curious one. It says that the basic properties of objects cannot discriminate between objects of equal trace. So, if Pierre believes that London is beautiful and Londres is not, we have two objects which happen to have the same trace in this world. Hence they must share all properties *in this world.* So, London and Londres can only be both beautiful or both ugly. This seems very plausible indeed. From a technical point of view, however, the fact that they cannot simply have different properties is a mere stipulation on our part. On the other hand, it is conceivable that there are basic predicates that are actually intensional, which would mean that they fail the substitution under (extensional) equality.

An alternative setup for strictly extensional basic predicates is the following. An interpretation is a function assigning to predicates in a world not tuples of objects but tuples of things. Then an object has a property if and only if its trace does. This approach is certainly more transparent because it attributes the fact that an object bears a property only to the fact that its trace does. Yet, technically it amounts to the same.

DEFINITION 13 (Coherence models). A *coherence model* is a triple $\langle \mathfrak{C}, \beta, w \rangle$, where $\mathfrak{C}$ is a coherence structure, $\beta : V \to U$ a valuation, $w \in W$ and ε as in Definition 4. We define the truth of a formula inductively as follows.

$$\begin{aligned}
\langle \mathfrak{C}, \beta, w \rangle \vDash P(\vec{t}) &:\Longleftrightarrow \varepsilon(\vec{t}) \in \mathfrak{I}_w(P), \\
\langle \mathfrak{C}, \beta, w \rangle \vDash s \doteq t &:\Longleftrightarrow \tau(\varepsilon(s), w) = \tau(\varepsilon(t), w), \\
\langle \mathfrak{C}, \beta, w \rangle \vDash \chi \wedge \varphi &:\Longleftrightarrow \langle \mathfrak{C}, \beta, w \rangle \vDash \chi; \varphi, \\
\langle \mathfrak{C}, \beta, w \rangle \vDash \neg\varphi &:\Longleftrightarrow \langle \mathfrak{C}, \beta, w \rangle \nvDash \varphi, \\
\langle \mathfrak{C}, \beta, w \rangle \vDash \bigvee x.\varphi &:\Longleftrightarrow \text{for some } \gamma \text{ with } \gamma \sim_x \beta : \langle \mathfrak{C}, \gamma, w \rangle \vDash \varphi, \\
\langle \mathfrak{C}, \beta, w \rangle \vDash \Diamond\varphi &:\Longleftrightarrow \text{there is } w' \text{ such that } w \lhd w' \text{ and } \langle \mathfrak{C}, \beta, w' \rangle \vDash \varphi.
\end{aligned}$$

$\mathfrak{C} \vDash \varphi$ if for all valuations β and all worlds $v : \langle \mathfrak{C}, \beta, v \rangle \vDash \varphi$.

It is a matter of straightforward verification to show that all axioms and rules of the minimal MPL are valid in a coherence frame. Moreover, the set of formulae valid in a coherence structure constitute a first-order MPL. Notice that the fourth postulate for equality holds in virtue of the special clause for equality and the condition that the interpretation must be equivalential. For if $\langle \mathfrak{C}, \beta, w \rangle \vDash y_i \doteq y_n$, then $\tau(\beta(y_i), w) = \tau(\beta(y_n), w)$. So, if $\langle \mathfrak{C}, \beta, w \rangle \vDash P(y_0, \ldots, y_{n-1})$ for $P \in \Pi$, then $\langle \beta(y_i) : i < n \rangle \in \mathfrak{I}(P)(w)$. Let $\beta' \sim_{y_i} \beta$ be such that $\beta'(y_i) = \beta(y_n)$. By equivalentiality, $\langle \beta'(y_i) : i < n \rangle \in \mathfrak{I}(P)(w)$. This means that $\langle \mathfrak{C}, \beta', w \rangle \vDash P(y_0, \ldots, y_{n-1})$, and so $\langle \mathfrak{C}, \beta, w \rangle \vDash [y_n/y_i]P(y_0, \ldots, y_{n-1})$. If $\mathfrak{F}$ is a coherence frame, put $\mathfrak{F} \vDash \varphi$ if $\langle \mathfrak{F}, \mathfrak{I} \rangle \vDash \varphi$ for all equivalential interpretations $\mathfrak{I}$. Evidently, $\{\varphi : \mathfrak{F} \vDash \varphi\}$ is a second-order MPL.

The difference with the counterpart semantics is that we have disentangled the quantification over objects from the quantification over worlds. Moreover, objects exist independently of worlds. Each object leaves a trace in a given world, though it need not exist there. Furthermore, two objects can have the same trace in any given world without being identical. However, identity of two objects holds in a world if and only if they have the same trace in it. If we also have function symbols, the clauses for basic predicates and equality will have to be generalized in the obvious direction.

To derive a completeness result for coherence structures we have to revise the construction from Section 3 only slightly, namely we have to define what the traces of objects are. To do this, let $\Delta \in W_{L^*}$ be a Henkin-world and c a constant. Then put $[c]_\Delta := \{d \in C_{L^*} : c \doteq d \in \Delta\}$. Now set

$$\tau_{L^*}(c, \Delta) := \langle [c]_\Delta, \Delta \rangle.$$

Then let

$$T_{L^*} := \{\langle [c]_\Delta, \Delta \rangle : \Delta \in W_{L^*},\ c \in C_{L^*}\}.$$

Finally, put

$$\mathfrak{Coh}_{L^*} := \langle\langle W_{L^*}, \lhd, C_{L^*}, T_{L^*}, \tau_{L^*} \rangle, \mathfrak{I}_{L^*} \rangle.$$

This is a coherence structure. For by Eq4, $\mathfrak{I}_{L^*}$ is equivalential, as is easily checked. Since $\langle [c]_\Delta, \Delta \rangle = \langle [d]_\Delta, \Delta \rangle$ iff $[c]_\Delta = [d]_\Delta$ iff $c \doteq d \in \Delta$, the following is immediate:

LEMMA 14. *Let φ be a sentence and Δ a Henkin-world. Then*

$$\langle \mathfrak{Coh}_{L^*}, \Delta \rangle \vDash \varphi \Longleftrightarrow \varphi \in \Delta.$$

THEOREM 15. *Every modal predicate logic with or without equality is complete with respect to coherence structures.*

In his [11], Gerhard Schurz introduced a semantics, called *worldline semantics*, in the context of analyzing Hume's is-ought thesis, i.e. the logical problem whether one may infer ethical value (normative) statements from factual (descriptive) statements. This semantics is very close to the coherence semantics defined in this paper. A *worldline frame* is a quintuple $\langle W, R, L, U, Df \rangle$, where W is a set of worlds, R the accessibility relation, $U \neq \varnothing$ a non-empty set of *possible objects*, $\varnothing \neq L \subseteq U^W$ a set of functions from possible worlds to possible objects (members of L are called *worldlines*), $Df : W \to \wp(U)$ a *domain function* such that $Df(w) =: D_w \subseteq U_w$, where $U_w := \{d \in U : \exists l \in L\ (l(w) = d)\}$ (the set of term extensions at world w) and $L_w = \{l \in L : \exists d \in D_w\ (l(w) = d)\}$ (the set of worldlines with extension in D_w). An interpretation V into a worldline frame is a function such that $V(t) \in L$ for any term t and $V_w(Q) \subseteq U_w^n$ for any n-ary predicate Q. If V is a worldline interpretation denote by $V[l/x]$ the interpretation that is like

V except that it assigns worldline l to the variable x. Since, unlike Schurz, we assume that free logical quantifiers are a defined notion, we suppose in the following that $D_w = U_w$ for all w. Then, in particular, $Df(w) = U_w$ and $L_w = L$ for all w, which means that Df can be omitted. The truth relation in worldline semantics can now be defined as follows:

DEFINITION 16 (Truth in worldline semantics). Let $\mathfrak{F} = \langle W, R, L, U\rangle$ be a worldline frame, V an interpretation and w a world, define

$$\begin{aligned}
\langle \mathfrak{F}, V, w\rangle \vDash P(\vec{t}) &:\Longleftrightarrow V_w(\vec{t}) \in V_w(P),\\
\langle \mathfrak{F}, V, w\rangle \vDash s \doteq t &:\Longleftrightarrow V(s)(w) = V(t)(w),\\
\langle \mathfrak{F}, V, w\rangle \vDash \chi \wedge \varphi &:\Longleftrightarrow \langle \mathfrak{F}, V, w\rangle \vDash \chi; \varphi,\\
\langle \mathfrak{F}, V, w\rangle \vDash \neg\varphi &:\Longleftrightarrow \langle \mathfrak{F}, V, w\rangle \nvDash \varphi,\\
\langle \mathfrak{F}, V, w\rangle \vDash \bigvee x.\varphi &:\Longleftrightarrow \text{for some } l \in L : \langle \mathfrak{F}, V[l/x], w\rangle \vDash \varphi,\\
\langle \mathfrak{F}, V, w\rangle \vDash \Diamond\varphi &:\Longleftrightarrow \text{there is } w' \text{ such that } w \lhd w' \text{ and } \langle \mathfrak{F}, V, w'\rangle \vDash \varphi.
\end{aligned}$$

It should be rather clear that the main difference between worldline and coherence semantics is terminological. While in worldline semantics one quantifies over worldlines and evaluates predicates and identity statements with respect to the value of a worldline at a particular world, in coherence semantics we quantify over modal individuals without specifying their internal structure, but assume a trace function that maps an individual at a world to its trace. So we can give the following translation. Given a coherence model $\langle \mathfrak{F}, \mathfrak{I}, \beta, w\rangle$ based on the coherence frame $\langle W, \lhd, U, T, \tau\rangle$, define a worldline model $\langle \mathfrak{G}, V, w'\rangle$ based on the worldline frame $\langle W', R, L, U'\rangle$ as follows. Set $W' := W$, $R := \lhd$, $U' := T$, and $w' := w$. Further, given $u \in U$, define $f_u : W' \to U'$ by letting $f_u(w) := t$ if $\tau(u, w) = t$ and set $L := \{f_u : u \in U\}$. Then, for $v \in W'$, we have

$$U'_v := \{t \in U' : \exists l \in L\ (l(v) = t)\} = \{t \in T : \exists u \in U\ (\tau(u, v) = t)\}.$$

Call $\mathfrak{G}$ the *worldline companion* of $\mathfrak{F}$.

PROPOSITION 17. *For every coherence frame $\mathfrak{F}$ and its worldline companion $\mathfrak{G}$ and for all φ:*

$$\mathfrak{F} \vDash \varphi \Longleftrightarrow \mathfrak{G} \vDash \varphi.$$

In particular, if a logic L is frame complete with respect to coherence frames, it is frame complete with respect to worldline frames.

PROOF. The proof is by a rather straightforward structural induction on φ. The only task is to define appropriate interpretations, so we only consider the atomic cases. Fix a coherence frame $\mathfrak{F}$ and its worldline companion $\mathfrak{G}$. Suppose $\varphi = P(x_0, \ldots, x_{n-1})$ and that $\langle \mathfrak{F}, \mathfrak{I}, \beta, w\rangle \nvDash \varphi$ for some equivalential interpretation $\mathfrak{I}$, valuation β and world w. Define

$$(*) \quad \langle t_0, \ldots, t_{n-1} \rangle \in V_w(P) \Longleftrightarrow t_i = \tau(w, u_i) \text{ and } \langle u_0, \ldots, u_{n-1} \rangle \in \mathfrak{I}_w(P)$$

and $\beta_w(x_i) = V_w(x_i)$. Then, clearly, $\langle \mathfrak{G}, V, w \rangle \nvDash P(x_0, \ldots, x_{n-1})$. Conversely, given an interpretation V and $\langle \mathfrak{G}, V, w \rangle \nvDash \varphi$, we can define an interpretation $\mathfrak{I}$ and a valuation β as in $(*)$ such that $\mathfrak{I}$ is equivalential and $\langle \langle \mathfrak{F}, \mathfrak{I} \rangle, \beta, w \rangle \nvDash \varphi$. The case $\varphi = (x_i \doteq x_j)$ is treated in the same way. ⊣

Notice that worldline frames require no condition of equivalentiality.

§5. **Coherence structures and counterpart structures.** Counterpart frames and structures were introduced in Kracht and Kutz [10]. They generalize the functor-semantics of Ghilardi. Call a relation $R \subseteq M \times N$ a *CE-relation* (*CE* stands for "counterpart existence") if for all $x \in M$ there exists a $y \in N$ such that $x \mathrel{R} y$ and, likewise, for all $y \in N$ there exists an $x \in M$ such that $x \mathrel{R} y$. This is a slight adaptation of the definition of that paper to take care of the fact that we now deal with possibilist quantifiers plus an existence predicate as opposed to "proper" free-logical quantifiers. Furthermore, we shall make more explicit the world dependence of the universes.

Definition 18 (Counterpart frames and structures). A *counterpart frame* is a quadruple $\langle W, T, \mathfrak{U}, \mathfrak{C} \rangle$, where $W, T \neq \varnothing$ are non-empty sets, $\mathfrak{U}$ a function assigning to each $v \in W$ a non-empty subset $\mathfrak{U}_v$ of T (its *domain*) and, finally, $\mathfrak{C}$ a function assigning to each pair of worlds v, w a set $\mathfrak{C}(v, w)$ of CE-relations from $\mathfrak{U}_v$ to $\mathfrak{U}_w$. A pair $\langle \mathfrak{W}, \mathfrak{I} \rangle$ is called a *counterpart structure* if $\mathfrak{W}$ is a counterpart frame and $\mathfrak{I}$ an interpretation, that is, a function assigning to each $w \in W$ and to each n-ary predicate letter a subset of $\mathfrak{U}_w^n$.

We say that v *sees* w in $\mathfrak{F}$ if $\mathfrak{C}(v, w) \neq \varnothing$. A *valuation* is a function η which assigns to every possible world v and every variable an element from the universe $\mathfrak{U}_v$ of v. We write η_v for the valuation η at v. A *counterpart model* is a quadruple $\mathfrak{M} = \langle \mathfrak{F}, \mathfrak{I}, \eta, w \rangle$, where $\mathfrak{F}$ is a counterpart frame, $\mathfrak{I}$ an interpretation, η a valuation and $w \in W$. Note that interpretations in counterpart frames differ from interpretations in coherence frames in that they do not assign values to constants. That is to say, unless otherwise stated, when working with counterpart frames we assume that the language does not contain constants.

Let $v, w \in W$ be given and ρ a CE-relation from $\mathfrak{U}_v$ to $\mathfrak{U}_w$. We write $\eta \xrightarrow{\rho} \widetilde{\eta}$ if for all $x \in V$: $\langle \eta_v(x), \widetilde{\eta}_w(x) \rangle \in \rho$. In the context of counterpart frames, $\widetilde{\eta} \sim_x^v \eta$ denotes a local x-variant at the domain of world v, i.e., $\widetilde{\eta}$ is a valuation that may differ from η only in the values that it assigns to the variable x at world v.

Definition 19 (Truth in a counterpart model). Let $\varphi(\vec{y})$ and $\chi(\vec{z})$ be modal formulae with the free variables $y_0, \ldots, y_{n-1}$ and $z_0, \ldots, z_{m-1}$, respectively.

Let $\mathfrak{C}$ be a counterpart structure, v a possible world and let η be a valuation. We define:

$$\begin{aligned}
\langle \mathfrak{C}, \eta, v \rangle \vDash x_i \doteq x_j &:\Longleftrightarrow \eta_v(x_i) = \eta_v(x_j), \\
\langle \mathfrak{C}, \eta, v \rangle \vDash R(\vec{y}) &:\Longleftrightarrow \langle \eta_v(y_0), \ldots, \eta_v(y_{n-1}) \rangle \in \mathfrak{I}_v(R), \\
\langle \mathfrak{C}, \eta, v \rangle \vDash \neg\varphi &:\Longleftrightarrow \langle \mathfrak{C}, \eta, v \rangle \nvDash \varphi, \\
\langle \mathfrak{C}, \eta, v \rangle \vDash \varphi \wedge \chi &:\Longleftrightarrow \langle \mathfrak{C}, \eta, v \rangle \vDash \varphi; \chi, \\
\langle \mathfrak{C}, \eta, v \rangle \vDash \Diamond\varphi(\vec{y}) &:\Longleftrightarrow \text{there are } w \in W,\ \rho \in \mathfrak{C}(v, w) \text{ and} \\
&\qquad \widetilde{\eta} \text{ such that } \eta \xrightarrow{\rho} \widetilde{\eta} \text{ and } \langle \mathfrak{C}, \widetilde{\eta}, w \rangle \vDash \varphi(\vec{y}), \\
\langle \mathfrak{C}, \eta, v \rangle \vDash \bigvee x.\varphi(x) &:\Longleftrightarrow \text{there is } \widetilde{\eta} \sim_x^v \eta \text{ such that } \langle \mathfrak{C}, \widetilde{\eta}, v \rangle \vDash \varphi(x).
\end{aligned}$$

Given a counterpart frame $\mathfrak{F}$, $\mathfrak{F} \vDash \varphi$ if for all interpretations $\mathfrak{I}$, all valuations η and worlds v, $\langle \langle \mathfrak{F}, \mathfrak{I} \rangle, \eta, v \rangle \vDash \varphi$.

The intuition behind counterpart frames is that objects do not exist; the only things that exist are the object traces (which belong to the domains of the worlds), and the counterpart relations. However, the notion of an object is still definable, even though it shall turn out that counterpart frames can have very few objects in this sense.

DEFINITION 20. Let $\mathfrak{F} = \langle W, T, \mathfrak{U}, \mathfrak{C} \rangle$ be a counterpart frame. An *object* is a function $f : W \to T$ such that (i) $f(v) \in \mathfrak{U}_v$ for all $v \in W$, (ii) for each pair $v, w \in W$ with $\mathfrak{C}(v, w) \neq \varnothing$ there is $\rho \in \mathfrak{C}(v, w)$ such that $\langle f(v), f(w) \rangle \in \rho$.

So, objects are constructed using the counterpart relation. If the trace b in world w is a counterpart of the trace a in world v, then there may be an object leaving trace a in v and trace b in w. If not, then not. However, there are frames which are not empty and possess no objects. Here is an example. Let $W = \{v\}$, $T = \{a, b\}$, $\mathfrak{U}_v = \{a, b\}$, and $\mathfrak{C}(v, v) = \{\rho\}$ with $\rho = \{\langle a, b \rangle, \langle b, a \rangle\}$. It is easy to see that this frame has no objects. The crux is that we can only choose one trace per world, but when we pass to an accessible world, we must choose a counterpart. This may become impossible the moment the frame is not a tree.

Counterpart frames show a different behaviour than coherence frames. As we have shown above, for each modal predicate logic there exists an adequate structure. However, counterpart structures satisfy a formula that is actually not generally valid when one thinks of the quantifiers as quantifying over intensional rather than extensional (trace-like) objects.

PROPOSITION 21. *Let $\mathfrak{F}$ be a counterpart frame, x and y variables not occurring in $\vec{z}$. Then for all formulae $\varphi(x, \vec{z})$*

$$\mathfrak{F} \vDash \bigwedge x. \bigwedge y.(x \doteq y) \to (\varphi(x, \vec{z}) \leftrightarrow \varphi(y, \vec{z})).$$

PROOF. Pick a valuation $\mathfrak{I}$, and let $\mathfrak{S} := \langle \mathfrak{F}, \mathfrak{I} \rangle$. It is clear that we can restrict our attention to formulae of the type $\varphi(x, \vec{z}) = \Diamond\chi(x, \vec{z})$. Let η be a valuation and v a world. Assume that $\langle \mathfrak{S}, \eta, v\rangle \vDash x \doteq y$. Then $\eta_v(x) = \eta_v(y)$. We will show that $\langle \mathfrak{S}, \eta, v\rangle \vDash \Diamond\chi(x, \vec{z}) \rightarrow \Diamond\chi(y, \vec{z})$. Suppose that $\langle \mathfrak{S}, \eta, v\rangle \vDash \Diamond\chi(x, \vec{z})$. Then there exists a world w, a $\rho \in \mathfrak{C}(v, w)$ and a valuation $\widetilde{\eta}$ such that $\eta \xrightarrow{\rho} \widetilde{\eta}$ and $\langle \mathfrak{S}, \widetilde{\eta}, w\rangle \vDash \chi(x, \vec{z})$. Now define η' by $\eta'_w(y) := \widetilde{\eta}_w(x)$, and $\eta'_{w'}(y') := \widetilde{\eta}_{w'}(y')$ for all w' and y' such that either $w' \neq w$ or $y' \neq y$. Then $\langle \mathfrak{S}, \eta', w\rangle \vDash \chi(y, \vec{z})$. Furthermore, for all variables $y' : \langle \eta_v(y'), \eta'_w(y')\rangle \in \rho$. For if $y' \neq y$ this holds by definition of η and choice of $\widetilde{\eta}$. And if $y' = y$ we have $\eta_v(x) = \eta_v(y)$, so that $\rho \ni \langle \eta_v(x), \widetilde{\eta}_w(x)\rangle = \langle \eta_v(y), \eta'_w(y)\rangle$. It follows that $\eta \xrightarrow{\rho} \eta'$ and therefore that $\langle \mathfrak{S}, \eta, v\rangle \vDash \Diamond\chi(y, \vec{z})$, as had to be shown. ⊣

From the previous theorem we deduce that also the following holds.

$$(\ddagger) \qquad \bigwedge x.\bigwedge y.(x \doteq y \wedge \Diamond\top) \rightarrow \Diamond(x \doteq y).$$

Namely, take

$$\bigwedge x.\bigwedge y.x \doteq y. \rightarrow .\Diamond(x \doteq z) \leftrightarrow \Diamond(y \doteq z).$$

By the above theorem, this is generally valid. Substituting x for z we get

$$\bigwedge x.\bigwedge y.x \doteq y. \rightarrow .\Diamond(x \doteq x) \leftrightarrow \Diamond(y \doteq x).$$

Applying standard laws of predicate logic yields (‡). We remark here that the logics defined in the literature (for example Ghilardi [6], Skvortsov and Shehtman [13] and Kracht and Kutz [10]), differ from modal predicate logics as defined here only in the additional laws of equality that they assume.

The modal Leibniz law of [10] allows for simultaneous substitution of all free occurrences of x by y in $\Diamond\chi$ (denoted by $\Diamond\chi(y//x)$), provided that $x \doteq y$ is true.

$$\bigwedge x.\bigwedge y.x \doteq y. \rightarrow .\Diamond\chi(x) \rightarrow \Diamond\chi(y//x).$$

Now, notice that in a modal predicate logic as defined above, the rule of replacing constants for universally quantified variables is valid. In counterpart frames this creates unexpected difficulties. For, suppose we do have constants and that they may be substituted for variables. Then we may derive from (‡), using the substitution of c for x and d for y:

$$(c \doteq d \wedge \Diamond\top) \rightarrow \Diamond(c \doteq d).$$

Since a constant has a fixed interpretation in each world, this means that if two constants are equal in a world and there exists some accessible world, then there will also be some accessible world in which they are equal. This is not generally valid. What is happening here is a shift from a *de re* to a *de dicto* interpretation. If we follow the traces of the objects, the formula is valid, but if we substitute intensional objects, namely constants, it becomes refutable. Notice that this situation is also reflected in the way non-rigid constants are treated in [5]. There, the two possible readings of the above formula, the de dicto and de re reading, are distinguished by actually binding the interpretation of the constants to the respective worlds by using the term-binding λ-operator.

Applied to Hesperus and Phosphorus, this means that if they are equal, then there is a belief world of George's in which they are equal. However, if George believes that they are different, this cannot be the case. So, the counterpart semantics cannot handle constants correctly — at least not in a straightforward way, i.e., without restricting the possible values of constants in accessible worlds. This paradox is avoided in Kracht and Kutz [10] by assuming that the language actually has *no* constants.

§6. Objectual counterpart structures. The connection between coherence frames and counterpart frames is not at all straightforward. Since the logic of a counterpart frame is a first-order modal predicate logic, one might expect that for every counterpart frame there is a coherence frame having the same logic. This is only approximately the case. It follows from Theorem 23 that for every counterpart *structure* there is a coherence structure having the same theory. This is not generally true for frames. However, adopting a modification of coherence frames proposed by Melvin Fitting in [4], namely balanced coherence frames (in [4] the corresponding frames are called Riemann FOIL frames), it can indeed be shown that for every counterpart frame there is a balanced coherence frame validating the same logic (under a translation).

Let us begin by elucidating some of the connections between counterpart and coherence frames. Note again that since counterpart structures as defined above do not interpret constants, we have to assume that the language does not contain constants.

First, fix a coherence structure $\mathfrak{C} = \langle W, \lhd, U, T, \tau, \mathfrak{I}\rangle$. We put $U_v := \{\tau(o, v) : o \in U\}$. This defines the domains of the world. Next, for $v, w \in W$ we put $\rho(v, w) := \{\langle \tau(o, v), \tau(o, w)\rangle : o \in U\}$ and $\mathcal{C}(v, w) := \varnothing$ if $v \lhd w$ does not obtain; otherwise, $\mathcal{C}(v, w) := \{\rho(v, w)\}$. Finally,

$$\langle \tau(a_i, w) : i < \Omega(P)\rangle \in \mathfrak{I}'(P)(w) \Longleftrightarrow \langle a_i : i < \Omega(P)\rangle \in \mathfrak{I}(P)(w).$$

Then $\langle W, T, \mathfrak{U}, \mathcal{C}, \mathfrak{I}'\rangle$ is a counterpart structure. We shall denote it by $CP(\mathfrak{C})$. Notice that there is at most one counterpart relation between any two worlds.

Conversely, let a counterpart structure $\mathfrak{N} = \langle W, T, \mathfrak{U}, \mathcal{C}, \mathfrak{I}\rangle$ be given. We put $v \lhd w$ iff $\mathcal{C}(v, w) \neq \varnothing$. $U := T$. Let O be the set of all objects $o : W \to T$. Further, $\tau(o, w) := o(w)$. This defines a coherence frame if the set of objects is nonempty.[1] Finally,

$$\langle o_i : i < \Omega(P)\rangle \in \mathfrak{I}'(P)(w) \Longleftrightarrow \langle o_i(w) : i < n\rangle \in \mathfrak{I}(P)(w).$$

It is easy to see that this is an equivalential interpretation. So, $\langle W, \lhd, O, U, \tau, \mathfrak{I}'\rangle$ is a coherence structure, which we denote by $CH(\mathfrak{N})$.

[1] Strictly speaking, we have to reduce the set U of traces to those elements $t \in T$ that actually are the trace of some object o, but this makes no difference semantically.

Unfortunately, the logical relation between these two types of structures is rather opaque, not the least since the notion of satisfaction in them is different. Moreover, the operations just defined are not inverses of each other. For example, as we have already seen, there exist counterpart structures with nonempty domains which have no objects. In this case $CP(CH(\mathfrak{N})) \not\cong \mathfrak{N}$. Also let $\mathfrak{C}$ be the following coherence frame. $W := \{v, w, x, y\}$, $T := \{1, 2, 3, 4, 5, 6\}$, $U = \{a, b\}$, $\lhd = \{\langle v, w\rangle, \langle w, x\rangle, \langle x, y\rangle\}$. Finally,

$$\tau(a, -) : v \mapsto 1, w \mapsto 2, x \mapsto 4, y \mapsto 5,$$
$$\tau(b, -) : v \mapsto 1, w \mapsto 3, x \mapsto 4, y \mapsto 6.$$

Generating the counterpart frame we find that 2 and 3 are counterparts of 1, and 5 and 6 are counterparts of 4. Hence, there are more objects in the counterpart frame than existed in the coherence frame, for example the function $v \mapsto 1, w \mapsto 2, x \mapsto 4, y \mapsto 6$.

DEFINITION 22 (Threads). Let $\mathfrak{N}$ be a counterpart frame. We call a sequence $\langle (w_i, t_i) : i < n\rangle$ a *thread* if (1) for all $i < n : w_i \in W$, $t_i \in \mathfrak{U}_{w_i}$, and (2) for all $i < n-1 : w_i \lhd w_{i+1}$ and $\langle t_i, t_{i+1}\rangle \in \rho$ for some $\rho \in \mathfrak{C}(w_i, w_{i+1})$. $\mathfrak{N}$ is *rich in objects* if for all threads there exists an object o such that $o(w_i) = t_i$ for all $i < n$.

Notice that if $\lhd$ has the property that any path between two worlds is unique then $\mathfrak{N}$ is automatically object rich. Otherwise, when there are two paths leading to the same world, we must be able to choose the same counterpart in it. This is a rather strict condition. Nevertheless, we can use unravelling to produce such a structure from a given one, which is then object rich. Additionally, we can ensure that between any two worlds there is at most one counterpart relation. We call counterpart frames that satisfy the condition $|\mathfrak{C}(v, w)| \leq 1$ for all worlds $v, w \in W$ *Lewisian counterpart frames*.

THEOREM 23. *For every counterpart structure $\mathfrak{N}$ there exists a Lewisian counterpart structure $\mathfrak{N}'$ rich in objects such that $\mathfrak{N}$ and $\mathfrak{N}'$ have the same theory.*

PROOF. Let $\mathfrak{N} = \langle\langle W, T, \mathfrak{U}, \mathfrak{C}\rangle, \mathfrak{I}\rangle$ be a counterpart structure. A *path* in $\mathfrak{N}$ is a sequence $\pi = \langle w_0, \langle\langle w_i, \rho_i\rangle : 0 < i < n\rangle\rangle$ such that $\rho_i \in \mathfrak{C}(w_{i-1}, w_i)$ for all $0 < i < n$. We let $e(\pi) := w_{n-1}$ and $r(\pi) = \rho_{n-1}$ and call these, respectively, the *end point* and the *end relation* of π. Let W' be the set of all paths in $\mathfrak{N}$ and $T' := T$. Further, let $\mathfrak{U}'_\pi := \mathfrak{U}_{e(\pi)}$ and for two paths π and μ put $\mathfrak{C}'(\pi, \mu) := r(\mu)$ if $r(\mu) \in \mathfrak{C}(e(\pi), e(\mu))$ and empty otherwise. Finally, let P be an n-ary predicate letter. Then $\mathfrak{I}'(P)(\pi) := \mathfrak{I}(P)(e(\pi))$. Now let $\mathfrak{N}' = \langle\langle W', T', \mathfrak{U}', \mathfrak{C}'\rangle, \mathfrak{I}'\rangle$. This is a Lewisian counterpart structure and clearly rich in objects. The following can be verified by induction. If β is a valuation on $\mathfrak{N}$, and w a world, and if β' is a valuation on $\mathfrak{N}'$ and π a path such that $e(\pi) = w$ and $\beta'_\pi(x_i) = \beta_w(x_i)$, then $\langle \mathfrak{N}, \beta, w\rangle \vDash \varphi$ iff $\langle \mathfrak{N}', \beta', \pi\rangle \vDash \varphi$ for all φ. The theorem now follows; for given β and w, β' and π satisfying these conditions can be found, and given β' and π also β and w satisfying these conditions can be found. ⊣

Notice by the way that in the propositional as well as the second-order case this theorem is false. This is so because the interpretation of a predicate in π must be identical to that of μ if the two have identical end points. If this is not the case, the previous theorem becomes false. However, if we are interested in characterizing MPLs by means of models, it follows from the above result that we can restrict ourselves in the discussion to Lewisian counterpart structures that are rich in objects.

But we can also strike the following compromise. Let us keep the counterpart semantics as it is, but interpret formulae in a different way. Specifically, let us define the following.

DEFINITION 24 (Objectual counterpart interpretations). We say that $\mathfrak{M} = \langle\langle\mathfrak{F},\mathfrak{I}\rangle,\beta,v\rangle$ is an *objectual counterpart model*, if $\mathfrak{F} = \langle W, T, \mathfrak{U}, \mathfrak{C}\rangle$ is a counterpart frame as before, $\mathfrak{I}$ is an *objectual interpretation*, that is, a counterpart interpretation that additionally assigns objects to constant symbols, β an *objectual valuation* into $\mathfrak{F}$, i.e., a function that assigns to each variable an object in a given a world. In this context, $\varepsilon_v(o) := \beta_v(o)$ if o is a variable and $\varepsilon_v(o) = \mathfrak{I}(o)(v)$ if o is a constant symbol.

Write $\beta \rightarrow^{\vec{y}}_{v,w} \beta$ if for some $\rho \in \mathfrak{C}(v,w)$ we have $\langle\beta_v(x_i),\beta_w(x_i)\rangle \in \rho$ for all $x_i \in \vec{y}$. Furthermore, write $\beta \rightarrow^{\vec{y}}_{v,w} \gamma$ if for some $\rho \in \mathfrak{C}(v,w)$ we have $\langle\beta_v(x_i),\gamma_w(x_i)\rangle \in \rho$ for all $x_i \in \vec{y}$, where γ is an objectual valuation. Terms t_i denote either variables or constants, $\vec{y}$ tuples of variables and $\vec{c}$ tuples of constants. The symbol $\vDash^*$ is called the *weak objectual truth-relation* and is defined thus (with $\mathfrak{M} := \langle\mathfrak{F},\mathfrak{I}\rangle$):

$$\begin{aligned}
\langle\mathfrak{M},\beta,v\rangle \vDash^* t_i \doteq t_j &:\Longleftrightarrow \varepsilon_v(t_i) = \varepsilon_v(t_j)\\
\langle\mathfrak{M},\beta,v\rangle \vDash^* R(\vec{t}) &:\Longleftrightarrow \langle\varepsilon_v(t_0),\dots,\varepsilon_v(t_{n-1})\rangle \in \mathfrak{I}_v(R)\\
\langle\mathfrak{M},\beta,v\rangle \vDash^* \neg\varphi &:\Longleftrightarrow \langle\mathfrak{M},\beta,v\rangle \nvDash^* \varphi\\
\langle\mathfrak{M},\beta,v\rangle \vDash^* \varphi\wedge\chi &:\Longleftrightarrow \langle\mathfrak{M},\beta,v\rangle \vDash^* \varphi;\chi\\
\langle\mathfrak{M},\beta,v\rangle \vDash^* \Diamond\varphi(\vec{y},\vec{c}) &:\Longleftrightarrow \text{there is } \beta \rightarrow^{\vec{y}}_{v,w} \gamma \text{ such that } \langle\mathfrak{M},\gamma,w\rangle \vDash^* \varphi(\vec{y},\vec{c})\\
\langle\mathfrak{M},\beta,v\rangle \vDash^* \bigvee y.\varphi(y,\vec{c}) &:\Longleftrightarrow \text{there is } \widetilde{\beta} \sim_y \beta \text{ such that } \langle\mathfrak{M},\widetilde{\beta},v\rangle \vDash^* \varphi(y,\vec{c}).
\end{aligned}$$

The *strong objectual truth-relation* $\vDash^\dagger$ is like $\vDash^*$ except for the clause for $\Diamond$ which is given by:

$$\begin{aligned}
\langle\mathfrak{M},\beta,v\rangle \vDash^\dagger \Diamond\varphi(\vec{y},\vec{c}) :\Longleftrightarrow{} & \text{there is } w \text{ such that } \beta \rightarrow^{\vec{y}}_{v,w} \beta\\
& \text{and } \langle\mathfrak{M},\beta,w\rangle \vDash^\dagger \varphi(\vec{y},\vec{c}).
\end{aligned}$$

These interpretations remove the asymmetry between variables and constants in the sense that constants and variables are now assigned the same kind of values. However, while the strong objectual interpretation brings us very close to coherence semantics, the weak interpretation still bears essential properties of counterpart semantics, namely that we may move via a counterpart relation to a new object. More precisely we have the following:

PROPOSITION 25. *The rule of substituting constants for universally quantified variables is valid in the strong objectual interpretation. More specifically, for every counterpart frame* $\mathfrak{F}$

$$\mathfrak{F} \vDash^{\dagger} \left(\bigwedge x.\varphi\right) \rightarrow [c/x]\varphi.$$

Furthermore, there is an objectual counterpart model $\mathfrak{M}$ *such that*

$$\mathfrak{M} \nvDash^{\dagger} \bigwedge x. \bigwedge y.(x \doteq y) \rightarrow \left(\varphi(x, \vec{z}) \leftrightarrow \varphi(y, \vec{z})\right).$$

Both claims are false for the weak objectual interpretation.

PROOF. For the first claim suppose that β is an objectual valuation, $\mathfrak{I}$ an objectual interpretation, v a world and that $\langle\langle\mathfrak{F}, \mathfrak{I}\rangle, \beta, v\rangle \vDash^{\dagger} \bigwedge x.\varphi$. We only need to consider the case where $\varphi = \Diamond\psi$. We then have that for every x-variant $\widetilde{\beta} : \langle\langle\mathfrak{F}, \mathfrak{I}\rangle, \widetilde{\beta}, v\rangle \vDash^{\dagger} \Diamond\psi(x)$. I.e., for every object $o = \widetilde{\beta}(x)$ there is $\widetilde{\beta} \rightarrow^{x}_{v,w} \widetilde{\beta}$ and $\langle\langle\mathfrak{F}, \mathfrak{I}\rangle, \widetilde{\beta}, w\rangle \vDash^{*} \psi(x)$. Now $\mathfrak{I}(c) = \widetilde{\beta}(x)$ for some x-variant $\widetilde{\beta}$ of β from which the claim follows immediately.

For the second claim, fix the following simple model $\mathfrak{M}$. Let $W = \{v, w\}$, $T = \{a, b, b'\}$, $\mathfrak{U}(v) = \{a\}$, $\mathfrak{U}(w) = \{b, b'\}$, $\mathfrak{C}(v, w) = \{\rho\}$ where $\rho = \{\langle a, b\rangle, \langle a, b'\rangle\}$ and $\mathfrak{I}(P)(w) = \{b\}$ and $\beta(x) = o$ with $o(v) = a$ and $o(w) = b$ and $\beta(y) = o'$ with $o'(v) = a$ and $o'(w) = b'$. o and o' are the only objects in this model. It should be obvious that $\mathfrak{M} \vDash^{\dagger} x \doteq y \wedge \Diamond P(x)$ while $\mathfrak{M} \nvDash^{\dagger} \Diamond P(y)$.

Consider now the weak objectual interpretation. Take the model just defined and assume furthermore that $\mathfrak{I}(c) = o'$. Then clearly $v \vDash^{*} \bigwedge x.\Diamond P(x)$ while $w \nvDash^{*} \Diamond P(c)$, which shows that the rule is not valid. That the formula in the second claim is still valid under the weak objectual interpretation should be clear. ⊣

The following is also straightforward.

THEOREM 26. *Let* $\mathfrak{N}$ *be a counterpart structure rich in objects,* v *a world and let* β *be an objectual valuation and* $\widetilde{\beta}$ *a counterpart valuation such that* $\beta_v(x_i) = \widetilde{\beta}_v(x_i)$ *for all variables. Then for all* φ:

$$\langle\mathfrak{N}, \beta, v\rangle \vDash^{*} \varphi \Longleftrightarrow \langle\mathfrak{N}, \widetilde{\beta}, v\rangle \vDash \varphi.$$

The proof is by induction on φ. The two relations differ only with respect to formulae of the form $\Diamond\chi$. Here, object richness assures that for each choice of counterparts in the successor worlds an object exists. (Actually, for that we only need that every element of a domain is the trace of some object.)

§7. Passing back and forth: balanced coherence frames. By the previous theorem we can introduce the notion of an object into counterpart frames, which then makes them rather similar to coherence frames. However, counterpart structures with object valuations are still different from coherence structures. A different approach is to translate $\Diamond$ in order to accommodate the truth relation $\models^*$ within the language of counterpart structures.

$$\begin{aligned}
(x \doteq y)^\gamma &:= x \doteq y, \\
P(\vec{y})^\gamma &:= P(\vec{y}), \\
(\neg\varphi)^\gamma &:= \neg\varphi^\gamma, \\
(\varphi \wedge \chi)^\gamma &:= \varphi^\gamma \wedge \chi^\gamma, \\
\left(\bigvee x.\varphi\right)^\gamma &:= \bigvee x.\varphi^\gamma, \\
(\Diamond\varphi(y_0, \dots, y_{n-1}))^\gamma &:= \bigvee z_0. \cdots \bigvee z_{n-1}. \bigwedge_{i<n} z_i \doteq y_i \wedge \Diamond\varphi(\vec{z}/\vec{y})^\gamma.
\end{aligned}$$

Here, y_i $(i < n)$ are the free variables of φ and z_i $(i < n)$ distinct variables not occurring in φ. This is actually unique only up to renaming of bound variables. Further, notice that $\bigwedge_{i<n}$ denotes a finite conjunction, not a quantifier. This translation makes explicit the fact that variables inside a $\Diamond$ are on a par with bound variables. (In linguistics, one speaks of $\Diamond$ in the context of counterpart frames as an *unselective binder*.) Notice now that

$$\begin{aligned}
&\left(\bigwedge x_0. \bigwedge x_1.x_0 \doteq x_1 \to (\Diamond\varphi(x_0) \leftrightarrow \Diamond\varphi(x_1))\right)^\gamma \\
&= \bigwedge x_0. \bigwedge x_1.x_0 \doteq x_1 \to ((\Diamond\varphi(x_0))^\gamma \leftrightarrow (\Diamond\varphi(x_1))^\gamma) \\
&= \bigwedge x_0. \bigwedge x_1.x_0 \doteq x_1 \to \left(\left(\bigvee x_3.x_3 \doteq x_0 \wedge \Diamond\varphi(x_3)^\gamma\right)\right. \\
&\qquad \left.\leftrightarrow \left(\bigvee x_3.x_3 \doteq x_1 \wedge \Diamond\varphi(x_3)^\gamma\right)\right).
\end{aligned}$$

This principle is actually valid in coherence structures. For it is a substitution instance of the following theorem of predicate logic.

$$\bigwedge x_0. \bigwedge x_1.x_0 \doteq x_1 \to \left(\left(\bigvee x_3.x_3 \doteq x_0 \wedge \varphi\right) \leftrightarrow \left(\bigvee x_3.x_3 \doteq x_1 \wedge \varphi\right)\right).$$

PROPOSITION 27. *Let $\mathfrak{N}$ be a counterpart structure and x a world. Then for any φ:*

$$\langle \mathfrak{N}, x\rangle \models \varphi^\gamma \iff \langle \mathfrak{N}, x\rangle \models \varphi.$$

In object rich structures also $\models$ and $\models^*$ coincide, which makes all four notions the same. So, while in counterpart structures the formulae φ and φ^γ are equivalent, they are certainly not equivalent when interpreted in coherence structures.

In [11] it is shown that worldline semantics provides for the same class of frame complete logics in the absence of extra equality axioms as standard constant domain semantics. It follows that the same holds for coherence frames. This means that while coherence frames allow for a more natural treatment of non-rigid designation for example, unlike counterpart frames, they do not enlarge the class of frame complete logics unless one moves to the full second-order semantics as we will do in Section 10. But there is a different approach to this problem. Instead of introducing algebras of admissible interpretations we can assume that certain worlds are isomorphic copies of each other. So, we add to the frames an equivalence relation between worlds and require that predicates are always interpreted in the same way in equivalent worlds. This idea is basically due to Melvin Fitting (see [4]). Here we use a slightly different approach. Namely, we add a first-order bisimulation to coherence frames. To be precise, let $\mathfrak{F}$ be a coherence frame. Call a relation $\mathcal{E} \subseteq W \times W$ a *world-mirror on* $\mathfrak{F}$ if $\mathcal{E}$ is an equivalence relation and whenever $v \ \mathcal{E} \ w$ and $v \lhd u_1$, there is a u_2 such that $w \lhd u_2$ and $u_1 \ \mathcal{E} \ u_2$. Intuitively, two mirrored worlds v and w may be understood as a situation seen from two different perspectives (because v and w may have "different histories", but have the "same future"). In [9] world-mirrors are called *nets*, and it is shown that an equivalence relation is a net if and only if it is induced by a p-morphism.

DEFINITION 28 (Balanced coherence frames). A *balanced coherence frame* is a pair $\langle \mathfrak{F}, \mathcal{E} \rangle$ where $\mathfrak{F} = \langle W, \lhd, U, T, \tau \rangle$ is a coherence frame and $\mathcal{E}$ is a world-mirror on $\mathfrak{F}$. An interpretation $\mathfrak{I}$ is called *balanced*, if it is equivalential and $\langle u_0, \ldots, u_{n-1} \rangle \in \mathfrak{I}_v(P)$ iff $\langle u_0, \ldots, u_{n-1} \rangle \in \mathfrak{I}_w(P)$ for all n-ary relations P and for all worlds v, w such that $v \ \mathcal{E} \ w$. A *balanced coherence model* is a triple $\langle \langle \mathfrak{B}, \mathfrak{I} \rangle, \beta, w \rangle$, where $\mathfrak{B}$ is a balanced coherence frame, $\mathfrak{I}$ a balanced interpretation, β a valuation and w a world.

The next theorem gives the connection between counterpart frames and balanced coherence frames.

THEOREM 29. *For every counterpart frame* $\mathfrak{F}$ *there exists a balanced coherence frame* $\mathfrak{F}^*$ *such that for all formulae* φ*:*

$$\mathfrak{F} \vDash \varphi \Longleftrightarrow \mathfrak{F}^* \vDash \varphi^{\gamma}.$$

PROOF. Fix a counterpart frame $\mathfrak{F} = \langle W, T, \mathfrak{U}, \mathfrak{C} \rangle$. Let $\mathfrak{F}' := \langle W', T', \mathfrak{U}', \mathfrak{C}' \rangle$ be the unravelled Lewisian counterpart frame defined as in Theorem 23. We define a balanced coherence frame $\mathfrak{F}^* = \langle \langle W^*, \lhd, U^*, T^*, \tau \rangle, \mathcal{E} \rangle$ from $\mathfrak{F}'$. Let $W^* = W'$, $T^* = T'$ and define $\pi \lhd \nu :\Leftrightarrow \mathfrak{C}'(\pi, \nu) \neq \varnothing$. Since $\mathfrak{F}'$ is rich in objects, there is an object $o : W' \to T$ for every thread in $\mathfrak{F}'$. Define U^* as the set of all objects in $\mathfrak{F}'$ and set $\tau^*(o, w) = t :\Leftrightarrow o(w) = t$. Finally, for $\pi, \nu \in W^*$, set $\pi \ \mathcal{E} \ \nu :\Leftrightarrow e(\pi) = e(\nu)$, where $e(\pi)$, $e(\nu)$ again denote the endpoints of $e(\pi)$ and $e(\nu)$, respectively. Clearly, $\mathcal{E}$ is a world-mirror for it is an equivalence relation and if $\pi \lhd \mu$ and $\pi \ \mathcal{E} \ \nu$, there is a path μ' such that $r(\mu) = r(\mu')$ and $\nu \lhd \mu'$, hence $\mu \ \mathcal{E} \ \mu'$.

Given a valuation β, an interpretation $\mathfrak{I}$ and a world v of $\mathfrak{F}$, Theorem 23 yields

$$\langle\langle\mathfrak{F},\mathfrak{I}\rangle,\beta,v\rangle \vDash \psi \iff \langle\langle\mathfrak{F}',\mathfrak{I}'\rangle,\beta',\nu\rangle \vDash \psi$$

for all ψ, where $\mathfrak{I}'(P)(\pi) := \mathfrak{I}(P)(e(\pi))$ for all worlds π, $\beta'_\pi(x_i) := \beta_w(x_i)$ if $e(\pi) = w$ and ν is a world in $\mathfrak{F}'$ such that $e(\nu) = v$.

Set

$$\mathfrak{I}^*(P)(\pi) := \{\langle o_0,\dots,o_{n-1}\rangle \in (U^*)^n : \langle o_0(\pi),\dots,o_{n-1}(\pi)\rangle \in \mathfrak{I}'(P)(\pi)\}$$

and choose an object valuation β^* on $\mathfrak{F}^*$ such that $\beta^*(x_i)(\nu) = \beta'_\nu(x_i)$. Such a valuation exists because $\mathfrak{F}'$ is rich in objects, whence there is an object leaving trace $\beta'_\nu(x_i)$ in world ν. Furthermore, $\mathfrak{I}^*$ is a balanced interpretation by definition, so $\mathfrak{M}^* := \langle\langle\mathfrak{F}^*,\mathfrak{I}^*\rangle,\beta^*,\nu\rangle$ defines a balanced coherence model. Finally note that every balanced interpretation in $\mathfrak{F}^*$ is of the form $\mathfrak{I}^*$ for some interpretation $\mathfrak{I}$ in $\mathfrak{F}$. Hence it suffices to show the following:

$$\langle\langle\mathfrak{F}',\mathfrak{I}'\rangle,\beta',\nu\rangle \vDash \psi \iff \langle\langle\mathfrak{F}^*,\mathfrak{I}^*\rangle,\beta^*,\nu\rangle \vDash \psi^\gamma$$

for all ψ. The claim is proved by induction. The atomic case follows from the definitions of $\mathfrak{I}^*$ and τ^* and the Boolean cases are trivial. The quantificational case follows again from object richness. So, consider the case $\psi = \Diamond\chi(y_0,\dots,y_{n-1})$ and assume first that $\langle\langle\mathfrak{F}',\mathfrak{I}'\rangle,\beta',\nu\rangle \nvDash \psi$. We have to show that $\mathfrak{M}^* \vDash \bigwedge z_0.\cdots\bigwedge z_{n-1}.(\bigwedge_{i<n} z_i \doteq y_i \to \Box\neg\chi(\vec{z}/\vec{y})^\gamma)$. Choose objects $o_0,\dots,o_{n-1}$ and an objectual $\vec{z}$-variant $\widetilde{\beta^*}$ such that $\widetilde{\beta^*}(z_i) = o_i$ and $o_i(\nu) = \beta^*(y_i)(\nu)$ for all $i < n$. Then $\langle\langle\mathfrak{F}^*,\mathfrak{I}^*\rangle,\widetilde{\beta^*},\nu\rangle \vDash \bigwedge_{i<n} z_i \doteq y_i$. Assume further that for some $\pi \in W^*$ we have $\nu \lhd \pi$, i.e. that there is a $\rho \in \mathfrak{C}'(\nu,\pi)$. Then, since $\mathfrak{F}'$ is a Lewisian counterpart frame, ρ is unique and hence $\langle o_i(\nu), o_i(\pi)\rangle \in \rho$ for all $i < n$. Then, by assumption, we have $\langle\langle\mathfrak{F}',\mathfrak{I}'\rangle,\widetilde{\beta},\pi\rangle \nvDash \chi$, where $\widetilde{\beta_\pi}(y_i) = \widetilde{\beta^*}(z_i)(\pi)$, so by induction it follows that $\langle\langle\mathfrak{F}^*,\mathfrak{I}^*\rangle,\widetilde{\beta^*},\pi\rangle \vDash \neg\chi(\vec{z}/\vec{y})^\gamma$.

Conversely, suppose that $\mathfrak{F}^* \nvDash \varphi^\gamma$. Again we consider only the case of

$$\varphi^\gamma = \bigvee z_0.\cdots.\bigvee z_{n-1}\left(\bigwedge_{i<n} z_i \doteq y_i \wedge \Diamond\psi(\vec{z}/\vec{y})^\gamma\right).$$

Pick a balanced interpretation $\mathfrak{I}^*$, an object valuation β^* and a world ν such that $\langle\langle\mathfrak{F}^*,\mathfrak{I}^*\rangle,\beta^*,\nu\rangle \nvDash \varphi^\gamma$. We need to show that $\langle\langle\mathfrak{F}',\mathfrak{I}'\rangle,\beta',\nu\rangle \vDash \Box\neg\psi(\vec{y})$. Let π be a world such that $\nu \lhd \pi$ in $\mathfrak{F}'$. Then $\delta \in \mathfrak{C}(e(\nu),e(\pi))$ and $\delta = r(\pi)$. Suppose $\widetilde{\beta}'$ is a counterpart valuation such that $\widetilde{\beta}' : \beta' \xrightarrow{\delta} \widetilde{\beta}'$. By object richness there are objects $u_i \in U^*$ $(i < n)$ such that $u_i(\nu) = \beta'_\nu(y_i)$ and $u_i(\pi) = \widetilde{\beta}'_\pi(y_i)$. Hence there is an objectual $\vec{z}$-variant $\widetilde{\beta^*}$ such that $\widetilde{\beta^*}(z_i)(\nu) = \beta^*(y_i)(\nu)$ and $\widetilde{\beta^*}(z_i)(\pi) = \widetilde{\beta}'_\pi(y_i)$ for all i. Then $\langle\langle\mathfrak{F}^*,\mathfrak{I}^*\rangle,\widetilde{\beta^*},\nu\rangle \vDash \bigwedge_{i<n} z_i \doteq y_i$ and thus $\langle\langle\mathfrak{F}^*,\mathfrak{I}^*\rangle,\widetilde{\beta^*},\pi\rangle \vDash \neg\psi^\gamma(\vec{z}/\vec{y})$. But then $\langle\langle\mathfrak{F}',\mathfrak{I}'\rangle,\widetilde{\beta}',\pi\rangle \vDash \neg\psi^\gamma(\vec{y})$ by induction, and the claim follows. ⊣

This result has interesting consequences. For example, since counterpart semantics is frame complete with respect to all first-order extensions QL of canonical propositional modal logics L (compare [6]), the same holds true for the translation QL^{γ} with respect to balanced coherence frames. Now we noted above that coherence frames per se characterize the same logics as standard constant domain semantics if no extra equality axioms are involved. But it is known that already rather simple canonical propositional logics have frame incomplete predicate extensions. In [2] it is shown that to complete frame incomplete MPLs by adding appropriate axioms, one needs mixed de re formulae rather than substitution instances of purely propositional formulae. So, the above result gives a hint on where the source for frame incompleteness with respect to standard semantics is to be found. In particular, note that the translation $.^{\gamma}$ leaves propositional formulae untouched, whereas de re formulae of the form $\Diamond\varphi(y_0, \ldots, y_{n-1})$ are transformed into formulae $\bigvee z_0. \cdots \bigvee z_{n-1}. \bigwedge_{i<n} z_i \doteq y_i \wedge \Diamond\varphi(\vec{z}/\vec{y})^{\gamma}$, which are de re formulae involving equality.

So what we need if we want to use standard possible worlds semantics to characterize a large class of logics via frame completeness are basically three things: firstly, the distinction between trace and object, secondly a different understanding of the modal operator as given by $.^{\gamma}$, and, thirdly, the assumption that certain worlds are copies of each other.

Let us make this claim more explicit. Given a standard constant domain frame $\langle W, \lhd, U\rangle$, we may add, as before, an equivalence relation $\mathcal{E}$ relating worlds. Furthermore, we technically do not need traces but can add a family of equivalence relations $(\mu_w)_{w\in W}$ interpreting equality at each world. Let us call frames of the form $\mathfrak{F} = \langle W, \lhd, U, (\mu_w)_{w\in W}, \mathcal{E}\rangle$ *balanced standard frames*. An interpretation $\mathfrak{I}$ is called *admissible*, if interpretations agree on worlds related by $\mathcal{E}$, and, moreover, they respect the equivalence relations μ_w in the sense that $\vec{a} \in \mathfrak{I}(w)(P)$ iff $\vec{b} \in \mathfrak{I}(w)(P)$ whenever $a_i \mu_w b_i$ for all i. We may think of objects being related by μ_w as indiscriminable with respect to world w and basic extensional predicates. Call a valuation $\widetilde{\gamma}$ a *w-ignorant $\vec{x}$-variant* of γ, if $\widetilde{\gamma}(x_i)\mu_w\gamma(x_i)$ for all $x_i \in \vec{x}$. The truth definition for balanced standard frames is as usual except for the equality and modal clauses, which are as follows:

$$\langle\mathfrak{F}, \mathfrak{I}, \gamma, w\rangle \models x \doteq y \iff \gamma(x)\mu_w\gamma(y);$$

$$\langle\mathfrak{F}, \mathfrak{I}, \gamma, w\rangle \models \Diamond\varphi(\vec{x}) \iff \text{there is a } w\text{-ignorant } \vec{x}\text{-variant } \widetilde{\gamma} \text{ and a world } v \rhd w \text{ such that } \langle\mathfrak{F}, \mathfrak{I}, \widetilde{\gamma}, v\rangle \models \varphi(\vec{x});$$

It should be rather clear that there is a bijective correspondence between balanced coherence frames and balanced interpretations on the one hand and balanced standard frames and admissible interpretations on the other. Furthermore, for every $\langle W, \lhd, U, T, \tau, \mathcal{E}\rangle$ there is a $\langle W, \lhd, U, (\mu_w)_{w\in W}, \mathcal{E}\rangle$ such that for all φ

$$\langle W, \lhd, U, T, \tau, \mathcal{E}\rangle \vDash \varphi^{\gamma} \Longleftrightarrow \langle W, \lhd, U, (\mu_w)_{w \in W}, \mathcal{E}\rangle \vDash \varphi.$$

Hence, the following theorem is an immediate corollary to Theorem 29.

THEOREM 30. *For every counterpart frame there is a balanced standard frame having the same logic.*

§8. Varieties of equality. Scott [12] proposes various kinds of identity. The first is the one we have discussed so far, namely *identity in trace*. The second, stronger notion, is the *inherent identity in trace*, which we shall denote by $\doteq^*$. Two objects satisfy this at a world if they are identical in trace at all subsequent worlds. The third is the *global identity in trace*, which we denote by $\approx$. Two objects are globally identical in trace at w if their traces are identical in all worlds that can be reached from w by either moving forward or backward along the relations. The fourth is *strong identity in trace*, denoted by $\approx^+$. Two objects are strongly identical in trace if they have identical trace in all worlds. The fifth is *identity as object*, denoted by $\equiv$. This is the numerical identity of objects. The semantics can be formally defined as follows. Denote by $T(w)$ the set of all worlds which are accessible from w in a series of steps. More formally, we define this as follows.

DEFINITION 31 (Transits). Let $\mathfrak{C} = \langle W, \lhd, U, T, \tau\rangle$ be a coherence frame. Then define $v \lhd^n w$ inductively by (a) $v \lhd^0 w$ iff $v = w$, (b) $v \lhd^{n+1} w$ iff there is a $u \in W$ such that $v \lhd^n u \lhd w$. Further, put $v \lhd^* w$ iff there is an n such that $v \lhd^n w$. Define $T(v) := \{w : v \lhd^* w\}$, and $Z(v) := \{w : v(\lhd \cup \lhd^{\smile})^* w\}$.

Here, $R^{\smile} := \{\langle y, x\rangle : \langle x, y\rangle \in R\}$ is the converse relation of R.

DEFINITION 32 (Equality in coherence models). Let $\langle \mathfrak{C}, \beta, v\rangle$ be a coherence model.

$$\begin{aligned}
\langle \mathfrak{C}, \beta, v\rangle \vDash x \doteq y &:\Longleftrightarrow \tau(\beta(x), v) = \tau(\beta(y), v),\\
\langle \mathfrak{C}, \beta, v\rangle \vDash x \doteq^* y &:\Longleftrightarrow \text{for all } w \in T(v) : \tau(\beta(x), w) = \tau(\beta(y), w),\\
\langle \mathfrak{C}, \beta, v\rangle \vDash x \approx y &:\Longleftrightarrow \text{for all } w \in Z(v) : \tau(\beta(x), w) = \tau(\beta(y), w),\\
\langle \mathfrak{C}, \beta, v\rangle \vDash x \approx^+ y &:\Longleftrightarrow \text{for all } w \in W : \tau(\beta(x), w) = \tau(\beta(y), w),\\
\langle \mathfrak{C}, \beta, v\rangle \vDash x \equiv y &:\Longleftrightarrow \beta(x) = \beta(y).
\end{aligned}$$

As it turns out, although all these notions are different semantically, we can only distinguish simple identity in trace from the other relations, that is to say, the latter four cannot be defined by means of modal axioms in the standard modal language using $\doteq$. Metatheoretically, the interrelations between $\approx$, $\approx^+$, $\doteq^*$ and $\equiv$ can — besides the usual axioms for identity (reflexivity, symmetry, transitivity) — be given by the following postulates.

$$\bigwedge x. \bigwedge y. x \equiv y \to x \approx^+ y,$$
$$\bigwedge x. \bigwedge y. x \approx^+ y \to x \approx y,$$
$$\bigwedge x. \bigwedge y. x \approx y \to x \doteq^* y,$$
$$\bigwedge x. \bigwedge y. x \doteq^* y \to x \doteq y,$$
$$\bigwedge x. \bigwedge y. x \doteq^* y \to \Box(x \doteq^* y).$$

Global identity in trace implies strong identity in trace if the frame is connected. A frame is called *cyclic* if for all $v : T(v) = Z(v)$. S5-frames are cyclic. Tense frames are also cyclic. (Notice that we have not defined $T(v)$ for polymodal frames. See however [9] for a definition.) Inherent identity in trace implies global identity in cyclic frames. Finally, notice that it is always possible to factor out the equivalence relation $\equiv$. That is to say, without changing the logic we can identify all objects that have the same trace function. Effectively, a coherence frame in which $\equiv$ is the identity is the same as a worldline frame. This is the content of Proposition 17.

§9. **Objects in metaframes.** Shehtman and Skvortsov have introduced in [13] the metaframe semantics and shown that it is complete for all modal predicate logics. Their results were stated and proved for superintuitionistic logics and extensions of S4. However, by removing some of the category theoretic definitions one can generalize these results easily to arbitrary modal predicate logic.

DEFINITION 33. Σ denotes the category of finite ordinals and functions between them.

DEFINITION 34 (Metaframes). A *general metaframe* M is a contravariant functor from the category Σ into the category of general frames. In particular, for every n, $M(n) = \langle F_n, \lhd_n, \mathbb{F}_n \rangle$ is a general frame, and for each $\sigma : m \to n$, $M(\sigma)$ is a p-morphism from $M(n)$ to $M(m)$. A *metaframe* is a contravariant functor from Σ into the category of Kripke-frames. We call the members of F_n *n-points*.

The idea is this. $M(0)$ represents the frame of possible worlds and $M(n)$ for $n > 0$ represents n-tuples over worlds. The arrows are needed to be able to identify the worlds and the tuples. For example, there is a unique map $0_n : 0 \to n$ for each n. Consequently, we have a map $M(0_n) : M(n) \to M(0)$. Thus, for each $a \in M(n)$, the *world* of a is $M(0_n)(a)$. Further, there is a (unique) natural embedding $i_{n,n+1} : n \to n+1 : i \mapsto i$. Hence, we define a *projection* of $a \in M(n+1)$ onto $M(n)$ by $M(i_{n,n+1})(a)$. We shall write

$a \downarrow b$ if $a \in M(n+1)$ for some $n \in \omega$ and $b = M(i_{n,n+1})(a)$. Further, write $p_i^n : 1 \to n$ for the unique map sending 0 to i and if $\sigma : m \to n$, write $x^\sigma := \langle x_{\sigma(0)}, x_{\sigma(1)}, \dots, x_{\sigma(m-1)} \rangle$.

DEFINITION 35 (Interpretations). Let M be a general metaframe. An *interpretation* on M is a function ξ assigning to each predicate letter P an internal set of $M(\Omega(P))$, i.e. $\xi(P) \in \mathbb{F}_{\Omega(P)}$.

For $a \in M(n)$ and $\sigma : m \to n$, $m = \Omega(P)$ we define

$$\begin{aligned}
\langle M, \xi, a \rangle \vDash P(x^\sigma) &:\Longleftrightarrow M(\sigma)(a) \in \xi(P), \\
\langle M, \xi, a \rangle \vDash x_i \doteq x_j &:\Longleftrightarrow M(p_i^n)(a) = M(p_j^n)(a), \\
\langle M, \xi, a \rangle \vDash \neg\chi &:\Longleftrightarrow \langle M, \xi, a \rangle \nvDash \chi, \\
\langle M, \xi, a \rangle \vDash \varphi \wedge \chi &:\Longleftrightarrow \langle M, \xi, a \rangle \vDash \varphi; \chi, \\
\langle M, \xi, a \rangle \vDash \Diamond\chi &:\Longleftrightarrow \text{exists } b \rhd_n a : \langle M, \xi, b \rangle \vDash \chi, \\
\langle M, \xi, a \rangle \vDash \bigvee x_n.\chi &:\Longleftrightarrow \text{exists } b \downarrow a : \langle M, \xi, b \rangle \vDash \chi.
\end{aligned}$$

We can identify $M(0)$ with the sets of worlds, $M(1)$ with the sets of objects, $M(2)$ with the sets of pairs of objects, and so on. Now, the definitions will not suffice to define a MPL from a metaframe unless it satisfies a further condition. Let $\sigma : m \to n$. Write σ^+ for the unique function from $m+1$ to $n+1$ such that $\sigma^+ \restriction m = \sigma$ and $\sigma(m) = n$. Further, for $m \leq n$, $i_{m,n} : m \to n : j \mapsto j$ is the unique inclusion.

DEFINITION 36 (Modal metaframes). A metaframe satisfies the *lift property* if for all $\sigma : m \to n$ and $a \in M(n)$, $b \in M(m+1)$ such that $M(\sigma)(a) = M(i_{m,m+1})(b) = d \in M(m)$ there exists a $c \in M(n+1)$ such that

$$a = M(i_{n,n+1})(c) \quad \text{and} \quad b = M(\sigma^+)(c).$$

A metaframe is a *modal metaframe* if it satisfies the lift property.

Shehtman and Skvortsov give a canonical procedure to obtain a modal metaframe from a modal predicate logic, see also Bauer [1]. Let L be given. We then let $M(n)$ be the set of all complete n-types $\Gamma = \Gamma(x_0, \dots, x_{n-1})$. They form a frame, where $\Gamma \lhd_n \Delta$ iff for all $\Box\varphi \in \Gamma$ we have $\varphi \in \Delta$. Further, if $\sigma : m \to n$, then $M(\sigma) : M(n) \to M(m)$ is defined by

$$M(\sigma)(\Delta) := \{\chi : [x_{\sigma(0)}/x_0, x_{\sigma(1)}/x_1, \dots, x_{\sigma(m-1)}/x_{m-1}]\chi \in \Delta\}.$$

Actually, the definition of truth in a model can be changed somewhat. First, it can be shown that if the free variables of χ are in the set $\{x_i : i < n\}$, we have

$$\langle M, \xi, a \rangle \vDash \chi \Longleftrightarrow \text{for all } b \downarrow a : \langle M, \xi, b \rangle \vDash \chi.$$

In this form we see that the truth of a formula is upward persistent. (The above-mentioned conditions on metaframes are such that this holds.)

Now write $a \sim_i c$ for the following. Let $\sigma : n \to n : i \mapsto n-1$, $n-1 \mapsto i$, $j \mapsto j$ $(j \notin \{i, n-1\})$. (If i is not less than $n-1$, this map is the identity.)

Then $a \sim_i c$ iff there exist b such that $M(\sigma)(a) \downarrow b$ and $M(\sigma)(c) \downarrow b$. Then we have

$$\langle M, \xi, a\rangle \vDash \bigvee x_i.\chi \iff \text{there is } c \sim_i a : \langle M, \xi, c\rangle \vDash \chi.$$

This is the form that we shall use later on. (It is closer in spirit to cylindrification.)

This construction is highly abstract. We shall illustrate it with a very simple example. Our language contains only equality. Suppose that we have a logic L which contains $\varphi \leftrightarrow \Box\varphi$ for every formula φ. Then we have $\Gamma \lhd_n \Delta$ iff $\Gamma = \Delta$ for all n-types Γ and Δ. So, the relations are trivial. Suppose also that the logic contains the sentence saying that there are exactly three objects. We shall calculate the cardinalities of the $M(n)$. There is exactly one 0-type Γ_0, since the logic is POST-complete. There exists exactly one 1-type, since all objects are indistinguishable. There exist 2 2-types, namely the type containing $x_0 \doteq x_1$ and the other containing $x_0 \not\doteq x_1$. The general formula is as follows. Let n be given. Choose a function $f : \{0, 1, \ldots, n-1\}$ into the set $\{0, 1, 2\}$. Then for this function the associated type is

$$t_f := \{x_i \doteq x_j : f(i) = f(j),\ i, j < n\} \cup \{\neg(x_i \doteq x_j) : f(i) \neq f(j),\ i, j < n\}.$$

Now let $f \approx g$ iff there is a permutation $\pi : \{0, 1, 2\} \to \{0, 1, 2\}$ such that $f = \pi \circ g$. Obviously, $t_f = t_g$ iff $f \approx g$. (Case 1) $f(i) = f(j)$ for all $i, j < n$. There are 3 functions, all representing the same type. (Case 2) The image of f has at least two members. There are $3^n - 3$ many such functions. Each type is represented by six functions. This gives $(3^{n-1} - 1)/2$ functions. In total we have $(3^{n-1} + 1)/2$ functions. The series is

$$1, 2, 5, 14, 41, 122, \ldots.$$

It is clear that the objects in such a frame are very difficult to recover. For this reason, Shehtman and Skvortsov define a *cartesian metaframe* (see below in Section 12 for a definition). This is a metaframe in which for each 0-type Γ the set of all n-types Δ with sentential reduct Γ is isomorphic to the n-fold cartesian product of the set of 1-types with sentential reduct Γ. Moreover, the projections are the maps $M(\iota_i^n)$, where $\iota_i^n : 1 \to n : 0 \mapsto i$. As Bauer [1] shows, each metaframe that allows to fuse types (a condition which we shall not define here) has a logically equivalent cartesian metaframe. The canonical metaframe defined above satisfies this condition. Thus, every logic is complete with respect to cartesian metaframes. Still, these proofs are very tedious. In the next section we shall show how the present results allow to prove completeness with respect to metaframe semantics using the previous completeness result.

§10. Going second order. In Kracht and Kutz [10] we have defined a notion of second order modal logics. Although they technically correspond to

Π^1_1-formulae, we shall nevertheless call them second order logics. To be precise we shall describe them as logics over a slightly different language. Namely, while before we had a set Π of predicate symbols, now we assume to have predicate variables of any given arity.

DEFINITION 37 (Symbols and languages). The languages of *second order modal predicate logic*, abbreviated collectively by $\mathbf{MPL}^2$, contain the following symbols.

1. A denumerable set $V := \{x_i : i \in \omega\}$ of *object variables.*
2. A denumerable set $C := \{c_i : i \in \omega\}$ of *constants.*
3. For each $n \in \omega$, a denumerable set $PV^n := \{P^n_i : i \in \omega\}$ of *predicate variables.*
4. *Boolean functors* $\bot, \wedge, \neg$.
5. *Quantifiers* $\bigvee, \bigwedge$.
6. A set $M := \{\Box_\lambda : \lambda < \kappa\}$ of *modal operators.*

Furthermore, $\Omega(P^n_i) = n$ for all $n, i \in \omega$.

This language does not contain an existence predicate constant, but this is only for convenience. There are no further complications in introducing predicate constants as well, but we have omitted them here (with the exception of equality) to keep the notation reasonably simple. As before, we deal with only one modal operator. The generalization is obvious. The following substitution principle has first been discussed by Steven Kleene in his [8].

DEFINITION 38 (Second order substitution). Let φ and χ be formulae of $\mathbf{MPL}^2$ and P be a predicate variable. Then $[\chi/P]\varphi$ denotes the formula that is obtained by replacing every occurrence of $[\vec{y}/\vec{x}]P$ by $[\vec{y}/\vec{x}]\chi$, where bound variables get replaced in the usual (first-order) way to prevent accidental capture.

We shall describe this substitution principle in a little more detail. Notice that χ can have free variables that are not among the variables $x_0, \dots, x_{\Omega(P)-1}$. Let $\vec{z}$ be these variables. Then we replace φ by a bound variant φ', in which all variables of φ occurring in $\vec{z}$ are replaced by suitable variables not occurring in either φ or χ. Next, we perform the replacement of any occurrence of $[\vec{y}/\vec{x}]P$ for some variables $\vec{y}$ by $[\vec{y}/\vec{x}]\chi$. This time, no bound variant needs to be chosen. For example, let

$$\varphi = \bigvee x_2.\bigwedge x_0.\bigwedge x_1.P^1_0(x_2) \wedge P^2_1(x_0, x_2) \rightarrow P^2_1(x_1, x_0).$$

Let $\chi = \bigvee x_1.P^3_0(x_2, x_1, x_0)$. Suppose that we want to replace P^2_1 by χ. Then since x_2 occurs free in χ, we shall replace bound occurrences of x_2 in φ by x_4. This gives

$$\varphi' = \bigvee x_4.\bigwedge x_0.\bigwedge x_1.P^1_0(x_2) \wedge P^2_1(x_0, x_4) \rightarrow P^2_1(x_1, x_0).$$

Finally, we have to replace $P_1^2(x_0, x_4)$ by $[x_4/x_1]\chi = \bigvee x_1.P_0^3(x_2, x_1, x_0)$ and also $P_1^2(x_1, x_0)$ by $[x_1/x_0, x_0/x_1]\chi = \bigvee x_3.P_0^3(x_2, x_3, x_1)$.

$$[\chi/P_1^2]\varphi = \bigvee x_4.\bigwedge x_0.\bigwedge x_1.P_0^1(x_2) \wedge \bigvee x_1.P_0^3(x_2, x_1, x_0) \\ \to \bigvee x_3.P_0^3(x_2, x_3, x_1).$$

The following definition is analogous to Skvortsov and Shehtman [13].

DEFINITION 39 (Second order MPLs). A *second order MPL* is a set L of **MPL**^{2}-formulae satisfying the following conditions.

1. L contains all instances of axioms of first-order logic.
2. L is closed under all rules of first-order logic.
3. L is closed under second-order substitution.
4. L contains all instances of axioms of the modal logic K.
5. L is closed under the rule $\varphi/\Box\varphi$.
6. $\Diamond \bigvee y.\varphi \leftrightarrow \bigvee y.\Diamond\varphi \in L$.

Clearly, a second order MPL can be viewed as a special sort of a first-order MPL, by taking Π to be the union of the sets PV^m. This allows us to speak, for example, of the canonical structure $\mathfrak{Can}_{L^*}$ for L. However, these languages are technically distinct, since the predicate variables are not interpreted in the structure. Their value is not fixed in the structure, just like the value of an object variable is not fixed in a first-order structure. This means that technically we get a different notion of structure. However, the way we get these structures is by abstracting them from the corresponding first-order structures. Thus, we begin with a second order MPL L and interpret it as a first-order MPL, also called L, for which we then build the canonical structure $\mathfrak{Can}_{L^*}$. Starting with this structure we shall define the second-order structure for L^*.

DEFINITION 40 (Second order coherence frames). A *second order coherence frame* is a triple $\langle W, \lhd, U\rangle$, where $\langle W, \lhd\rangle$ is a Kripke-frame and U a set.

Given a second order coherence frame, we call a member of $W \times U^n$ an *n-point* and a subset of $W \times U^n$ an *n-set*. Let $p = \langle v, \vec{a}\rangle$ and $q = \langle w, \vec{c}\rangle$ be n-points. We write $p \sim_k q$ if $a_i = c_i$ for all $i \neq k$.

$$\mathsf{V}_k(A) := \{q : \text{exists } p \in A : p \sim_k q\}.$$

(If $k < n$ does not obtain, then we may put $\mathsf{V}_k(A) := A$.) Next, let $\sigma : m \to n$. Then we define $\widehat{\sigma}$ on n-points as follows.

$$\widehat{\sigma}(\langle v, \vec{a}\rangle) := \langle v, \langle a(\sigma(i)) : i < m\rangle\rangle.$$

If p is an n-point, $\widehat{\sigma}(p)$ is an m-point. So, $\widehat{-}$ is a contravariant functor from Σ into the set of functions from points to points. This is also directly verified.

If $\tau : \ell \to m$ then

$$\begin{aligned}\widehat{\sigma \circ \tau}(p) &= \langle v, \langle a(\sigma \circ \tau(i)) : i < \ell \rangle\rangle \\ &= \widehat{\sigma}(\langle v, \langle a(\tau(i)) : i < \ell \rangle\rangle) \\ &= \widehat{\tau}(\widehat{\sigma}(p)) \\ &= \widehat{\tau} \circ \widehat{\sigma}(p).\end{aligned}$$

For an m-set A we put

$$\mathsf{C}_\sigma(A) := \{p : \widehat{\sigma}(p) \in A\}.$$

It is directly verified that C is covariant, that is, $\mathsf{C}_{\sigma\circ\tau} = \mathsf{C}_\sigma \circ \mathsf{C}_\tau$. For we have for an ℓ-set A:

$$\begin{aligned}\mathsf{C}_{\sigma\circ\tau}(A) &= \{p : \widehat{\sigma \circ \tau}(p) \in A\} \\ &= \{p : \widehat{\tau}(\widehat{\sigma}(p)) \in A\} \\ &= \{p : \widehat{\sigma}(p) \in \mathsf{C}_\tau(A)\} \\ &= \mathsf{C}_\sigma(\mathsf{C}_\tau(A)) \\ &= \mathsf{C}_\sigma \circ \mathsf{C}_\tau(A).\end{aligned}$$

And finally we set

$$\blacklozenge A := \{\langle w, \vec{a}\rangle : \text{exists } v \rhd w : \langle v, \vec{a}\rangle \in A\}.$$

DEFINITION 41 (n-complexes and towers). An *n-complex* over a 2nd order coherence frame is a set $\mathbb{C}_n$ of n-sets closed under intersection, complement, the operations V_k, $\blacklozenge$ and C_σ for every $\sigma : n \to n$. A *tower* is a sequence $\langle \mathbb{C}_n : n \in \omega\rangle$ such that $\mathbb{C}_n$ is an n-complex for every n, and for every $\sigma : n \to m$, $\mathsf{C}_\sigma : \mathbb{C}_n \to \mathbb{C}_m$.

DEFINITION 42 (Generalized second order coherence frames). A *generalized second order coherence frame* is a quadruple $\mathfrak{S} = \langle W, \lhd, U, \mathfrak{T}\rangle$, where $\langle W, \lhd, U\rangle$ is a second order coherence frame and $\mathfrak{T} = \langle \mathbb{C}_i : i \in \omega\rangle$ a tower over it. A *valuation* into $\mathfrak{S}$ is a pair ξ and β of mappings, where ξ is defined on all predicate variables, and $\xi(P_i^m) \in \mathbb{C}_m$ for all $m, i \in \omega$ and β assigns to each $x_i \in V$ a member of U.

$$\begin{aligned}\langle \mathfrak{S}, \xi, \beta, v\rangle \vDash P_i^m(\vec{y}) &:\Longleftrightarrow \langle \beta(y_i) : i < m\rangle \in \xi(P_i^m), \\ \langle \mathfrak{S}, \xi, \beta, v\rangle \vDash \neg\chi &:\Longleftrightarrow \langle \mathfrak{S}, \xi, \beta, v\rangle \nvDash \chi, \\ \langle \mathfrak{S}, \xi, \beta, v\rangle \vDash \varphi \wedge \chi &:\Longleftrightarrow \langle \mathfrak{S}, \xi, \beta, v\rangle \vDash \varphi; \chi, \\ \langle \mathfrak{S}, \xi, \beta, v\rangle \vDash \bigvee y.\chi &:\Longleftrightarrow \text{for some } \beta' \sim_y \beta : \langle \mathfrak{S}, \xi, \beta', v\rangle \vDash \chi, \\ \langle \mathfrak{S}, \xi, \beta, v\rangle \vDash \Diamond\chi &:\Longleftrightarrow \text{for some } w \rhd v : \langle \mathfrak{S}, \xi, \beta, w\rangle \vDash \chi.\end{aligned}$$

We write $\mathfrak{S} \vDash \varphi$ if for all valuations ξ, β and all worlds v : $\langle \mathfrak{S}, \xi, \beta, v\rangle \vDash \varphi$.

Notice that Shehtman and Skvortsov define the truth of a formula at an n-point. We can do the same here. Namely, we set

$$\langle \mathfrak{S}, \xi, \langle v, \vec{a} \rangle \rangle \vDash \varphi$$

iff for any valuation β such that $\beta(x_i) = a_i$ for all $i < n$ we have

$$\langle \mathfrak{S}, \xi, \beta, v \rangle \vDash \varphi.$$

An inductive definition can be given as well. We can also define a valuation on a metaframe in the following way. We say that a function $\beta : V \to \bigcup_n M(n)$ is a *valuation* if for every $n \in \omega$ (a) $\beta(x_n) \in M(n)$, and (b) $\beta(x_{n+1}) \downarrow \beta(x_n)$. Thus, $\beta(x_n)$ is an n-point which is the projection of the $n+1$-point $\beta(x_{n+1})$.

We shall show that second order MPLs are complete with respect to this semantics. So, let L be a second order MPL. We understand it as a first-order MPL, which we denote by the same letter. Then we can construct the canonical first-order coherence structure $\mathfrak{Can}_{L^*} = \langle W_{L^*}, \lhd, C_{L^*}, \mathfrak{I}_{L^*} \rangle$. We shall now define a second order canonical *frame* from it. This construction is completely general. First, observe that we can transport the notion of n-point as well as the satisfaction of a formula at an n-point to frames (and first-order coherence frames).

Take a modal (first-order) structure $\mathfrak{S} = \langle W, \lhd, U, \mathfrak{I} \rangle$. Let φ be a formula such that the free variables occurring in it are contained in $\{x_i : i < n\}$. Then we write $[\varphi]_n$ for the set of n-points satisfying φ. Formally,

$$[\varphi]_n := \{p : \langle \mathfrak{S}, p \rangle \vDash \varphi\}.$$

Now set

$$\mathbb{C}_n := \{[\varphi]_n : fvar(\varphi) \subseteq \{x_i : i < n\}\}.$$

Finally, we put $\mathfrak{T} := \langle \mathbb{C}_n : n \in \omega \rangle$. Now let

$$\mathfrak{S}^2 := \langle W, \lhd, U, \mathfrak{T} \rangle.$$

LEMMA 43. *$\mathfrak{S}^2$ is a second order generalized coherence frame. Furthermore, with $\xi(P) := \mathfrak{I}(P)$ we have for every n-point p and every formula φ with free variables in $\{x_i : i < n\}$:*

$$\langle \mathfrak{S}, p \rangle \vDash \varphi \iff \langle \mathfrak{S}^2, \xi, p \rangle \vDash \varphi.$$

PROOF. We need to verify that $\mathfrak{T}$ is a tower. This follows from the following equations. (In the last clause, $\sigma : n \to m$.)

$$\begin{aligned} [\neg\chi]_n &= -[\chi]_n, \\ [\varphi \wedge \chi]_n &= [\varphi]_n \cap [\chi]_n, \\ \Big[\bigvee x_k.\chi\Big]_n &= \mathsf{V}_k([\chi]_n), \\ [\Diamond\chi]_n &= \blacklozenge[\chi]_n, \\ [[\vec{x}^\sigma/\vec{x}]\chi]_n &= \mathsf{C}_\sigma([\chi]_m). \end{aligned}$$

Only the last clause needs comment.

$$\begin{aligned}[[\vec{x}^{\sigma}/\vec{x}]\chi]_n &= \{p \in U \times W^n : \langle \mathfrak{S}, p\rangle \vDash [\vec{x}^{\sigma}/\vec{x}]\chi\} \\ &= \{p \in U \times W^n : \langle \mathfrak{S}, \widehat{\sigma}(p)\rangle \vDash \chi\} \\ &= \{p \in U \times W^n : \widehat{\sigma}(p) \in [\chi]_m\} \\ &= \mathsf{C}_\sigma([\chi]_m).\end{aligned}$$

The second claim is immediate to verify. ⊣

THEOREM 44 (Second order completeness). *Let L be a second order modal logic without equality with the canonical structure $\mathfrak{Can}_{L^*}$ and φ a formula. Then $\mathfrak{Can}^2_{L^*} \vDash \varphi$ iff $\varphi \in L$. It follows that L is complete with respect to second-order generalized coherence frames.*

PROOF. We pass to the first-order canonical structure $\mathfrak{Can}_{L^*}$ of the (first-order) MPL L. Let ξ_{L^*} be defined by

$$\xi_{L^*}(P_i^m) := [P_i^m(x_0, \dots, x_{m-1})]_m.$$

Then, by first order completeness and Lemma 43 we get

$$\langle \mathfrak{Can}^2_{L^*}, \xi_{L^*}\rangle \vDash \varphi \Longleftrightarrow \varphi \in L.$$

We have to show that if $\varphi \in L$ then for every valuation ξ into $\mathfrak{Can}^2_{L^*}$ we have

$$\langle \mathfrak{Can}^2_{L^*}, \xi\rangle \vDash \varphi.$$

For this establishes $\mathfrak{Can}^2_{L^*} \vDash \varphi$ in case $\varphi \in L$. If $\varphi \notin L$ then we have anyway

$$\langle \mathfrak{Can}^2_{L^*}, \xi_{L^*}\rangle \nvDash \varphi$$

by first-order completeness and Lemma 43. Now for the proof of the claimed fact. Assume that ξ is given. By definition of the tower $\mathfrak{T}_{L^*}$ there exists for every predicate variable P a formula $\pi_P(\vec{x})$ such that

$$\xi(P) = \xi_{L^*}(\pi_P(\vec{x})).$$

Let $pvar(\varphi)$ denote the set of predicate variables occurring in φ. Define

$$\varphi_* := [\pi_P/P : P \in pvar(\varphi)]\varphi.$$

This formula is unique up to renaming of bound variables. Then, by induction, it is verified that

$$\langle \mathfrak{Can}^2_{L^*}, \xi, \Delta\rangle \vDash \varphi \Longleftrightarrow \langle \mathfrak{Can}^2_{L^*}, \xi_{L^*}, \Delta\rangle \vDash \varphi_*.$$

Since $\varphi \in L$ and L is closed under second order substitution we have $\varphi_* \in L$ as well. Hence the right-hand side obtains, and therefore the left-hand side is true as well. This is what we had to prove. ⊣

This construction of retracting the valuation ξ and adding the tower of definable sets is applicable to any first order coherence structure.

§11. A logic that is incomplete with respect to coherence frames. In this section we will give an axiomatization of a 2nd order modal predicate logic that is the logic of a single counterpart frame having two distinct counterpart relations. This logic will also turn out not to be valid on any coherence frame. Define the following counterpart frame $\mathfrak{F}$: let $W = \{w\}$ be the set of worlds, $U_w = \{a, b\}$ be the universe of w and $\mathcal{C}(w, w) = \{f, g\}$ the set of counterpart relations from w to itself, where $f : a \to a, b \to b$ and $g : a \to b, b \to a$. Notice that this frame is not object rich, for $\langle a, b\rangle$ is a thread in $\mathfrak{F}$ but there is no object that leaves both a and b as its trace in w.

Call Λ the second order MPL axiomatized as follows:

(a) B: $p \to \Box\Diamond p$, T: $p \to \Diamond p$, $\mathsf{4}$: $\Diamond\Diamond p \to \Diamond p$,
(b) $\mathsf{alt}_2 : \Diamond p \wedge \Diamond q \wedge \Diamond r \to \Diamond(p \wedge q) \vee \Diamond(p \wedge r) \vee \Diamond(q \wedge r)$,
(c) $\bigwedge x_0, x_1, x_2.(x_0 \doteq x_1 \vee x_1 \doteq x_2 \vee x_0 \doteq x_2)$,
(d) $\bigvee x_0, x_1.\neg(x_0 \doteq x_1)$,
(e) $\bigwedge x_0, x_1.x_0 \doteq x_1 \to \Box(x_0 \doteq x_1)$,
(f) $\bigwedge x_0, x_1.\neg(x_0 \doteq x_1) \to \Box\neg(x_0 \doteq x_1)$,
(g) $\bigwedge x_0, x_1.P(x_0) \wedge \neg(x_0 \doteq x_1). \to .\Box(P(x_0) \vee P(x_1))$.

Clearly, $\mathfrak{F} \vDash \Lambda$. In counterpart frames, the axioms (c) and (d) together state that there exist exactly 2 things in each world. When interpreted in coherence frames, they state that there are exactly two object traces in each world, but there may still be infinitely many objects. However, by equivalentiality, this implies that at most two objects are discriminable in each world.

Furthermore, (e) and (f) state that identity and difference are necessary. Finally, (g) states that whatever applies to a given object either applies to it in a successor world, or to the other object. It may be arguable from a philosophical point of view whether Λ is a genuine logic, because it makes *existence assumptions* about e.g., the number of objects or object traces in the world. Nevertheless, it surely is a logic in the technical sense of Definition 2.

THEOREM 45. *There is no class $\mathcal{K}$ of Lewisian counterpart frames such that Λ is the second order modal logic of $\mathcal{K}$. There is no class $\mathcal{C}$ of coherence frames such that Λ is the second order modal logic of $\mathcal{C}$.*

PROOF. The propositional reduct of this logic is the logic $\mathsf{S5.alt}_2$, which possesses exactly two nonisomorphic Kripke frames, namely the 1-point reflexive frame and the 2-point frame, where accessibility is universal. Furthermore, by necessity of identity and distinctness and the fact that the number of objects is constant and finite, counterpart relations are bijective functions. Suppose now that we have two worlds, v and w. Suppose further that there is a formula φ that is true of a but not of b in v, where a and b are the objects of the domain of v. Then, by virtue of (g), φ can be true only of one of the objects in w. There is however nothing that guarantees this if $v \neq w$. This argument is valid both for counterpart frames and coherence frames. So, we can have only one world. It remains to show that if there is only one world then the logic of the frame is stronger than Λ. Now, if $\mathfrak{C}$ is a coherence frame containing only one self-accessible world or if it is a counterpart frame containing only one world with one counterpart relation, then $\mathfrak{C} \models \Diamond p \to \Box p$, which is not a theorem of Λ, since $\mathfrak{F} \nvDash \Diamond p \to \Box p$. $\dashv$

It follows that whereas Λ is characterized by its canonical coherence model, compare Theorem 15, there is no coherence frame in which Λ is valid, that is to say, Λ is coherence frame incomplete. Evidently, if we add the right towers, completeness is regained (of course, with respect to generalized frames). On the other hand, we have shown in Section 7 how the notion of coherence frame can be modified to gain the same expressive power as counterpart semantics without moving to the full second-order semantics.

§12. **Cartesian metaframes.** In this section we shall use the previous completeness proof to derive a very simple completeness proof for the metaframe semantics. Shehtman and Skvortsov give the following definition.

DEFINITION 46 (Cartesian metaframes). A metaframe M is called *cartesian* if the following holds.

1. There is a set U and a family $\{W_u : u \in U\}$ of nonempty and pairwise disjoint sets such that $M(0) = U$ and $M(n) = \bigcup_{u \in U}(W_u)^n$ for every n. We write $a \lhd_1 b$ iff there are $u, v \in U$ such that $a \in W_u$, $b \in W_v$, $u \lhd_0 y$ and $\langle u, a\rangle \lhd_1 \langle v, b\rangle$.
2. $\langle v, \vec{a}\rangle \lhd_n \langle w, \vec{c}\rangle$ iff
   (a) $v \lhd_0 w$
   (b) $a_i \lhd_1 c_i$ for all $i < n$ and
   (c) for all $i < j < n$: if $a_i = a_j$ then also $c_i = c_j$.
3. For every $\sigma : n \to m$, $m, n > 0$, $M(\sigma) = \widehat{\sigma} : \vec{a} \mapsto \langle a(\sigma(i)) : i < n\rangle$. For $\sigma : 0 \to n$, $n > 0$, $M(\sigma) : \vec{a} \mapsto u$, where u is such that $\vec{a} \in (W_u)^n$.

For $\sigma : n \to 0$, there is no definition of $M(\sigma)$ given. One possibility is to choose an object $u^* \in W_u$ for every $u \in U$ and then let $M(\sigma) : u \mapsto (u^*)^n$. It is an approximation of the idea that the elements of the nth frame are n-tuples. While cartesian metaframes assume that the n-tuples are tuples of things, we shall offer another variant, where the idea is that the tuples are in fact tuples of objects.

DEFINITION 47 (Cubic metaframes). A metaframe M is called *cubic* if the following holds.

1. There are sets U and W such that $M(0) = U$ and $M(n) = U \times M^n$ for every n.
2. $\langle u, \vec{a}\rangle \lhd_n \langle v, \vec{c}\rangle$ iff $\vec{c} = \vec{a}$ and $u \lhd_0 v$.
3. For every $\sigma : n \to m$, $M(\sigma) = \widehat{\sigma} : \langle u, \vec{a}\rangle \mapsto \langle u, \langle a(\sigma(i)) : i < n\rangle\rangle$.

It is first of all to be checked that the above requirements define a contravariant functor from Σ to the class of generalized frames. (a) $M(n)$ is a general frame, as is easily seen. (b) for each $\sigma : m \to n$, $M(\sigma)$ is a p-morphism from $M(n)$ to $M(m)$. Namely, suppose that $p = \langle v, \vec{a} \rangle \lhd_n \langle w, \vec{c} \rangle = q$. Then $\vec{c} = \vec{a}$ and $v \lhd w$. Hence $\langle a(\sigma(i)) : i < m \rangle = \langle c(\sigma(i)) : i < m \rangle$, and so $\widehat{\sigma}(p) \lhd_m \widehat{\sigma}(q)$. Second, suppose that $\widehat{\sigma}(p) \lhd_m q'$. Then $q' = \langle w', \langle c'(i) : i < m \rangle \rangle$ for some w' such that $v \lhd w'$ and $c'(i) = a(\sigma(i))$ for each $i < m$. So, put $q := \langle w', \vec{a} \rangle$. Then $p \lhd_n q$ and $M(\sigma)(q') = q$. Third, let $A \in \mathfrak{T}(m)$. Then $\mathrm{C}_\sigma(A) \in \mathfrak{T}(n)$, by definition of towers. This proves that $M(\sigma)$ is a p-morphism. (c) For each $\sigma : m \to n$ and $\tau : n \to q$, $M(\tau \circ \sigma) = M(\sigma) \circ M(\tau)$. But by previous calculations, $M(\tau \circ \sigma)(p) = \widehat{\sigma} \circ \widehat{\tau}(p) = M(\sigma) \circ M(\tau)(p)$.

PROPOSITION 48. *For every cubic metaframe M there exists a semantically equivalent cartesian metaframe N.*

PROOF. Let M be a cubic metaframe. Put $N(0) := M(0)$ and $W_u := \{u\} \times W$ for all $u \in U$ and $\langle u, a \rangle \lhd_1 \langle v, b \rangle$ iff $u \lhd_0 v$ and $a = b$. Then $N(n) := \{\langle \langle u, a_i \rangle : i < n \rangle : a_i \in W\}$ for all n and $\vec{a} \lhd_n \vec{c}$ iff there are u and v such that $u \lhd_0 v$ and $a_i = \langle u, o_i \rangle$, $c_i = \langle v, o_i \rangle$ for some $o \in W$. The p-morphisms $M(\sigma)$ are straightforwardly defined. ⊣

The reason that this works is a construction that we have used before: the trace of an object at u may be the pair consisting of u and the object itself. It is easy to establish a bijective correspondence between second order generalized coherence frames and cubic generalized metaframes. Given a second order coherence frame $\langle W, \lhd, U, \mathfrak{T} \rangle$, we simply define $M(n) := \langle W \times U^n, \lhd_n, \mathfrak{T}(n) \rangle$, where $\langle v, \vec{a} \rangle \lhd_n \langle w, \vec{c} \rangle$ iff $\vec{a} = \vec{c}$ and $v \lhd w$. This is a cubic generalized metaframe. Conversely, let M be a cubic generalized metaframe, with $M(n) = \langle W \times U^n, \lhd_n, \mathbb{M}_n \rangle$ for every n. Then let $\lhd := \lhd_0$ and $\mathfrak{T}(n) := \mathbb{M}_n$. Then it is easily checked that $\langle W, \lhd, U, \mathfrak{T} \rangle$ is a second order general coherence frame. As an immediate consequence we get the following

THEOREM 49. *Every second order MPL without equality is complete with respect to cubic generalized metaframes.*

Notice that the way this result has been obtained is by abstraction from the first-order case, rather than the first-order case being an application of the second-order case. Completeness with respect to cartesian generalized metaframes now follows, given that cubic metaframes are special cartesian metaframes.

§13. **Equality.** Let us turn to the treatment of equality in coherence frames and metaframes. We have argued earlier that two objects may be equal in one world and different in another. Equality in a world has been regulated by the trace function. The most direct way to account for equality between objects is therefore to add the trace function into the generalized coherence frame. Another way is to add equality as a predicate constant whose interpretation is a equivalence relation on U in each individual world. Thus, we add a constant $\Delta \in \mathbb{C}_2$ such that the following holds. Write $a\ \Delta_w\ b$ iff $\langle w, a, b\rangle \in \Delta$.

1. $a_0\ \Delta_w\ a_0$ for all $w \in W$, $a_0 \in U$.
2. If $a_0\ \Delta_w\ a_1$ then $a_1\ \Delta_w\ a_0$, for all $w \in W$, $a_0, a_1 \in U$.
3. If $a_0\ \Delta_w\ a_1$ and $a_1\ \Delta_w\ a_2$ then $a_0\ \Delta_w\ a_2$ for all $w \in W$, $a_0, a_1, a_2 \in U$.

Additionally, a valuation must satisfy the following property. Call $A \in \mathbb{C}^n$ *equivalential* if for all $p = \langle w, \vec{a}\rangle \in A$, $q = \langle w, \vec{b}\rangle$ such that $a_i\ \Delta_w\ b_i$ for all $i < n$ then $q \in A$. Then we require that for every predicate P, $\xi(P)$ must be an equivalential set. However, notice that equivalential sets are not closed under $\blacklozenge$!

It may be disappointing to see that we have not been able to reduce $\doteq$ to simple identity. However, there is to our knowledge no semantics under which this is so (and for reasons given below it is not to be expected either). Let us look for example at metaframes. In a metaframe, $\langle M, \xi, a\rangle \vDash x_i \doteq x_j$ if $pr_{n,i}(a) = pr_{n,j}(a)$. Furthermore, a frame interpreting $\doteq$ (called an $m^=$-*metaframe*) must satisfy the following requirement:

$$(0^\sharp) \qquad \text{For all } n \text{ and } a, b \in M(n)\ a = b \Longleftrightarrow pr_{n,i}(a) = pr_{n,i}(b).$$

This condition effectively eliminates the distinction between object and trace. However, in general metaframes the possibility of distinct developments for identical objects still exists: let a be in $M(2)$. Think of a as the pair $\langle a_0, a_1\rangle$. If $pr_{2,0}(a) = pr_{2,1}(a)$, then $a_0 = a_1$. Now, accessibility is a relation between pairs, so if $a \lhd_2 b = \langle b_0, b_1\rangle$, we may or may not have $b_0 = b_1$. If we move to cartesian metaframes, the situation is different, however. For now, if $a_0 \lhd_1 b_0$ and $a_1 \lhd_1 b_1$ then from $a_0 = a_1$ we expect $a \lhd_2 \langle b_0, b_0\rangle, \langle b_1, b_1\rangle, \langle b_1, b_0\rangle$ as well. Shehtman and Skvortsov make some maneuvers to avoid this consequence.

First, let us look at a definition of cartesian metaframes and assume that the clause that $a_i = a_j$ implies $c_i = c_j$ was not there. Then the following principle is valid. If $\varphi(y, \vec{z})$ is a formula such that x_0 and x_1 do not occur in $\vec{z}$ then

$$M \vDash \bigwedge x_0. \bigwedge x_1. x_0 \doteq x_1 \to (\Diamond[x_0/y]\varphi(y, \vec{z}) \leftrightarrow \Diamond[x_1/y]\varphi(y, \vec{z})).$$

For let ξ be a valuation and $a \in M(n)$. Assume that $\langle M, \xi, a\rangle \vDash x_0 \doteq x_1$. Then we have $pr_{n,0}(a) = pr_{n,1}(a)$. Let $\alpha_0 : n-1 \to n : i \mapsto i+1$ and $\alpha_1 : n-1 \to n : 0 \mapsto 0,\ i \mapsto i+1\ (i \neq 0)$. Put $a_0 := M(\alpha_0)(a)$, $a_1 := M(\alpha_1)(a)$. Intuitively, a_0 is a reduced by its 0th coordinate, a_1 is a reduced by its 1st coordinate. From this it follows that $a_0 = a_1$. Assume next that $\langle M, \xi, a\rangle \vDash \Diamond[x_0/y]\varphi(y, \vec{z})$. Then there is a b such that $a \lhd_n b$ and $\langle M, \xi, b\rangle \vDash [x_0/y]\varphi(y, \vec{z})$. Put $b_0 := M(\alpha_0)(b)$ and $b_1 := M(\alpha_1)(b)$. We have $a_0 \lhd_{n-1} b_0$, since $M(\alpha_0)$ is a p-morphism, and likewise $a_1 \lhd_{n-1} b_1$. Since x_0 is not free in $[x_1/y]\varphi(y, \vec{z})$ we have $\langle M, \xi, b_1\rangle \vDash [x_0/y]\varphi(y, \vec{z})$. From this follows

$\langle M, \xi, a_1\rangle \vDash \Diamond[x_0/y]\varphi(y, \vec{z})$, and so $\langle M, \xi, a\rangle \vDash \Diamond[x_1/y]\varphi(y, \vec{z})$. Likewise for the other direction.

So we find, as indicated, that without the clause, metaframes imitate counterpart semantics. However, Shehtman and Skvortsov have added it. Thereby they avoid counterpart semantics, but there is a price to be paid.

LEMMA 50. *Let M be a cartesian $m^{=}$-metaframe. Then*

$$M \vDash \bigwedge x_0. \bigwedge x_1.x_0 \doteq x_1 \rightarrow \Box(x_0 \doteq x_1).$$

So, neither of the alternatives is completely general. It turns out that metaframe semantics could have been saved in the same way as coherence semantics, namely by adding a constant interpreting equality. This seems to be necessary. If we do not treat equality in this way, we must assume that the interpretation of identity is an equivalence relation. Shehtman and Skvortsov have shown that the condition ($0^\sharp$) makes the semantics less general: there are formulae which are not generally valid but valid in all metaframes satisfying ($0^\sharp$). This indicates that equating objects and object traces even done in metaframes à la Shehtman and Skvortsov cannot eliminate the problems of identity.

§14. Conclusion. In this paper we have defined a new semantics for modal predicate logic, namely coherence frames. Coherence frames differ from counterpart frames in that variables are interpreted in the same way as constants, namely by objects. We have shown completeness both for first-order and for second-order MPLs with respect to generalized coherence frames. From this we have derived a completeness theorem for second order MPLs with respect to generalized metaframe semantics. In fact, completeness with respect to cubic generalized metaframes is obtained rather directly.

The proposal of distinguishing between an object and its trace is certainly a very far reaching one but not without justification. Many philosophers have argued that there may exist different identity criteria for objects (see van Leeuwen [15] for a review of these ideas). A statue is not the same as the material it is made of. Hence, though perhaps trace identical, the two are not the same objects. There are predicates that are sensitive to this difference (again see [15]). These predicates reject the postulate Eq4. It goes beyond the scope of this paper to review the possibilities that coherence structures offer in this respect.

Acknowledgement. We would like to thank Kit Fine, Melvin Fitting, Greg Restall, Gerhard Schurz and Frank Wolter for various helpful discussions and remarks as well as an anonymous referee for her or his comments and corrections.

REFERENCES

[1] SEBASTIAN BAUER, ***Metaframes, Typen und Modelle der modalen Prädikatenlogik***, Master's thesis, Fachbereich Mathematik, Humboldt–Universität zu Berlin, 2000.

[2] MAX J. CRESSWELL, *How to complete some modal predicate logics*, ***Advances in Modal Logic. Volume 2*** (Michael Zakharyaschev et al., editors), CSLI Publications, Stanford, 2001.

[3] MELVIN FITTING, ***Types, Tableaus and Gödel's God***, Trends in Logic, vol. 12, Kluwer Academic Publishers, Dordrecht, 2002.

[4] ———, *First-order intensional logic*, ***Annals of Pure and Applied Logic***, vol. 127 (2004), no. 1-3, pp. 171–193, Provinces of logic determined. Essays in the memory of Alfred Tarski. Parts IV, V and VI. Edited by Z. Adamowicz, S. Artemov, D. Niwinski, E. Orlowska, A. Romanowska and J. Wolenski.

[5] MELVIN FITTING and RICHARD L. MENDELSOHN, ***First–Order Modal Logic***, Kluwer Academic Publishers, Dordrecht, 1998.

[6] SILVIO GHILARDI, *Quantified extensions of canonical propositional intermediate logics*, ***Studia Logica***, vol. 51 (1992), pp. 195–214.

[7] GEORGE E. HUGHES and MAX J. CRESSWELL, ***A New Introduction to Modal Logic***, Routledge, London, 1996.

[8] STEVEN C. KLEENE, ***Introduction to Metamathematics***, Wolters-Noordhoff Publishing, Groningen, 1971.

[9] MARCUS KRACHT, ***Tools and Techniques in Modal Logic***, Studies in Logic and the Foundations of Mathematics, vol. 142, Elsevier Science Publishers, Amsterdam, 1999.

[10] MARCUS KRACHT and OLIVER KUTZ, *The Semantics of Modal Predicate Logic I. Counterpart–Frames*, ***Advances in Modal Logic. Volume 3*** (Frank Wolter, Heinrich Wansing, Maarten de Rijke, and Michael Zakharyaschev, editors), World Scientific Publishing Company, 2002.

[11] GERHARD SCHURZ, ***The Is-Ought Problem: An Investigation in Philosophical Logic***, Trends in Logic, vol. 1, Kluwer Academic Publishers, Dordrecht, 1997.

[12] DANA SCOTT, *Advice on Modal Logic*, ***Philosophical Problems in Logic. Some Recent Developments*** (Karel Lambert, editor), Reidel, Dordrecht, 1970, pp. 143–174.

[13] DIMITER P. SKVORTSOV and VALENTIN B. SHEHTMAN, *Maximal Kripke–Type Semantics for Modal and Superintuitionistic Predicate Logics*, ***Annals of Pure and Applied Logic***, vol. 63 (1993), pp. 69–101.

[14] RICHMOND H. THOMASON, *Some completeness results for modal predicate calculi*, ***Philosophical Problems in Logic. Some Recent Developments*** (Karel Lambert, editor), Reidel, Dordrecht, 1970, pp. 56–76.

[15] JACQUES VAN LEEUWEN, ***Individuals and Sortal Concepts. An Essay in Logical Metaphysics***, Ph.D. thesis, University of Amsterdam, 1991.

DEPARTMENT OF LINGUISTICS, UCLA
3125 CAMPBELL HALL
405 HILGARD AVENUE
BOX 951543
LOS ANGELES, CA 90095-1543, USA
E-mail: kracht@humnet.ucla.edu

DEPARTMENT OF COMPUTER SCIENCE
THE UNIVERSITY OF MANCHESTER
KILBURN BUILDING, OXFORD ROAD
MANCHESTER, M13 9PL, UK
E-mail: kutz@cs.man.ac.uk

INTENSIONALITY AND COERCION

MICHIEL VAN LAMBALGEN AND FRITZ HAMM

Abstract. Frege informally characterized intension or *Sinn* as *Art des Gegebenseins*, that is, 'mode of presentation' of an object or event. The linguistic phenomenon of coercion, as exemplified by the shift in aspectual class of the verb 'love' from state in (a) to activity in (c):

(a) I love her.

(b) * I am loving her.

(c) I am loving her more and more every day, the more I get to know her.

thus falls in the domain of intensional phenomena. We present a fully computational theory of aspectual coercion, based on the event calculus of Aritificial Intelligence, as reformulated in constraint logic programming. The cognitive background of this formalism, as well as other applications to tense and aspect, can be found in the authors' *The Proper Treatment of Events* (Blackwell 2004).

§1. Aspectual coercion as an intensional phenomenon. Ever since Vendler's famous classification of verbs with respect to their time schemata, linguists have distinguished at least four aspectual classes or Aktionsarten. These classes comprise states, activities, accomplishments, achievements and points[1], which are exemplified in (1) to (5).

(1) *States*: know, love, be beautiful.

(2) *Activities*: run, push a cart, draw.

(3) *Accomplishments*: cross the street, build a house, draw a letter.

(4) *Achievements*: begin, reach, arrive.

(5) *Points*: flash, spot, blink.

Several linguistic tests were developed to distinguish aspectual classes. For example, adverbial modification with *for two hours* is possible with activities (John ran for two hours) but not with accomplishments (*John built a house for two years) or achievements. Temporal modification with the adverbial *in two hours* shows the reverse pattern. Usually only activities and accomplishments can occur in the progressive. Expressions like **knowing the answer* are

[1]This category was introduced by Moens and Steeedman, see [20].

Intensionality
Edited by Reinhard Kahle
Lecture Notes in Logic, 22

ungrammatical. But there is also a crucial difference in the behaviour of activities and accomplishments with respect to their progressivised forms. From *John was pushing a cart* we can infer that John pushed a cart. However, the inference from *John was crossing the street* to *John crossed the street* is invalid. This phenomenon was dubbed the "imperfective paradox" by Dowty[2].

Note, however, that the aspectual properties introduced above cannot be *lexical* properties of the respective verbs alone. For instance, *John drank* is certainly an activity and *John drank beer* is an activity too. However, *John drank a glass of beer* is not an activity but an accomplishment. Therefore, the choice of the object NP (a mass term versus an indefinite NP) is a determining factor for aspectual classes. The next example from [20] shows that Aktionsart is not fully determined prior to the sentence level. *Chapman arrived* is an achievement, as can be seen from the ungrammaticality of **Chapman arrived all night*. But if we choose a bare plural subject instead of a proper name as subject NP, we get the grammatical sentence *Visitors arrived all night*.

Therefore the choice of the object and subject NPs can force verbs to be interpreted in different aspectual classes. This reinterpretation process was dubbed *coercion* in [14]. Unfortunately, this term does not refer to a coherent class of phenomena. Consider example (6):

(6) Pollini played the sonata for two days.

Pollini played the sonata is an accomplishment and accomplishments are generally bad with for-adverbials as pointed out above. Nevertheless sentence (6) can mean that Pollini played the respective sonata repeatedly within a timespan of two days. Thus the accomplishment *Pollini played the sonata* is coerced to an iterative reading. But this time we cannot identify a linguistic constituent of the sentence which is responsible for this reinterpretation. The expression *for two days* does not force an iterative reading in *John slept for two days* and neither does the accomplishment *Pollini played the sonata*.

Worse, even cases of metonymy are often considered as instances of coercion. Thus a sentence like *The pianists are on the top shelf* interpreted with respect to a scene within a CD-shop means that the CDs of pianists are on the top shelf. The NP *The pianists* is in this context reinterpreted as *CDs of pianists* on the basis of a threatening type conflict with the VP *be on the top shelf*.

In this paper we will certainly not do justice to all phenomena which are grouped together under the linguistic term *coercion*, but we will concentrate on those which pertain to Vendler's classification outlined above. Moreover, we will not comment on the vast literature on reinterpretation but refer the interested reader to a selection of more recent work[3].

---

[2] Many tests which we cannot mention here were devised by linguists. An almost comprehensive list is in [6].

[3] See especially [14], [20], [16], [17], [5], [4], [8].

In several quarters of the semantic world, Vendler's classification of verbs into states, achievements, activities and accomplishments is often conceived of as pertaining to inherent properties of a verb (or, rather verb phrase). By contrast, various forms of cognitive grammar treat aspect as a way to *impose* temporal structure on events, or a way to conceptualise the temporal constitution of events. In the words of Croft [3, p. 70]:

> The aspectual grammatical constructions determine in part the temporal structure of the event it describes via conceptualization.

The difference between these ways of treating aspect is subtle but important, as can be made clear by means of an example due to Croft (*op. cit.*). If the verb *love* is considered to be inherently stative, then

(7) *I am loving her

is ungrammatical, and cognitive linguists would agree; here the intended content has to be expressed by

(8) I love her.

But if (7) is provided with a context, it may become grammatical, as in

(9) I am loving her more and more every day, the more I get to know her.

This phenomenon is hard to explain if stativity is taken to be a property of the verb *love*. It seems more profitable to explain the contrast by saying that the events referred to in (8) and (9) are conceptualised differently, so that the event taken as an atom in (8) is considered to have stages and phases in (9). Thus, the phenomenon of *aspectual coercion* comes to the fore: the potential of grammatical constructions, such as the progressive, to "move" a verb from one aspectual category to another.

Note that in general we cannot expect to arrive at the result of such a "move" in a compositional way. To see this, consider again example (6). As pointed out, the linguistic material itself does not give any clue that an iterative reading is prominent here. We have to rely on extralinguistic knowledge such as the timespan it usually takes to play a sonata to derive this meaning. A theory of coercion effects should therefore allow to combine very different aspects of meaning within a unified formal framework.

Traditional accounts of intensionality have been very much concerned with objects (Morning Star/Evening Star) and propositions (φ in "I believe that φ"). The above considerations suggest that intensionality also plays a prominent role when talking about events. In fact, Frege's view of intentionality as *Art des Gegebenseins* fits the phenomena outlined above very well.

In modern formal semantics, Frege's notion of *Art des Gegebenseins* is reconstructed via possible worlds. Therefore, intensional phenomena in semantics are those where possible worlds play an essential role for the description of their meaning. A quick glance at natural language constructions analysed with possible worlds in modern semantics reveals that this again is an extremely heterogenous domain. Here are two examples which demonstrate this diversity quite clearly. The first are propositional attitudes like *John believes that Pythagoras was looking for the rational* $\sqrt{2}$. Here the set of worlds which support the sentence *Pythagoras was looking for the rational* $\sqrt{2}$ and are compatible with John's belief are used to models John's belief state.

For a progressivised accomplishment like *Pollini was playing the sonata* which is in general analysed as an intensional construction, the role of worlds is completely different. In Dowty's by now classical analysis, worlds are used to model *inertia*.

Furthermore, the standard tests for identifying intensional constructions emphasise the observed difference. According to the first, a construction is *extensional* if equivalent expressions can be substituted *salva veritate*. An expression is intensional if there are exceptions to this substitutivity principle. Clearly propositional attitudes are intensional according to this principle. The progressive, on the other hand, is not intensional according to this test. From *Bill was painting Dan* (*Kavanaugh*) we can intuitively infer *Bill was painting Julian* (*Barnes*), given that Dan (Kavanaugh) and Julian (Barnes) are the same person.

The second criterion says that intensional constructions are those for which the rule of existential generalisation fails. Again, propositional attitudes are intensional according to this criterion[4]. On the basis of this principle, the progressive, too, is classified as intensional. From *Mary was drawing a circle* we cannot infer that there exists a circle that Mary was drawing.

Other intensional constructions in natural languages, such as modals, infinitives, verbs like *seek* versus at first sight extensional verbs like *fetch* clearly demonstrate the empirical heterogenity of the intensional domain.

The conclusion is that according to the practice of possible world semantics, *coercion* would certainly be considered an intensional phenomenon. However, given the empirical diversity on the one hand and the rather one-dimensional methodological approach of possible world semantics on the other, this seems to be a rather weak reason. Therefore, the clearest conceptual motivation for treating *coercion* as an intensional phenomenon rests on Frege's description of his concept *Art des Gegebenseins*.

---

[4]Neither principle is watertight. This was observed for the first principle in [18]. Consider the sentence

Clark Kent went into the phone booth, and Superman came out.

Traditionally this is not thought of as an intensional construction; nevertheless *Clark Kent* cannot be substituted for *Superman* salva veritate in this sentence. The counterexample to the second principle is from [2]. The sentence *The rational* $\sqrt{2}$ *does not exist* is considered extensional, but existential generalisation applied to it results in the nonsense *There exists something such that it does not exist*.

§2. A brief introduction to the event calculus. We need a formal system in which all the conceptualisations of events as they occur in natural language can be defined. To have an idea of the kind of structure that is necessary, it is useful to consider one of the most complex classes of verbs, called *accomplishments* by Vendler, of which examples are "draw a circle", "write a letter", "cross the street". Events representing such verbs have an elaborate internal structure. On the one hand there is an activity taking place (draw, write, cross), on the other hand an "object" is being "constructed": the circle, the letter, or the path across the street. Furthermore, each such event is associated with a natural consequent state, where a (complete) circle has been drawn, a letter has been written, or a street has been crossed. We may also remark that the activities involved are volitional and so come with events marking the beginning or end of the activity. The literature contains several formalisms for reasoning about events, which have their roots in planning systems in artificial intelligence. It has been suggested several times that such formalisms might be useful for the semantics of natural language, although [10] seems to be the first paper where the actual computations are done. We borrow the basic format from [19], although the computational tools will be different.

The above considerations suggest that at least the following ontology is necessary for talking about events:

1. individuals
2. real numbers, both to represent time and to code "stages" of partial objects
3. time-dependent properties, such as activities
4. changing partial objects
5. events, marking the beginning and end of time-dependent properties.

Both time-dependent properties and changing partial objects can be brought under the heading of a *fluent*. A fluent is a function which may contain variables for individuals and reals and which is interpreted in a model as a set of time points. Events will be taken in the sense of event types, from which event tokens are obtained by anchoring the event type to a time point (sometimes also to an interval).

Before going on to a discussion of the primitive predicates, we must fix our usage of the term "event". The attentive reader will have noticed that we have used "event" in two senses: as a generic term, as when we said that events can be conceptualised in different ways, and in a more specific sense, as marking the beginning and end of time-dependent properties. In the context of the formal theory, events in the second sense are taken to be unstructured objects (possibly paramatrised). Henceforth, events in the first sense will be called *eventualities*, following Bach's usage in [1][5].

[5]This distinction is not meant to be absolute. It could very well be that an event in the second sense does have interesting internal structure. The distinction is meant to be relative to the choice

Given this ontology, the following choice of basic predicates seems natural. We want to be able to say that events happen, that fluents are initiated and terminated by events, or that a fluent was true at the beginning of time. If f is a variable over fluents, e a variable over events, and t a variable over time points, we may write the required predicates as

1. $\mathit{Initially}(f)$
2. $\mathit{Happens}(e, t)$
3. $\mathit{Initiates}(e, f, t)$
4. $\mathit{Terminates}(e, f, t)$

These predicates are to be interpreted in such a way, that if $\mathit{Happens}(e, t) \wedge \mathit{Initiates}(e, f, t)$, then f will begin to hold after (but not at) t; if $\mathit{Happens}(e, t) \wedge \mathit{Terminates}(e, f, t)$, then f will still hold at t. Strictly speaking, this convention makes sense only for events which are not extended in time. For the general case, one needs some axioms additional to those presented below; we will omit these for the sake of simplicity.

The possibility of having changing partial objects requires its own special predicates, namely

5. $\mathit{Trajectory}(f_1, t, f_2, d)$
6. $\mathit{Releases}(e, f, t)$

We have seen above that an activity, while it is going on, may change a partial object. The first predicate in the above list embodies this. Here, one should think of f_1 as an activity and of f_2 as a certain stage of a partial object. The predicate then expresses that if f_1 holds from t until $t + d$, then at $t + d$, f_2 holds. In applications, f_2 will have a real argument and will be of the form $f_2(g(t + d))$ for some continuous function g. The predicate *Releases* is necessary to cancel the effect of those axioms of the event calculus which intend to express the so-called "principle of inertia"[6]. These axioms have the form: if there are no f-relevant events between t_1 and t_2, then the truth value of f at t_1 is the same as that at t_2. We introduce two special predicates for f-relevant events. The first predicate expresses the idea that there is a terminating or releasing event between t_1 and t_2; the second predicate states that there is an initiating or releasing event between t_1 and t_2.

7. $\mathit{Clipped}(t_1, f, t_2)$
8. $\mathit{Declipped}(t_1, f, t_2)$

Lastly, we need the truth predicate

9. $\mathit{HoldsAt}(f, t)$

---

of a level of granularity, and the formal theory applies to a specific level only. It is a different matter entirely to model a transition between levels.

[6]Which says: "unless there is explicit information to the contrary, we may assume an action does not affect a fluent".

In the usual set-up of the event calculus, it is only *said* that *HoldsAt* is a truth predicate; the defining axioms for the truth predicate are lacking. In planning applications of the event calculus, fluents are typically derived from first order formulas in a language disjoint from that of the event calculus itself. "Derive" here refers to an operation which transforms formulas into terms, for example, Gödelisation, or what in AI is termed reification. It is of course easy to declare a two-valued truth predicate applying to such fluents only. However, once the fluents are codes of formulas which may also contain predicates from the event calculus, this becomes problematic, and in that case one needs to add a logic program for the truth predicate to the event calculus. The proper logic for such a truth predicate is Kleene's strong three-valued logic. We do not know of any planning application which needs this generality, but, as explained in [10], for a semantics of natural language this generality is essential. Gödelisation or reification is the formal counterpart of the syntactic operation of nominalisation, and when this procedure is iterated, as in

(10) My father objecting to my not going to church

the truth conditions for sentences involving this expression involve nested occurrences of *HoldsAt* as well. These observations notwithstanding, in the interest of brevity and legibility we shall not add a truth theory, and we shall assume some external interpreter available which relates a fluent to the formula (not containing predicates of the event calculus) from which it derives.

All in all, we then have

DEFINITION 1. An EC-structure is a many-sorted structure of the form[7] $(R \uplus D \uplus E \uplus F; 0, 1, +, \cdot; a_1 \ldots a_k, f_1 \ldots f_n, e_1 \ldots e_m, \mathit{Basic})$, where

1. R denotes a real closed field, D a set of individual objects, E a set of events, F a set of fluents.
2. $a_1 \ldots a_k$ are elements of D and interpret individual constants.
3. $f_1 \ldots f_n$ are the interpretations of fluent functions (with variables for reals and individuals).
4. $e_1 \ldots e_m$ are the interpretations of event functions (with variables for reals and individuals).
5. *Basic* is an interpretation of the basic predicates introduced above.

DEFINITION 2. The predicates *Initially*, *Happens*, *Initiates*, *Terminates*, *Trajectory* and *Releases* are called the *primitive predicates* of the event calculus (We consider *HoldsAt* to belong to the truth theory.).

[7] $\uplus$ denotes disjoint union.

The axioms of the event calculus given below are modified from [19], the difference being due to the fact that we prefer a constraint logic programming approach, whereas Shanahan uses circumscription. In the following, all variables are assumed to be universally quantified.

AXIOM 1. $\mathit{Initially}(f) \wedge \neg \mathit{Clipped}(0, f, t) \rightarrow \mathit{HoldsAt}(f, t)$.

AXIOM 2. $\mathit{Happens}(e, t) \wedge \mathit{Initiates}(e, f, t) \wedge t < t' \wedge \neg \mathit{Clipped}(t, f, t') \rightarrow \mathit{HoldsAt}(f, t')$.

AXIOM 3. $\mathit{Happens}(e, t) \wedge \mathit{Terminates}(e, f, t) \wedge t < t' \wedge \neg \mathit{Declipped}(t, f, t') \rightarrow \neg \mathit{HoldsAt}(f, t')$.

AXIOM 4. $\mathit{Happens}(e, t) \wedge \mathit{Initiates}(e, f_1, t) \wedge t < t' \wedge t' = t + d \wedge \mathit{Trajectory}(f_1, t, f_2, d) \wedge \neg \mathit{Clipped}(t, f_1, t') \rightarrow \mathit{HoldsAt}(f_2, t')$.

AXIOM 5. $\mathit{Happens}(e, s) \wedge t < s < t' \wedge (\mathit{Terminates}(e, f, s) \vee \mathit{Releases}(e, f, s)) \rightarrow \mathit{Clipped}(t, f, t')$.

AXIOM 6. $\mathit{Happens}(e, s) \wedge t < s < t' \wedge (\mathit{Initiates}(e, f, s) \vee \mathit{Releases}(e, f, s)) \rightarrow \mathit{Declipped}(t, f, t')$.

The set of axioms of the event calculus will be abbreviated by EC. In the absence of further axioms, one can construct a model for EC by interpreting fluents as finite unions of halfopen intervals (of the form $(a, b]$) and assuming that each event either initiates or terminates a fluent and that fluents are initiated or terminated by events only. Interestingly, such models are also available in the presence of further sentences, laying down preconditions on *Happens* and the other primitive predicates. Indeed, this property will play an important role in the proposed aspectual calculus.

2.1. Scenarios. While the above axioms state what is generally true about fluents and events, the structure of specific eventualities must be described by micro-theories. For example, in the case of "draw a circle" the eventuality is composed (at least) of an activity ("draw") and a changing partial object (the circle in its various stages of completion); the micro-theory should specify how the activity "draw" is related to the amount of circle constructed. This is done by means of three definitions.

DEFINITION 3. A state $S(t)$ at time t is a complex body (cf. Definition 5) involving only

1. literals of the form $(\neg)$ $\mathit{HoldsAt}(f, t)$, for t fixed and possibly different f.
2. equalities between fluent terms, and between event terms.
3. formulas in the constraint language $\mathcal{L}$.

DEFINITION 4. A *scenario* is a conjunction of statements of the form

1. $\mathit{Initially}(f)$, or
2. $\mathit{HoldsAt}(f, t_0)$ (t_0 a constant), or

3. $\forall t\ (S(t) \rightarrow \mathit{Initiates}(e, f, t))$, or
4. $\forall t\ (S(t) \rightarrow \mathit{Terminates}(e, f, t))$, or
5. $\forall t, d\ (S(f_1, f_2, t, d) \rightarrow \mathit{Trajectory}(f_1, t, f_2, d))$, or
6. $\forall t\ (S(t) \rightarrow \mathit{Releases}(e, f, t))$, or
7. $\forall t, s\ (S(t) \rightarrow \mathit{Happens}(e, t))$

where $S(t)$ is a state in the sense of Definition 3. These formulas may contain additional constants for objects[8], reals or time points and can be prefixed by universal quantifiers over time points, reals and objects.

§3. Minimal models. When it comes to making inferences, the event calculus as presented so far is very weak, and definitely not a model for human reasoning about events. Common sense reasoning seems to be governed by the principle of inertia: "normally, an action does not affect a property of the world". This requires a non-classical notion of validity, in which one does not look at *all* models of a set of premises, but only at *minimal* models, that is models in which the interpretations of the primitive predicates of the event calculus are as small as is consistent with those premises. In such models there are no spurious happenings of events which may nonetheless affect classical validity. For example, we would like to derive from a scenario for "cross the street" that, barring unforeseen circumstances, one reaches the other side of the street. It is impossible to derive such a conclusion classically, because there will be a model in which lightning *happens* and has the effect of *terminating* the crossing. There are several ways of formalising this intuitive idea. Shanahan chose circumscription, we opt for constraint logic programming with (a version of) negation as failure, mainly because we want to have the reals as the underlying structure of time.

3.1. Logic programming with real-closed fields. Constraint logic programming is concerned with the interplay of two languages. Let $\mathcal{L}$ be the language $\{0, 1, +, \cdot, <\}$, $\mathcal{T}$ the complete theory of $(\mathbb{R}, 0, 1, +, \cdot, <)$ in $\mathcal{L}$, i.e. the theory of real-closed fields. Let $\mathcal{K}$ be another language, consisting of programmed predicate symbols. The constraint programming language CLP($\mathcal{T}$) consists on the one hand of *constraints*, which are first order formulas from the language $\mathcal{L}$, and on the other hand of formulas from $\mathcal{K}$, whose terms come from $\mathcal{L}$[9]. Constraint logic programs differ from logic programs by allowing constraints in the bodies of rules and in queries. For example, primitive constraints in CLP($\mathcal{T}$) include formulas of the form $s < t$ and $s = t$, where s, t are terms from $\mathcal{L}$. *Definite* constraint logic programs have the form

$$B_1 \wedge \cdots \wedge B_m \wedge c \longrightarrow A,$$

[8]We do not allow functions on objects.

[9]We shall slightly relax this condition later.

where B_i and A are atoms in $\mathcal{K}$ and c is a constraint. Likewise, a query has the logical form

$$B_1 \wedge \cdots \wedge B_m \wedge c \longrightarrow \bot.$$

We shall use the notation

$$?c, B_1 \ldots B_m$$

for queries, always with the convention that c denotes the constraint, and that the remaining formulas come from $\mathcal{K}$. The words "query" and "goal" will be used interchangeably.

The aim of a constraint computation is to express a programmed predicate symbol entirely in terms of constraints. Thus, unlike the case of ordinary logic programming, the last node of a successful branch in a computation tree contains a constraint instead of the empty clause.

For our purposes, definite constraint logic programs are not yet expressive enough. We follow [21] in allowing (classical equivalents of) arbitrary first order formulas in the body of program clauses. More precisely, we define a *complex subgoal* recursively to be

1. an atom in $\mathcal{K}$, or
2. $\neg\exists\overline{x}\,(B_1 \wedge \cdots \wedge B_m \wedge c)$, where c is a constraint and each B_i is a complex subgoal.

DEFINITION 5. A *complex body* is a conjunction of complex subgoals. A *normal program* is a formula $\phi \to A$ of CLP($\mathcal{T}$) such that ϕ is a complex body and A is an atom.

If, in the second clause of the above definition, we take $\overline{x}$ to be empty, we obtain ordinary goals (with constraints), which as indicated will be written as $?c, B_1 \ldots B_m$.

The interpretation of negation most congenial to constraint logic programming is *constructive* negation [21]. In the customary negation as failure paradigm, negative queries differ from positive queries: the latter yield computed answer substitutions, the former only the answers "true" or "false". Constructive negation tries to make the situation more symmetrical by also providing computed answer substitutions for negative queries. Applied to constraint logic programming, this means that both positive and negative queries can start successful computations ending in constraints. As in the case of negation as failure, the proper logic for negation in constraint logic programming is Kleene's strong three-valued logic with truth values $\{\boldsymbol{t}, \boldsymbol{f}, \boldsymbol{u}\}$. Semantic consequence in this logic will be denoted by $\models_3$.

The fundamental technical tool is likewise the *completion* of a program:

DEFINITION 6. Let $\mathcal{P}$ be a normal program, consisting of clauses

$$\overline{B}^1 \wedge c_1 \longrightarrow p^1(\overline{t}^1), \ldots, \overline{B}^n \wedge c_n \longrightarrow p^n(\overline{t}^n),$$

where p^i are atoms. The completion of $\mathcal{P}$, denoted by comp($\mathcal{P}$), is computed by the following recipe:

1. choose a predicate p that occurs in the head of a clause of $\mathcal{P}$
2. choose a sequence of new variables $\overline{x}$ of length the arity of p
3. replace in the ith clause of $\mathcal{P}$ all occurrences of a term in $\overline{t}_i$ by a corresponding variable in $\overline{x}$ and add the conjunct $\overline{x} = \overline{t}_i$ to the body; we thus obtain $\overline{B}^i \wedge c_i \wedge \overline{x} = \overline{t}_i \rightarrow p^i(\overline{x})$
4. for each i, let $\overline{z}_i$ be the set of free variables in $\overline{B}^i \wedge c_i \wedge \overline{x} = \overline{t}_i$ not in $\overline{x}$
5. given p, let $n_1, \ldots, n_k$ enumerate the clauses in which p occurs as head
6. define Def(p) to be the formula

$$\forall \overline{x}\, \left(p(\overline{x}) \longleftrightarrow \exists \overline{z}_{n_1} \left(\overline{B}^{n_1} \wedge c_{n_1} \wedge \overline{x} = \overline{t}_{n_1}\right) \vee \cdots \vee \exists \overline{z}_{n_k} \left(\overline{B}^{n_k} \wedge c_{n_k} \wedge \overline{x} = \overline{t}_{n_k}\right)\right).$$

7. comp($\mathcal{P}$) is then obtained as the formula $\bigwedge_p$ Def(p), where the conjunction ranges over predicates p occurring in the head of a clause of $\mathcal{P}$[10].

We see that the axioms form a normal logic program in the sense of CLP($\mathcal{T}$), with constraints of the form $s < t$[11]. In the logic program, the programmed predicates are ($\neg$)*HoldsAt*, *Clipped* and *Declipped.* The programs for the remaining primitive predicates, i.e. the scenario and the dynamics, also form a normal logic program. We then see that applying program-completion (cf. Definition 6) to a scenario in the sense of Definition 4 entails that the predicates *Happens*, *Initiates* and *Terminates* are definable in terms of *HoldsAt*.

The fixed points of the three-valued consequence operator associated to the normal logic program then coincide with the three-valued models of the completion. The minimal fixed point is a good candidate for a minimal model in the sense discussed above. In fact, it turns out that in the cases of interest to us the consequence operator has a unique fixed point.

3.2. Structure of the minimal models. In natural language semantics it is a contested issue whether the fundamental temporal entities are points or intervals. The event calculus neatly sidesteps this issue, by taking the basic entities to be events and fluents, which are not explicit functions of time and which can be interpreted on structures with very different ontologies for time. Even if we take the structure underlying time to be $\mathbb{R}$, that does not constitute an ontological commitment to points. Ontological commitment is generated rather by representation theorems, which correlate the events and fluents with point sets in a given structure. It may then very well turn out that, even when time is taken to be $\mathbb{R}$, fluents and events can be represented as sets of intervals, so that points have no role of their own to play.

[10]By the definition of complex subgoal, $\leftrightarrow$ does not occur in normal programs. The occurrence of $\leftrightarrow$ in the completion is interpreted in the manner of Łukasiewicz: $\varphi \leftrightarrow \psi$ is assigned $\boldsymbol{t}$ when φ, ψ are assigned the same truth value in $\{\boldsymbol{t}, \boldsymbol{f}, \boldsymbol{u}\}$, and $\boldsymbol{f}$ otherwise.

[11]It is technically convenient to replace Axiom 3 by its contraposition, whose head is *Declipped*.

The situation is slightly more complicated in the case of fluents admitting real parameters, for example fluents representing possibly changing partial objects. Again speaking intuitively, one would expect change to be continuous, with at most a finite number of jumps. One may thus surmise that a fluent with a real parameter should be interpreted by a semialgebraic set:

DEFINITION 7. A subset of $\mathbb{R}^n$ is *semialgebraic* if it is a finite union of sets of the form $\{x \in \mathbb{R}^n \mid f_1 = \cdots = f_k = 0,\ g_1 > 0, \ldots,\ g_l > 0\}$, where f_i and g_j are polynomials.

Intuitively at least, fluents corresponding to natural language expressions (e.g., verbs) are semialgebraic, and we would like this to fall out of the set-up, without further stipulations. The kind of change it is possible to program depends on the one hand on the constraint language chosen and on the other hand on the constraint logic program. Now in RCF only semialgebraic sets can be defined; but would it be possible to extend the range of definable sets by a constraint logic program to RCF?

The main mathematical question is thus to determine what the structure of a fixed point of the consequence operator is; or, more precisely, how the scenario, the dynamics and the axioms of the event calculus jointly affect the structure of the fixed point.

The next few theorems give some pertinent results. For proofs, see [23].

DEFINITION 8. A finite branch in a computation tree is *successful* if its last node contains a constraint only. A finite branch is *finitely failed* if its last node is of the form $?c'$, G' (G' nonempty) with no more resolution steps possible. A query $?c$, G is *finitely evaluable* if all branches in a derivation tree starting from $?c$, G end either in success or in finite failure.

THEOREM 1. *Let RCF be the theory of real-closed fields. Let $\mathcal{P}$ =* EC + SCEN + DYN *and $?G$ a finitely evaluable query in the language of the event calculus. Then there exists a semialgebraic set c such that RCF* +comp($\mathcal{P}$) $\models \forall(G \leftrightarrow c)$.

COROLLARY 1. *Let $\mathcal{P}$* = EC + SCEN + DYN. *If for all fluents $f(\overline{x})$ in the scenario and dynamics, the query $?HoldsAt(f(\overline{x}), t)$ is finitely evaluable, the theory RCF* + comp($\mathcal{P}$) *has a unique model (modulo the underlying real-closed field), which is obtained after finitely many iterations of the consequence operator. In this model the primitive predicates are also represented by semialgebraic sets.*

COROLLARY 2. *Let $\mathcal{P}$* = EC + SCEN + DYN. *The following are equivalent:*

a. *any model of RCF* + comp($\mathcal{P}$) *is completely determined by its restriction to RCF;*
b. *$HoldsAt(f(\overline{x}), t)$ is semialgebraic.*

The hypothesis of finite evaluability of $?HoldsAt(f(\overline{x}), t)$ is rather strong. In principle, one can determine $HoldsAt(f(\overline{x}), t)$ completely by starting up derivations from both $?HoldsAt(f(\overline{x}), t)$ and $?\neg HoldsAt(f(\overline{x}), t)$ and stop when the collected answer constraint in one tree is the complement of that in the other; the trees may then still contain branches which are neither successful nor failed. Let us call this notion *essential evaluability*.

THEOREM 2. *Let RCF be the theory of real-closed fields. Let* $\mathcal{P}$ = EC + SCEN + DYN *and* ?G *a query in the language of the event calculus such that* ?G *is essentially evaluable. Then there exists a semialgebraic set* c *such that* $RCF + \mathrm{comp}(\mathcal{P}) \models \forall(G \leftrightarrow c)$.

There is also a corresponding completeness result.

THEOREM 3. *Let* $\mathcal{P}$ = EC + SCEN + DYN. *The query* ?$HoldsAt(f(\overline{x}), t)$ *is essentially evaluable if* $RCF + \mathrm{comp}(\mathcal{P}) \models \forall t(HoldsAt(f, t) \leftrightarrow c(t))$ *for a constraint* c.

§4. From VPs to fluents and events. In order to use the event calculus for a semantics of natural language, we need a device which codes formulas as objects or functions. We shall borrow some machinery from [9], which was applied in [10] to the semantics of nominalisation.

Let L_0 be some first order language and S_0 a theory formulated in L_0. We briefly review the requirements on L_0 and S_0 which allow such coding.

First, one requires that L_0 contains an individual constant $\bar{0}$, a binary function symbol π and two unary function symbols π_1 and π_2. We shall often write (τ_1, τ_2) for $\pi(\tau_1, \tau_2)$. Secondly, we assume that S_0 proves the following statements concerning these functions

(11a) $(x, y) \neq \bar{0}$,

(11b) $\pi_1(x, y) = x \wedge \pi_2(x, y) = y$.

If $M_0 \models S_0$, $(\cdot, \cdot)$ is a pairing function in M_0, and π_1 and π_2 are the corresponding projection functions.

One may now define tuples inductively by putting: $(\tau) = \tau$ and $(\tau_1, \ldots, \tau_{k+1}) = ((\tau_1, \ldots, \tau_k), \tau_{k+1})$. Similarly, one may define the corresponding projection operations π_i^k $(1 \leq i \leq k)$ such that: $\pi_i^k(x_1, \ldots, x_k) = x_i$.

These constructs suffice to define an abstract form of Gödel numbering.

DEFINITION 9. Let L be some extension of L_0. Then we may code formulas of L as terms in L_0. We write $\ulcorner\varphi\urcorner$ for the Gödel number in L_0 of φ in L. This notation will be used interchangeably both for the term in L_0 and for the object denoted by that term in a model M_0.

We will now put this machinery to work. Let φ be a formula with free variables among $x_1, \ldots, x_k, y_1, \ldots, y_n$. The L_0-term $(\ulcorner\varphi\urcorner, y_1, \ldots, y_n)$ contains $x_1, \ldots, x_k$ as bound variables and $y_1, \ldots, y_n$ as free variables. Since $x_1, \ldots, x_k$ are bound by abstraction, the following notation makes sense:

DEFINITION 10. $\Delta_n\ \varphi[\hat{x}_1,\ldots,\hat{x}_n,y_1,\ldots,y_m] = (\ulcorner\varphi\urcorner,y_1,\ldots,y_m)$. For $n=1$ we will use standard set theoretical notation $\Delta_1\{x \mid \varphi(x,y_1,\ldots,y_n)\} = \varphi[\hat{x},y_1,\ldots,y_n]$. If both m and n are equal to 0, we write $\ulcorner\varphi\urcorner$.

Finally, we are able to explain what all this has to do with the event calculus. Assume verbs always have a parameter for time, as in run(t). Thus, unlike event semantics in the Davidsonian tradition, we do not assume that verbs have an event parameter; rather, various kinds of eventualities will be constructed from the verb with its time parameter. Of particular importance are the *event type*[12], given formally by $\ulcorner\exists t.\,\text{run}(t)\urcorner$, and the *fluent*[13] run$[\hat{t}]$. Note the difference between the two abstractions: the latter is a function of time (to truth values), while the former is an object.

§5. Eventualities. The event calculus provides a good starting point for discussing the ontology required for verb semantics. Our general picture owes much to the pioneering work of Moens and Steedman (cf. [20]), but the apparatus of the event calculus allows us to add formal precision.

An eventuality is a structure composed of four parts

1. activity (in the strict sense)
2. changing, partial object
3. culminating event
4. consequent state

This is almost like the "event nucleus" of Moens and Steedman, except that we have added the "changing, partial object", which plays a prominent role in the treatment of accomplishments and of the progressive. Accordingly, each VP is denoted by a quadruple, each element of which is of the form "-", "e" (third argument only) or "f" (first, second and fourth argument)[14]. The elements of the quadruple are related in several ways. If the VP is represented by a structure of the form (f_1, f_2, e, f_3), then f_1 is a fluent which corresponds to an activity, f_2 is a fluent which corresponds to a partial object which changes under the influence of f_1 (thus f_2 will in general contain variables), e is both the culminating event of the activity and an initiating event for the resulting consequent state, f_3. The structure thus satisfies the properties *Initiates*(e, f_3, t) and *Terminates*(e, f_1, t). There is no distinguished role for an event initiating f_1, since there may be many of them, for example if the activity f_1 is interrupted from time to time. It is of course implicit that the

[12] Also called *perfect nominal*.

[13] Also called *imperfect nominal*.

[14] For simplicity we have treated only the case where the direct *object* is partial and changing. A similar phenomenon can occur in subject position however, as in the well-known causative/inchoative alternation

(i) a. The cook thinned the gravy.
 b. The gravy thinned.

activity f_1 is nontrivial, so that the scenario must contain a reference to an initiating event, perhaps in the form *Initially*.

The fluents in this structure play different roles in a scenario and in the dynamics. f_1 is not allowed to occur in *Releases* and in the third argument of *Trajectory*, but may occur in the latter's first argument. This is because activities are volitional and must hence be initiated and terminated by actions. For f_2 it is the other way around; since states are causally inert, they are not allowed as the first argument in *Trajectory*. Usually, the consequent state f_3 will be an instantiation of f_2, so it inherits the latter's syntactic restrictions. The relationship between the fluents f_1 and f_2 is given by the dynamics, i.e. a statement of the form $S(f_1, f_2, t, d) \rightarrow \mathit{Trajectory}(f_1, t, f_2, d)$, where $S(f_1, f_2, t, d)$ is a state in the sense of Definition 3.

Note that an eventuality *determines* the scenario and dynamics in the sense that these should at least fix the meaning of the positive components of the quadruple. Thus, for a quadruple of the form $(f_1, f_2, -, -)$ nothing needs to be said about a culminating event, but this changes when the remaining slots are filled, for example when an activity (*draw*) is coerced into an accomplishment (*draw a circle*). Often the scenario will need to contain more, for example a reference to an initiating event. As will be seen in the following sections, different *Aktionsarten* correspond to the various ways in which the slots in the quadruple can be filled.

5.1. Formal definition of aspectual classes. We will define aspectual classes as specific types of eventualities, not as verb classes, because we believe, as mentioned in the introduction, that verbs can by and large be coerced into any aspectual class. As we have seen an eventuality can contain an event and up to three fluents; one may then distinguish types of eventualities according to the slots filled in the quadruple.

State	$(-,-,-,+)$
Activity (strict)	$(+,-,-,-)$
Activity (wide)	$(+,+,-,-)$
Accomplishment	$(+,+,+,+)$
Achievement	$(-,-,+,+)$
Points	$(-,-,+,-)$

We will add a few comments on this table.

1. – indicates that a slot is empty, + that it is filled with an object of the appropriate category.
2. The distinction between states and activities is given by the different syntactic roles they play in the event calculus. The following table shows the argument slots that can be filled by states and activities ($\mathit{Traj}._i$ refers to the ith argument of *Trajectory*)

	Releases	*Traj.*1	*Traj.*3
States	+	−	+
Activity (strict)	−	+	−

There are no restrictions on occurrences in the other primitive predicates.

3. The difference between an activity in the strict sense (cf. *push*) and in the wide sense (*push a cart*), is that in case of the latter what Dowty [7] calls the "incremental theme" has been added not the cart itself, but the changing position of the cart.
4. The quadruples occurring in the above list can all be described by VPs. It is an interesting question whether some of the other quadruples also characterise VPs. Observe that there are quadruples which characterise eventualities corresponding to *sentences.* Here are a few examples. To this end, it is useful to give a wider interpretation to the second component, which we may also take to denote a gradable state, such as "love to a certain degree". With this stipulation, a quadruple like $(-,+,-,+)$ is exemplified by the sentence

(12) For a while she remembered him dimly, but she soon forgot him completely.

In the same vein, the quadruple $(-,+,+,+)$ corresponds to

(13) She felt some affection for him for a while, but after that incident, her love was dead.

In these examples we have used quasi-Boolean operations such as "but", the question is whether there are simple VPs which need to be characterised by such quadruples.

5.2. Example of an accomplishment: "build a house". To fix ideas, we give a scenario for the accomplishment "build a house", thus illustrating the role of the various components in the quadruple. An accomplishment is characterised by a quadruple $(+,+,+,+)$, so that the scenario to be written must contain a reference to all these components.

With this in mind, we need the following terms in the language of the event calculus.

1. *build* is an activity fluent.
2. *house*(x) is a parametrised fluent representing the construction stage (x) of a house.
3. c is a real constant indicating a construction stage at which the house is considered finished; thus *house*(c) is the fluent representing the consequent state.

4. *start* is any event initiating building.
5. *finish* is the canonical event terminating building, namely when the house is finished.
6. g is a real-valued function relating the building activity to the construction stage.

Note that so far we have not introduced an object *house*; all we have is a fluent $house(c)$. However, nothing prevents us from stipulating that $House(house(c))$, thus effectively turning $house(c)$ into an object as well.

1. $Initiates(start,build,t)$
2. $Initiates(finish,house(c),t)$
3. $Terminates(finish,build,t)$
4. $HoldsAt(build,t) \wedge HoldsAt(house(c),t) \rightarrow Happens(finish,t)$
5. $Releases(start,house(x),t)$
6. $HoldsAt(house(g(t)),t) \rightarrow Trajectory(build,t,house(g(t+d)),d)$.

The two statements 5 and 6 are jointly called the *dynamics*; this is characteristic of both activities (in the wide sense) and accomplishments. Accomplishments are distinguished from activities (in the wide sense) by statements describing the behaviour of the canonical terminating event, here statements 2–4.

The preceding statements should be thought as being part of the lexical entry of "build a house"; of course, the scenario for "build" itself is much more complex than that presented here. In the context of a sentence, episodic information will be added, such as

7. $Happens(start,t_0)$
8. $Initially(house(a))$,

where a is a constant representing a contextually determined initial construction stage of the house.

In [10] we have explained at length how verbs and fluents are related. Briefly, the fluent *build* results from the transitive verb x *builds* y by existentially quantifying the object position (y) and then nominalising. One may furthermore view the participle *built* as arising from *build* by existentially quantifying the subject position (x). The consequent state then satisfies $\exists x\ builds(f(c))$; thus there is a pleasant symmetry between the first and last component of the quadruple.

5.3. Example of an achievement: "reach the top". Here we need a terminating event[15] *reach* and a fluent *be-at-the-top*, related by

1. $Initiates(reach,be\text{-}at\text{-}the\text{-}top,t)$.

Episodic information can now only be of the form

2. $Happens(reach,t_0)$.

[15]"Terminating" with respect to an activity not explicitly mentioned.

Note that negation as failure will ensure that *Initially*(*be-at-the-top*) will be false.

5.4. The progressive. The following sentence makes perfect sense

(14) John was building a house last year, but he had to stop when his credit failed.

Upon reflecting, there is something paradoxical about this sentence. Whereas it belongs to the meaning of the accomplishment *build a house* that the activity ("build") is directed toward the consequent state of a finished ("built") house, the actual occurrence of that consequent state can be denied without contradiction. So how can a seemingly essential component of the meaning be denied, without affecting the meaning itself? This is known as the imperfective paradox'. The literature is replete with attempted resolutions of the paradox, ranging from explaining the problem away (cf. the recent Michaelis [13]) to various invocations of possible worlds (see Dowty [6], Landman [12] or de Swart [4]). Possible worlds solutions are based upon the idea that

> The progressive picks out a stage of a process/event which, if it does not continue in the real world, has a reasonable chance of continuing in some other possible world [4, p. 355].

but differ in the (largely informal) descriptions of the possible worlds used. For example, [6] uses so-called "inertia worlds", worlds which are identical with the present world until "now", but then continue in a way most compatible with the history of the world until "now". Thus these approaches are intensional in the formal sense of using possible worlds. In fact, most authors (but not all) would agree that the progressive creates an intensional context: even though John may have stopped building at a stage when it was unclear whether he was building a house or a barn, still only one of

(15a) John was building a house.

(15b) John was building a barn

can be true of the situation described.

Our solution (first proposed in [10]) will use the event calculus, but before we go into this, we discuss Michaelis' attempt in [13] to explain the problem away; this will then show why an elaborate machinery is necessary.

Explicitly denying that the progressive creates an intensional context, Michaelis writes

> Under the present proposal, the Progressive sentence *She is drawing a circle* denotes a state which is a subpart not of the accomplishment type *She- draw a circle* but of the activity type which is entailed by the causal representation of the accomplishment type. Since this activity can be identified with the preparatory activity that circle

> drawing entails, circle drawing can in principle be distinguished from square drawing etc. within the narrow window afforded by the Progressive construal [and] does not require access to culmination points either in this world or a possible world ... [13, p. 38].

We find this rather doubtful. Without access to a person's intention it may be very hard to tell initially whether she is drawing a circle or a square, building a barn or building a house. But that person's intention in performing an activity is characterised precisely by the associated consequent state, even though the latter cannot yet be inferred from the available data.

Here the event calculus comes to our rescue. In the event calculus, an activity comes with a theory (the scenario) which describes the intended consequences of that activity; this is precisely the function of the dynamics. However, unlike approaches such as Parsons' [15], where one quantifies existentially over events, the scenario is a universal theory and does not posit the occurrence of the intended consequences. Their existence is guaranteed only in minimal models of the scenario combined with the axioms for the event calculus, in which no unforeseen obstacles occur. Thus, the meaning of an accomplishment (as embodied in the scenario) involves a culminating event *type* (which therefore must exist); but there are no existential claims about the corresponding event *token*[16]. These are handled by different mechanisms. The use of the progressive indicates that the appropriate mechanism is default inference. By contrast, tense can be defined via ordinary existential quantification, so stays within the realm of monotonic inference.

We will now make the above considerations slightly more formal. Consider the sentence

(16) John is building a house.

Viewed purely denotationally, this is a statement about an activity, formally *HoldsAt(build,now)*, conjoined with the scenario describing the lexical meaning of "build a house". We claim however, that the progressive also involves a particular way of looking at the culminating event: that its occurrence is to be expected. We will see presently that the set of statements associated to (16) nonmonotonically implies that the house will be finished. If we apply the progressive not to an accomplishment but to an activity, there is no canonical culminating event. But the crucial presupposition for the use of the progressive remains the presence of a dynamics.

THEOREM 4. *Let $\mathcal{P}$ be the constraint logic program (w.r.t. a real-closed field) consisting of EC, the scenario* 1–6, *and the episodic statements*

7. *HoldsAt(build,now)*
8. *Initially(house(a))*.

---

[16]And similarly for the state consequent upon a culminating event type.

Suppose $\lim_{t \to \infty} g(t) \geq c$. *Then* comp($\mathcal{P}$) *has a unique model, and in this model there is a time* $t \geq now$ *for which* $HoldsAt(house(c), t)$. *By virtue of the stipulation that* $House(house(c))$, *there will be a house at time* t.

The effect of taking the completion is that in a model, only those events occur which are forced to happen due to the scenario; similarly, only those influences of events are considered which are explicitly stated in the scenario. In this particular case, since the scenario does not mention the event "going broke", the completion excludes this possibility; and similarly for other possible impediments to completing the construction. It is then a consequence of the general result Theorem 1 that the completion actually has a unique model.

Notice, however, that the existence of a time at which the house is finished is only guaranteed in minimal models. Thus, this inference is nonmonotonic: if we obtain more episodic information, the conclusion may fail.

§6. Explaining coercion. The default eventuality associated to, say, "reach the top", is an achievement. As such, the application of the progressive is excluded, because the achievement lacks the requisite dynamics. Nevertheless, in some contexts the progressive is appropriate, as in

(17) They were reaching the top when a blizzard forced them to go back.

In this case the achievement "reach the top" is *coerced* into an accomplishment, to which the progressive can be applied felicitously. The purpose of this section is to give a formal treatment of this operation, or rather, family of operations. We will distinguish three forms: additive coercion, subtractive coercion, and cross-coercion.

6.1. Additive coercion. This is the simplest kind of coercion, which consists in elaborating a scenario.

Activity $\rightsquigarrow$ accomplishment. The verb "build" is an activity, which is transformed into an accomplishment by adding a direct object such as "a house". In the simple set up outlined above, the only property governing "build" is 1. The effect of adding the direct object is that this scenario is extended with a dynamics, and a set of statements describing the behaviour of the culminating event. Hence the name "additive coercion".

Achievements $\rightsquigarrow$ accomplishments. Above we remarked that sentence (17) requires the transformation of "reach the top" into an accomplishment. That is, since the progressive requires for its application the presence of a dynamics, this must first be added to the scenario. This can be done by adding an activity *climb*, a changing partial object *height*(x) representing the height gained during climbing, a dynamics, and a statement relating the event *reach* to the activity, say by means of $HoldsAt\ (climb, t) \wedge HoldsAt\ (height(8.850m), t) \rightarrow Happens(reach, t)$.

6.2. Subtractive coercion. We now discuss several instances of the converse transformation, in which parts of a scenario are deleted.

Accomplishment $\rightsquigarrow$ activity. The most interesting case of this form of coercion occurs when we consider bare plurals: "drink a glass of wine" is an accomplishment, but "drink wine" is an activity (in the wide sense). What the activity and the accomplishment have in common is a dynamics, the crucial sentence of which has the following form

1. $\mathit{HoldsAt}(f_2(g(t)), t) \rightarrow \mathit{Trajectory}(\mathit{drink}, t, f_2(g(t+d)), d)$.

Here, f_1 is the activity, and f_2 is the changing partial object. Now activities and accomplishments are differentiated from each other by the fact that the scenario for an accomplishment requires sentences of the type

2. $\mathit{Terminates}(e(c), f_1, t)$
3. $\mathit{HoldsAt}(f_2(c), t) \wedge \mathit{HoldsAt}(f_1, t) \rightarrow \mathit{Happens}(e(c), t)$.

Here, $e(c)$ is a culminating event, dependent upon a constant c. An example of such a constant would be the quantity determined by a glass of wine; then $e(c)$ is the event "finish a glass of wine". Looking at the last sentence, one sees that, in a minimal model, the culminating event will only be activated if for some t, $g(t) = c$ can be derived.

Thus, the effect of "a glass of –" would be to add 2. and 3. Conversely, taking away "a glass of –" from "drink a glass of wine" would lead to the deletion of sentences 2. and 3.

6.3. Cross–coercion. We now discuss several difficult cases of coercion which cannot be viewed as adding or deleting sentences from a scenario. This kind of coercion occurs for instance when a state (which is of the form $(-,-,-,+)$) is coerced into an activity (of the form $(+,+,-,-)$) under the influence of the progressive. Of course a combination of the addition and subtraction operations would also yield this transformation, but presumably something else is going on.

State $\rightsquigarrow$ activity. Croft [3] gives the following convincing examples:

(18) She is resembling her mother more and more every day.

(19) I am loving her more and more, the more I get to know her.

Consider example (18). Let f be a fluent denoting *resemble*, conceived of as a state. Then the sentence

(20) She resembles her mother.

is represented as $\mathit{HoldsAt}(f, t)$.

The phrase "more and more" introduces a new fluent $f'(x)$ denoting *resemblance to degree* x. Furthermore, the application of the progressive introduces a dynamics consisting of the general form

1. $\mathit{Initiates}(e, f_1, t)$

2. $Releases(e, f_2, t)$
3. $HoldsAt(f_2, t) \rightarrow Trajectory(f_1, t, f_2, d)$.

Here, the f_1, f_2, e are *variables*, which have to be unified with terms provided by the discourse, namely f and $f'(x)$, and a contextually determined starting event e_0 which satisfies

1. $Happens(e_0, t_0)$
2. $t_0 < now$.

We claim that the observed coercion is due to the unification of f with the variable f_1. Suppose that f is substituted for f_2 in *Releases* and *Trajectory* predicate, then $Clipped(s, f, r)$ will be true of f if $s < t_0 < r$. This means that the query $?\neg Clipped(s, f, r)$ will fail for these s, r. By contrast if f is substituted for the first argument of the *Trajectory* predicate then $f'(x)$ can be substituted in the third argument and in *Releases* and successful computations can be constructed as usual. This implies that the state *resemble* is coerced into an activity via unification.

It should be noted that some native speakers of English find this example more problematic than Croft appears to do. From the formal side we may observe that this example involves much more processing than all the previous examples; it is possible that this difference is related to the difference in acceptability judgments.

The same analysis can then be applied to

(21) I am loving her more and more every day.

Examples (21) and (19) are distinguished by the fact that for (19) there are two candidate activity fluents, *love* and *get to know her* and accordingly two possible unifications. The case were "love" is coerced to be the activity is similar to example (18). When *get to know her* is interpreted to be the activity, this means that a new changing partial object is introduced, namely the changing degree of knowledge $g'(x)$ which is related to the degree of loving $f'(x)$ be means of some real-valued function h as in formula (22).

(22) $\forall x \, \forall t \, (HoldsAt(g'(x), t) \longrightarrow HoldsAt(f'(h(x)), t))$.

As before the progressive introduces a dynamics. Let the fluent g represent *get to know her*. By an argument similar to the one above, g is unified with the first variable of the *Trajectory*–predicate and $g'(x)$ with the third variable. This then yields a prediction for $f'(x)$ by means of the formula (22).
Points $\rightsquigarrow$ activity. Here we must consider a transformation from a quadruple $(-, -, +, -)$ to $(+, -, -, -)$ (or perhaps $(+, +, -, -)$). Consider

(23a) The light flashed.

(23b) The light was flashing all night.

"Flash" is a point; for this example we will therefore consider only the event *flash*. Sentence (23a) can then be formalised as

1. $Happens(flash, t_0)$
2. $t_0 \leq now$

In sentence (23b) "flash" has become an activity, hence formally a fluent, or pair of fluents. In [10] we have indicated (for the case of coerced nominals) a procedure which achieves just this. Namely, coercion is represented by the mapping $flash \mapsto Happens[flash, \hat{t}]$. The latter expression is a fluent. We propose that this fluent is identified with the changing partial object, and that an unspecified activity f_1 is added to which $Happens[flash, \hat{t}]$ is related via a dynamics. The only constraint this f_1 has to satisfy is that its corresponding finite set of intervals and/or points (which exists due to Theorem 1) coincides with that of $Happens[flash, \hat{t}]$. The dynamics may then simply be defined by $HoldsAt(Happens[flash, \hat{t}], t) \rightarrow Trajectory(f_1, t, Happens[flash, \hat{t}], d)$.

A non-instance: the progressive. The progressive is often described as a stativizer, so that the main clause in (24a) below would denote a state. The basis of this judgment is a comparison of (24a) to (24b) and (24c):

(24a) He was crossing the street when the truck hit him.

(24b) He was asleep when his friend came in.

(24c) He got up when his friend came in.

(24d) John drank wine when his son came home from school.
John drank whisky when his wife returned from work.
John's family thought he must be an alcoholic.

Sentence (24a) is like (24b), which has a stative main clause, in that the event mentioned in the subordinate clause takes place within the interval corresponding to the main clause. This is in contrast to (24c), where the main clause denotes an event which takes place after the event denoted by the subordinate clause.

This argument does not suffice to classify the progressive as an operator which yields states as values. The main clause in (24a) still refers to a period in which change is taking place, unlike the main clause of (24b). Even without the progressive, the activities in (24d) can be interpreted as in (24b)[17], and one would not want to say that *they* are stativized. The main clauses of (24a), (24b) and (24d) have in common that they refer to intervals, unlike (24c). Furthermore, the meaning of *when* is sensitive to the temporal profile of its

[17] There is also an event reading comparable to (24c).

first argument[18]. But it should be clear by now that temporal profile is only a small, derivative part of the meaning of a verb.

Activity/accomplishment $\rightsquigarrow$ state. To conclude the section on coercion we briefly discuss two examples about which much more should be said, in particular with regard to the syntax-semantics interface.

Negation. It has often been held (see for example [24]) that negating an activity results in a stative predicate. This intuition can to a certain extent be reproduced in the present framework. An activity (in the wide sense) consists of a fluent representing the activity proper, and a changing partial object. In a minimal model, the partial object changes only when the activity is "on"; outside of those intervals, the partial object does not change. It follows that the set of intervals complementary to that representing the activity has a state-like character. It differs from a true state in our sense by being initiated and terminated by events, since of course a terminating event for an activity is an initiating event for its negation.

Passive. In German and Dutch there is a form of the passive, the "Zustandspassiv" (indicated by a special auxiliary), which transforms an activity or an accomplishment into a (consequent) state[19].

(25a) Johann baut ein Haus. Es ist weiß.

(25b) Das Haus ist von Johann gebaut. Es ist weiß.

(25c) Das Haus wird von Johann gebaut. *Es ist weiß.

In (25b), the auxiliary *sein* indicates that the house is finished; by contrast, in (25c) the auxiliary *werden* refers to the process of building. This explains the * in (25c).

In English the passive is apparently ambiguous between the two readings. Compare

(26a) John is building a shed in his garden. This causes his neighbours much distress [because of the tremendous noise].

(26b) The shed that is built in John's garden causes his neighbours much distress [because it spoils their view].

(26c) A shed is built in John's garden. This causes his neighbours much distress.

"This" in sentence (26c) can refer both to an activity, as in (26a), and the result of that activity, as in (26b). If one conceives of the passive syntactically as movement of the object NP into subject position, an interesting tension between syntax and semantics comes to the fore. The NP *a house* in *build a house* is an incremental theme, and need not denote an object in the ontological

[18] An extensive analysis of *when*–clauses, tense, reference time and related phenomena is contained in the forthcoming book [22].

[19] For further discussion see [11].

sense. In our set up, the denotation of *a house* is as it were distributed over the changing partial object, the canonical terminating event, and the consequent state whose relations are governed by the scenario. Similarly, the verb *build* is not a two-place predicate, but a fluent with one (subject) parameter. So what happens semantically when the NP is moved into subject position?

One possibility is that nothing happens, in which case a passive sentence retains the process reading of the corresponding active sentence. Another possibility is that the NP is re-interpreted as a real object, to which an adjective can be applied. Indeed, in English grammar this form of the passive is known as "adjectival passive". For example, *built* would now be an adjective, obtained from the verb *build* by existentially quantifying the subject position. But the upshot of this is that "a house is built" now corresponds to the consequent state of the accomplishment, as indicated in Section 5.2.

§7. Conclusion. We can now explain in more detail why coercion is considered an intensional phenomenon in this paper.

The following formal reconstruction of Frege's notion *sense* is from [23].

DEFINITION 11. The *sense* of an expression is the constraint logic program (in the sense of CLP(T)) representing the lexical component, viewed as an algorithm which transforms an episode into the denotation of the expression in a model, using the axioms of the event calculus.

Consider again the explanation for the coercion effect which turns the stative sentence *She resembles her mother* into the activity expressed by *She resembles her mother more and more every day*.

The explanation is based on two possible unifications for fluent variables in the *Releases* and *Trajectory* predicates of the enriched scenario. However under the first substitution successful computation is blocked because $Clipped(s, f, r)$ is derivable under this unification. This failure is reached in a *finite* number of steps. By contrast the other possible substitution leads to a successful computation of a minimal model.

We can thus describe *coercion* as a map from the sense of an expression to a different but related sense of an expression. The relationship in determined both by linguistic and non-linguistic factors depending on the type of coercion involved.

REFERENCES

[1] E. BACH, *The algebra of events*, ***Linguistics and Philosophy***, vol. 9 (1986), pp. 5–16.

[2] G. BEALER and U. MÖNNICH, *Property theories*, ***Handbook of philosophical logic. Vol. IV*** (D. Gabbay and F. Guenthner, editors), Reidel, Dordrecht, 1989.

[3] W. CROFT, *The structure of events*, ***The new psychology of language*** (M. Tomasello, editor), Lawrence Erlbaum Associates, Mahwah, NJ, 1998.

[4] H. de Swart, *Aspect shift and coercion*, ***Natural Language and Linguistic Theory***, vol. 16 (1998), pp. 347–385.

[5] J. Dölling, *Polysemy and sort coercion in semantic representation*, ***Discourse and lexical meaning*** (P. Bosch and P. Gerstl, editors), Arbeitspapiere des SFB 340, Nr. 30, Stuttgart/Tübingen, 1993.

[6] D. Dowty, ***Word meaning and montague grammar***, Reidel, Dordrecht, 1979.

[7] ———, *Thematic proto–roles and argument selection*, ***Language***, vol. 67 (1991), pp. 547–619.

[8] M. Egg, ***Flexible semantic construction: the case of reinterpretation***, Habilitationsschrift, Universität des Saarlandes, 2000.

[9] S. Feferman, *Toward useful type-free theories I*, ***The Journal of Symbolic Logic***, vol. 49 (1984), pp. 75–111.

[10] F. Hamm and M. van Lambalgen, *Event calculus, nominalisation and the progressive*, Research report, ILLC, Amsterdam, December 2000, 76 pp. to appear in *Linguistics and Philosophy*. Available at `http://www.semanticsarchive.net`.

[11] A. Kratzer, *Building statives*, University of Massachusetts at Amherst, 2002.

[12] F. Landman, *The progressive*, ***Natural Language Semantics***, vol. 1 (1992), pp. 1–32.

[13] L. Michaelis, *A unification-based model of aspectual type-shifting*, University of Colorado at Boulder, 2001.

[14] M. Moens and M. Steedman, *Temporal ontology and temporal reference*, ***Computational Linguistics***, vol. 14 (1988), pp. 15–28.

[15] T. Parsons, ***Events in the semantics of English***, MIT University Press, Cambridge, MA, 1990.

[16] S. Pulman, *Aspectual shift as type coercion*, ***Transaction of the Philological Society***, vol. 95 (1997), pp. 279–317.

[17] J. Pustejovsky, *Type coercion and lexical selection*, ***Semantics and the lexicon*** (J. Pustejovsky, editor), Kluwer, Dordrecht, 1993.

[18] J. Saul, *Substitution and simple sentences*, ***Analysis***, vol. 57 (1997), pp. 114–118.

[19] M.P. Shanahan, ***Solving the frame problem***, The M.I.T. Press, Cambridge, MA, 1997.

[20] M. Steedman, *Temporality*, ***Handbook of logic and language*** (J. van Benthem and A. ter Meulen, editors), Elsevier, Amsterdam, 1997, pp. 895–938.

[21] P.J. Stuckey, *Negation and constraint logic programming*, ***Information and Computation***, vol. 118 (1995), pp. 12–33.

[22] M. van Lambalgen and F. Hamm, ***Events in time: Logic, semantics and cognition***, Blackwell, Malden, 2004.

[23] ———, *Moschovakis's notion of meaning as applied to linguistics*, ***Logic Colloquium '01*** (M. Baaz, S. Friedman, and J. Krajíček, editors), AK Peters, 2005, pp. 255–280.

[24] H. Verkuyl, ***A theory of aspectuality. the interaction between temporal and atemporal structure***, Cambridge University Press, Cambridge, 1993.

FGW / ILLC
UNIVERSITEIT VAN AMSTERDAM
NIEUWE DOELENSTRAAT 15
1012 CP AMSTERDAM, THE NETHERLANDS
E-mail: vanlambalgen@hum.uva.nl

SEMINAR FÜR SPRACHWISSENSCHAFT
UNIVERSITÄT TÜBINGEN
WILHELMSTR. 113
72074 TÜBINGEN, GERMANY
E-mail: friedrich.hamm@uni-tuebingen.de

INTENSIONALITY IN PHILOSOPHY AND METAMATHEMATICS

KARL-GEORG NIEBERGALL

Abstract. The relation between intensionality as understood in philosophy and in metamathematics (in particular in early writings of S. Feferman) is explored in this paper. It investigates whether the latter can be interpreted as an instance of the former and presents metamathematical examples of "serious intensionality".

Truth and *reference* (i.e., *extensions*) on the one side and *meaning* (i.e., *intensions*) on the other side are closely related: for example, principles like *every true sentence is meaningful* and *expressions with the same meaning refer to the same things* (if they refer) seem to be correct from an intuitive point of view. Of course, this does not mean that, e.g., truth or falsity are the meanings of (declarative) sentences. Nonetheless, were it not because of contexts taken from ordinary language — like those containing modalities and propositional attitudes — expressions like "meaning" could perhaps be explained by employing terminology taken solely from referential semantics. Thus, it has been claimed that for scientific purposes one may well get on with purely extensional languages.[1] In fact, this seems to be "obviously" true for the mathematical discourse.

It is mainly due to the work of S. Feferman that the topic of *intensionality* has nevertheless gained some relevance, perhaps even popularity, in the field of metamathematics (cf. primarily [12]).[2] But what is the relation between "intensionality" as understood in philosophy and as understood in metamathematics? Are these two concepts actually the same? If not, is it at least possible to make them fruitful for each other, e.g., by construing one of them as a special case of the other? This paper addresses these questions both

This paper is an elaboration of a talk held at the workshop *Intensionality* in Munich, October 2000. Many thanks to its organizer Dr. Reinhard Kahle, to Prof. Godehard Link and to its participants for comments. Special thanks to the referee for his helpful remarks and suggestions.

[1]This claim is a version of a so-called *thesis of extensionality*; see the writings of W.V.O. Quine and R. Carnap (e.g., [7], p. 245). It seems, however, that theses of extensionality have lost much of their philosophical attractiveness since the times of Wittgenstein's *Tractatus* and the Vienna Circle.

[2]G. Kreisel's and, for approaches having a more recursion theoretic flavour, Y. Moschovakis' work should also be mentioned; cf. [27].

Intensionality
Edited by Reinhard Kahle
Lecture Notes in Logic, 22

from a conceptual perspective and by presenting relevant metamathematical results.

§1. Intensionality in philosophy: an overview.

1.1. Intensionality: basic phrases. When dealing with *intensionality*, a natural starting point is G. Frege's distinction (see his seminal [16]) between "Bedeutung" (i.e., reference) and "Sinn" (i.e., meaning): expressions s, t may refer to the same entity while, at the same time, presenting that entity in different ways — whence s and t have different meanings. Natural as that distinction may be, Frege's choice of the word "Bedeutung" for reference has usually been regarded as an unhappy one and the details of his theory have never been widely accepted. Historically, it was superseded by Carnap's "method of extension and intension" (see [8]), which, in turn, lead to the presently reigning paradigms of formal semantics for non-extensional contexts — to possible worlds semantics and to approaches employing structured propositions (see [9] and [8]; situation semantics may also be mentioned (see [1])). Actually, these and other types of semantics dealing with non-extensional contexts are quite different from each other, but Carnap's preferred terminology of "intensions" and "extensions" has been retained.

In analytic philosophy, the contexts in which "extension" and "intension", words built up from them — like "extensional" and "intensional" — and related expressions — like "reference" and "meaning" — typically occur are:

(I.i) "(expression) s is meaningful", or "(expression) s has a meaning";

(I.ii) "(expression) s has the same extension as (expression) t"; and "(expression) s has the same intension as (expression) t", or "(expression) s is synonymous with (expression) t";

(I.iii) "the extension of (expression) s is a"; and "the intension of (expression) s is a", or "s expresses a", or "s expresses that a".

(II.i) "context C is extensional"; and "context C is intensional".

(II.ii) "axiom (or: principle) of extensionality".

(II.iii) "thesis of extensionality".

From the point of view of this paper's topic, contexts of type (II) are distinguished (when compared to the first group of phrases) in containing the expressions "extensional" and "intensional". Yet, I will first make some comments on the entries of type (I), for these enter into explanations of phrases of the second group.[3]

First, it is well known that phrases of type (I.i) can be defined using those of type (I.ii):

$$s \text{ is meaningful} :\Longleftrightarrow \exists x \, (s \text{ is synonymous with } x).$$

[3] My presentation is not a "neutral" sketch of intensionality in philosophy, though, but stresses the importance of phrases of type (I.ii); for more on this, see, e.g., [36].

Moreover, phrases of type (I.ii) are definable by those of type (I.iii):

$$s \text{ is synonymous with } t :\Longleftrightarrow$$
$$\exists x\ (\text{the meaning of } s = x \wedge \text{the meaning of } t = x).$$

In general, definitions going into the converse direction are hard to envisage. But contexts of type (I.iii) can be explained using those of type (I.ii) if *additional* — in particular set-theoretical — means are at hand; for that may make it possible to give "definitions by abstraction": for sentences ψ of L, e.g.,

$$a \text{ is the meaning of } \psi :\Longleftrightarrow$$
$$a = \{\varphi \mid \varphi \text{ is a sentence in L} \wedge \varphi \text{ is synonymous with } \psi\}.$$

Although it may not be quite satisfactory from an intuitive point of view that the meaning of a sentence, i.e. a proposition, should turn out to be a set of sentences,[4] this definition does its job. By this I mean that it yields (if synonymy is an equivalence relation)

$$\varphi \text{ is synonymous with } \psi \Longleftrightarrow \text{ the meaning of } \varphi = \text{the meaning of } \psi.$$

Alternatively, the phrase "a is the meaning of (expression) s" may be construed as

$$s \text{ is synonymous with (expression) "}a\text{"}.$$

This interpretation is probably also not the intended one. It seems particularly plausible, however, when s is a sentence: for one may wonder what "s expresses (the proposition) that a" could mean more than "s is synonymous with sentence 'a' ".[5]

Second, both from a systematic and an intuitive point of view, it need and should not be assumed that statements of the form "a has the same extension/intension as b" carry any *ontological commitment* — to say it in Quine's words — to extensions or intensions *qua* entities.[6] That is, those statements do not imply "$\exists x$ (x is the extension/intension of a)".[7] Thus, although "a is synonymous with b" *may* be defined by reference to meanings (as done above), there does not exist any necessity to define or understand phrases of type (I.ii) by employing those of type (I.iii) (or to define phrases of type (I.i) by those of type (I.ii)). Moreover, at least for those philosophers who admit ontological questions and claims to be meaningful and have ontological scruples (in a Quinean tradition, say), it should be preferable to treat the

---

[4]In fact, this definition is taken into consideration in, e.g., [7], [8], [40] and [34], but it is not the one the authors prefer in the end.

[5]Cf. [10] for more on this idea.

[6]To free myself from any seeming reference to intensions as entities, I will usually choose the formulation "a is synonymous with b" instead of "a has the same intension (or: meaning) as b".

[7]This has been emphasized by Quine, but is also noted, among others, in [8] and [40].

predicate "is synonymous with" as a primitive in order to avoid superfluous ontological commitments.

Furthermore, in texts from the founding era of intensional semantics (before possible worlds semantics took over), in particular, it can be observed that phrases of type (I.ii) have been taken to be intuitively more basic then those of type (I.iii). Frege, e.g., starts and closes [16] with the observation that sentences "$a = a$" and "$a = b$" sometimes have different cognitive values ("Erkenntniswerte"), even if both are true. That is, for Frege the *linguistic data* are that certain true sentences fail to be synonymous with each other (in some sense); meanings *as entities* are postulated only afterwards in order to give an explanation for the differing cognitive values. In [8], Carnap procedes similarly: he first introduces "equivalence" and "L-equivalence" (which are supposed to provide explications of "having the same extension" and "having the same intension") before the extension of a predicate is stipulated to be a class and the intension to be a property.[8]

In the rest of this section, entries of group (II) will be discussed. First, I present several (different) *principles of extensionality*, *theses of extensionality* and explications of "Context C is extensional" (or: "intensional"). Let me note here that in philosophy, it is *contexts* to which the predicates "extensional" and "intensional" are usually ascribed.[9] In the following subsection, I will analyze the interrelation between principles of extensionality and the extensionality of contexts somewhat more deeply. Roughly put, a context C is extensional if and only if a principle of extensionality is valid for C. Yet, it is not easy to describe the correspondence between extensional contexts and principles of extensionality in both general and precise terms: apart from the vagueness and lack of uniqueness of their formulations, the situation is complicated because of the various semantical principles relating them with each other which may be adopted.[10] This subsection should also illustrate how both the relevance and the validity of these principles and of theses of extensionality depend to a considerable degree on their exact formulation.

The *principle of extensionality* is most often formulated for quantificational languages: given terms s, t and formulas ψ of L, it is the schema

$$(Ext^{=}) \quad s = t \longrightarrow (\psi(s) \leftrightarrow \psi(t)).$$

As it is usually understood, this formulation presupposes that L is a language containing terms and the identity-predicate. In particular, $(Ext^{=})$ is not suitable for purely propositional languages L. Here, the principle of extensionality

[8]Note in this connection that it is merely formulations of the sort "A has the same extension/intension as B" which enter into the *definientia* of phrases of type (II).

[9]There are exceptions: in [6], Carnap also classifies *methods* as being extensional.

[10]See below for several examples.

is more properly stated as the schema

$(Ext^{\leftrightarrow})\quad (A \leftrightarrow B) \longrightarrow (\psi(A) \leftrightarrow \psi(B))$

(for all formulas A, B, ψ in L).[11] Further versions of principles of extensionality will be studied later on.

Coming to *extensionality* and *intensionality of contexts*, I take it that a context is just an expression or a set or sequence of expressions. For such a context C, it is common to define

C is extensional $:\Longleftrightarrow \forall A, B$ (A is a subexpression of $C \wedge A$ is coextensional with $B \Rightarrow C$ is coextensional with $C[B/A]$).[12]

Here, $C[B/A]$ is the result of the replacement of at least one occurence of A by B in C. The phrases "A is a subexpression of C" and "A occurs in C" are rather vague and can be made precise in various ways;[13] see, e.g., [8], par. 11, for details of one decision. A language L may then be called "extensional" if and only if all formulas C of L are extensional.

Now, "C is intensional" can been explained as (cf. [7], p. 242)

$$\neg(C \text{ is extensional}),$$

but also as (see more or less [8], p. 48).

$\neg$(C is extensional) $\wedge\, \forall A, B$ (A is subexpression of $C \wedge A$ is synonymous with $B \Rightarrow C$ is synonymous with $C[B/A]$).[14]

The second definition is primarily motivated by the wish to distinguish intensional from hyperintensional contexts (cf. [9]) and, thus, by the topic of propositional attitudes. These themes, however, will play no important role in this text. Let me point out that I will assume that the contexts C considered in the rest of this section do not contain means to express propositional attitudes.

Let me close this part with two typical examples for *theses of extensionality*, both of which are taken from early writings of Carnap:[15]

(TE.1) In every statement about a propositional function, the latter may be replaced by its extension symbol ([6], p. 72).

---

[11] The principles of extensionality studied in this paper — like $(Ext^{=})$ and $(Ext^{\leftrightarrow})$ — are principles of substitutivity. Though highly important, the *axiom of extensionality* from, e.g., set theory — i.e. "$\forall xy\ (\forall z\ (z \in x \leftrightarrow z \in y) \rightarrow x = y)$" — is not as relevant for the aim of this paper and thus will not be dealt with in it.

[12] Quantifications over expressions are always assumed to be relativized to the set of formulas, terms, predicates (or whatever is appropriate) of some language L.

[13] There are cases where I prefer to speak of an occurence of A in C when, actually, "A" is occuring in C.

[14] Actually, in [8], "B is extensional/intensional with respect to an occurence of A in C" is explained first; this supplies further possibilities to define "C is extensional/intensional".

[15] A thorough treatment of this topic is beyond the scope of this paper and not necessary for what I will discuss later on, anyway.

(TE.2) For every given intensional language S_1, an extensional language S_2 may be constructed such that S_1 can be translated into S_2 ([7], p. 245; see also the 2nd foreword to [6]).

(TE.1) *may* be construed as a quite strong claim, stating that the language L considered is extensional; let's call this (ThE.1) (for L). But since the predicates "propositional function" and "replace" occuring in (TE.1) are quite vague, and precise and univocal definitions for them are not forthcoming in [6], it is not easy to determine whether this interpretation is adequate.[16] Nevertheless, at least it seems that (TE.1) and (TE.2) are not equivalent with each other. As a matter of fact, Carnap proposed (TE.2) as an improvement over (TE.1), which he took to be refuted by the existence of languages in which modal notions — like "$\Box\psi$" for "ψ is necessary" — or propositional attitudes are expressible.[17] In what follows, I will discuss these "refutations" more thoroughly.

1.2. The correctness of principles and theses of extensionality. To start with, let me recall two examples which are commonly taken to establish that in general principles of extensionality are not sound. The first one (put forward by Quine; see e.g., [37]) is supposed to undermine $(Ext^{=})$; the second is directed against $(Ext^{\leftrightarrow})$. Both examples are formulated in languages whose vocabularies contain the expression "$\Box$" — standing for "is necessary" — as a *sentential operator*.

EXAMPLE 1. The following statements are true:
$9 =$ the number of planets (in our solar system), $\Box(9 > 7)$, $\neg\Box$(the number of planets > 7).

EXAMPLE 2. If we choose some tautology "$\top$" for "B" and "$\Box A$" for "$\psi(A)$" in $(Ext^{\leftrightarrow})$, the validity of $(Ext^{\leftrightarrow})$ and of "$\Box\top$" yield the validity of

$$A \longrightarrow \Box A.$$

Let's rephrase these considerations on (the failure of) principles of extensionality in terms of the (lack of) extensionality of contexts for Quine's example.

EXAMPLE 1, CONTINUED: The terms "9" and "the number of planets" are coextensional. Thus, if "$\Box(x > 7)$" ($\equiv C_0(x)$) were an extensional context, "$\Box(9 > 7)$" and "$\Box$(the number of planets > 7)" should be coextensional, too. But then, "$\Box(9 > 7)$" and "$\Box$(the number of planets > 7)" should have the same truth value — which is not the case.

---

[16]For example, it is not clear whether (TE.1) is equivalent to that version of a thesis of extensionality which states "that all statements about any propositional function are extensional (i.e., that there are no intensional statements)" ([6], p. 73): for the explanation "A statement is called *extensional* if it can be transformed into an extension statement . . . " from [6], p. 73, suffers from the vagueness of "transform" (i.e. "umformen").

[17]But let me add that (TE.2) is not without difficulties of its own. The problem is that if too many relations are admitted as *translations*, it may become trivialized.

It is of some interest to make explicit that this reformulation of *Example* 1 rests on two *semantical principles* which are widely accepted. Given terms s, t and formulas φ, ψ from some language L, they are

(SemP$_1$) $\ulcorner s = t \urcorner$ is true $\Leftrightarrow$ s is coextensional with t,

and

(SemP$_2$) $\ulcorner \varphi \leftrightarrow \psi \urcorner$ is true $\Leftrightarrow$ φ is coextensional with ψ.[18]

But do the examples presented above really establish the claims they were invented for? With respect to the theme of extensionality of contexts, recall Frege's proposal (see [16], and, for a similar position, [6], p. 77) that the reference of a name in an oblique context ("ungerade Rede") is what usually (in "gerader Rede") is taken to be its sense. Let's apply this idea to Quine's example (continued) and specialize the *definiens* of "C_0 is extensional" to "9" and "the number of planets", resp.; we obtain

(∗) "9" is coextensional with "the number of planets" $\Rightarrow$ "$\Box(9 > 7)$" is coextensional with "$\Box$(the number of planets > 7)".

When understood as usual, the terms "9" and "the number of planets" are not synonymous with each other — that at least I will assume; thus, their *occurences in context* C_0 will not refer to the same object. But in the *definiens* of "C is extensional", it is certainly presupposed that all occurences of the expression "A" considered there are coextensional with each other (if this is not so, there is no reason whatsoever to assume that there are any extensional contexts). Hence, (∗)'s antecedent "'9' is coextensional with 'the number of planets'" turns out to be false, whence (∗) becomes true — and we no longer have an argument for the nonextensionality of the context C.

With this type of reasoning, even the strong thesis of extensionality (ThE.1) is saved. It is natural to object, however, that this defense rests on a reinterpretation of the predicate "is coextensional with" as "is synonymous with": i.e., it is the claim

(ThI/E) If s is synonymous with t, then
$C(s)$ is true if and only if $C(t)$ is true

(for formulas C and terms s, t of L) which really has been dealt with rather than (ThE.1). Since C is not allowed to be hyperintensional, (ThI/E) may well be true; in fact, I will assume that this is the case.

As convincing as this reply may be, there is a residual problem with it: it rests on the assumption that we are able to distinguished rather strictly between synonymy and coextensionality; and this in turn may presuppose a decent understanding of "a is synonymous with b". Now, you need not be a Quinean sceptic — and doubt the very possibility of giving *any* explication of

[18]In this section, I am using the corners "$\ulcorner\ \urcorner$" as a quasi-quotation in Quine's sense; see [35]. In arithmetical formulas, they will stand for the device of Gödel numbering.

that expression — to notice that in the course of the history of philosophy, one had to learn that it is difficult to actually provide an explanation of it which is both clear, general and adequate to our pretheoretic understanding.

Let me nonetheless recall two important *explicanta* of "synonymous" that have actually been proposed in philosophy and logic.

(i) *Possible worlds semantics* (see [21]): this constitutes the commonly accepted framework for interpreting modal notions. Formulas φ from some modal language L (with "$\Box$") are evaluated in possible worlds models; i.e., "$\langle W, R\rangle, w \models \varphi[V]$" gets explained (with $W \neq \emptyset$, the set of "possible worlds", $R \subseteq W^2$, $w \in W$, V a value assignment for the sentential parameters). For sentences φ, ψ of L, "synonymy" relative to W is definable as follows:

$$\varphi \text{ is synonymous with } \psi :\Longleftrightarrow \{w \in W \mid \langle W, R\rangle, w \models \varphi\} = \{w \in W \mid \langle W, R\rangle, w \models \psi\}.$$

(ii) *The "provability" interpretation of necessity* (see [25], and also [7], [17], [23] and texts from provability logic, like [44], [3]): here, "$\Box\psi$" — more explicitly, "$\Box_T\psi$" — is taken to represent "ψ is provable in T", T being some theory given in advance. That is,

$$\varphi \text{ is synonymous}_T \text{ with } \psi$$

is explained as

$$T \vdash \varphi \longleftrightarrow \psi.$$

For reasons of space, I will not enter into a more detailed general discussion of these approaches. Incidentally, for this paper's purpose the question which is relevant is mainly whether they provide convincing explications of "synonymy" as applied to sentences from mathematics. On this, more will be said in the next section. The remaining part of this section is an investigation of counterexamples to principles of extensionality.

To begin with, I think it is remarkable that Frege's procedure for avoiding non-extensional contexts does not by itself yield the general truth of principles of extensionality. In fact, if "$C(x)$ is extensional" is understood as in (ThI/E), $C_0(x)$ is extensional; but this is compatible with the failure of $(Ext^{=})$, if that scheme is evaluated by employing (SemP_1) and (SemP_2). Yet, $(Ext^{=})$ can be made true for contexts containing modalities, too: one merely has to adopt the principle

(SemP_F) $\qquad \ulcorner s = t \urcorner \text{ is true } \Longleftrightarrow s \text{ is synonymous with } t$

(s, t terms in L) instead of (SemP_1), and $(Ext^{=})$ gets evaluated as (ThI/E).

But even without the assumption of the problematic (SemP_F), there may be principles of extensionality that can be maintained. For consider the schemata

$(Ext^{\Box=})$ $\qquad \Box(s = t) \longrightarrow (\psi(s) \leftrightarrow \psi(t))$

(with terms s, t and formulas ψ of L) and

$$(Ext^{\Box\leftrightarrow}) \qquad \Box(A \leftrightarrow B) \longrightarrow (\psi(A) \leftrightarrow \psi(B))$$

(with formulas A, B, ψ of L). Both of them are valid in rather weak modal logics; the truth of $(Ext^{\Box=})$, in particular, does not rest on (SemP_F), but is rather already guaranteed by the — certainly plausible — schema

$$\ulcorner\Box(s = t)\urcorner \text{ is true } \Longleftrightarrow s \text{ is synonymous with } t$$

(for terms s, t of L).[19]

Alongside $(Ext^{=})$ and $(Ext^{\leftrightarrow})$, $(Ext^{\Box=})$ and $(Ext^{\Box\leftrightarrow})$ are instances of the following schema (for some language L):

$$(Ext^{\equiv_1,\equiv_2}) \qquad A \equiv_1 B \longrightarrow \psi(A) \equiv_2 \psi(B).$$

Here, ψ is a formula of L, and A, B and $\equiv_1$, $\equiv_2$ are expressions of L such that A and B are of the same "category" — i.e., both are, e.g., terms, predicates or formulas — and $\equiv_1$, $\equiv_2$ satisfy the conditions on equivalence relations.[20]

If $(Ext^{\equiv_1,\equiv_2})$ is accepted as the general, abstract form of principles of extensionality,[21] $(Ext^{\Box=})$ and $(Ext^{\Box\leftrightarrow})$ *are* principles of extensionality. Yet, possibly because of such instances, the schema $(Ext^{\equiv_1,\equiv_2})$ may be criticized as being too wide. It may even be open to trivializing interpretations: consider, e.g., the case that "'A' = 'B'" is admitted as an instance of "$A \equiv_1 B$";[22] then "$A \equiv_1 B$" will certainly make "$\psi(A) \leftrightarrow \psi(B)$" true (at least if "$A$" is used, but not mentioned, in $\psi(A)$). Now, I do not want to defend $(Ext^{\equiv_1,\equiv_2})$ come what may. Still, I think there are contexts in which, say, $(Ext^{=})$ can be formulated, but other schemata are nonetheless taken to be more relevant as principles of extensionality — schemata which are, in addition, instances of

---

[19]This equivalence may, in turn, be reduced to two further often accepted semantical principles: for sentences A and terms s, t of L,

$$(\mathrm{SemP}_3) \qquad (A \text{ is analytic } \Leftrightarrow) A \text{ is necessarily true } \Longleftrightarrow \ulcorner\Box A\urcorner \text{ is true}$$

and

$$(\mathrm{SemP}_4) \qquad \ulcorner s = t\urcorner \text{ is necessarily true } \Longleftrightarrow s \text{ is synonymous with } t.$$

When dealing with $(Ext^{\Box\leftrightarrow})$, one should instead add

$$A \text{ is synonymous with } B \Longleftrightarrow \ulcorner A \longleftrightarrow B\urcorner \text{ is analytic}$$

to (SemP_3).

[20]For "$\alpha \equiv_2 \beta$", take the formula "$\alpha \leftrightarrow \beta$" (with formulas α, β); and choose for "$A \equiv_1 B$" the formula "$\Box(A = B)$" (with terms A, B) for Example 1 and "$\Box(A \leftrightarrow B)$" (with formulas A, B) for Example 2.

[21]This approach to extensionality is inspired by [26].

[22]I think this is intended and should be allowed; I consider this example and two similar ones later on.

$(Ext^{\equiv_1,\equiv_2})$.[23] The next example and those of Section 2.1 should make this assessment plausible.

Let's represent "is necessary" — possibly by regarding it as a shorthand for "is necessarily true" — by a (1-place) *predicate* "Nec" in a first-order language L. Being an instance of $(Ext^{=})$ (formulated in L), for all terms s, t of L,

$$s = t \longrightarrow (\mathrm{Nec}(s) \leftrightarrow \mathrm{Nec}(t))$$

is already provable in predicate logic. As a special case, we have the provability of

$$\text{“}A\text{”} = \text{“}B\text{”} \longrightarrow (\mathrm{Nec}(\text{“}A\text{”}) \leftrightarrow \mathrm{Nec}(\text{“}B\text{”}))$$

for formulas "A" and "B". Note that this instance of $(Ext^{=})$ is a metatheoretical sentence which, by assumption, belongs to L. Since, as usually construed, a quotation mark name of some formula "A" from a language L_0 is a term formulated in some metalanguage ML_0 of L_0, it may be assumed in our case that L ist just that metalanguage and "A" and "B" belong to L_0.

Now, although I see no problem in treating "is necessary" as a predicate and accepting $(Ext^{=})$ for it, I am reluctant to conclude that, after all, "x is necessary" is an extensional context; and I think most would agree. In my opinion, it is not $(Ext^{=})$ which is the most interesting instance of $(Ext^{\equiv_1,\equiv_2})$ in this case, but rather a predicate-logical reformulation of $(Ext^{\leftrightarrow})$ which can be stated approximately as follows

$$(Ext^{\mathrm{Nec},\leftrightarrow}) \qquad (A \leftrightarrow B) \longrightarrow (\mathrm{Nec}(\text{“}A\text{”}) \leftrightarrow \mathrm{Nec}(\text{“}B\text{”}))$$

for formulas "A", "B" of L_0. I write "approximately" because $(Ext^{\mathrm{Nec},\leftrightarrow})$ becomes well-formed only if the formulas "A" and their quotation-mark names ""A"" occuring in it belong to the same language. This is so if L_0 is included in ML_0 or L_0 is homophonically translated into ML_0, but need not always be the case. In general, a schema such as

$$(Ext^{\mathrm{T,Nec},\leftrightarrow}) \qquad \mathrm{True}(\ulcorner \varphi \leftrightarrow \psi \urcorner) \longrightarrow (\mathrm{Nec}(\varphi) \leftrightarrow \mathrm{Nec}(\psi))$$

(with φ, ψ of L_0 and "True(x)" plus "Nec(x)" belonging to ML_0) seems to be a reasonable variant of $(Ext^{\mathrm{Nec},\leftrightarrow})$.[24] Another point worth mentioning is that there exists, again, a thesis of extensionality which is equivalent to $(Ext^{\mathrm{Nec},\leftrightarrow})$. Under the assumption of (SemP_2), it says

(ThE.Nec) If "A" is coextensional with "B", then
"Nec('A')" is coextensional with "Nec('B')"

for sentences "A" and "B" of L_0.

---

[23]It is partly because of this flexibility that I regard $(Ext^{\equiv_1,\equiv_2})$ to be helpful in studying extensionality.

[24]Names of expressions from some L may also be construed as terms of L itself: this is usually achieved by encoding, e.g., Gödel numbering, but axiomatic object-language quotation theories are possible, too. In such a case, $(Ext^{\mathrm{Nec},\leftrightarrow})$ may be conceived of as a well-formed formula of some object language without further ado. — I will return to this remark in the next section.

I began this subsection with a reminder of examples which are intended as refutations of theses of extensionality and the general soundness of principles of extensionality. I then continued with some suggestions for taking away the force from these examples and with rejections of those suggestions, accusing them of turning the original theses and principles into trivialities, and finally closed with supplying nontrivial versions of those principles and theses. This undecidedness of mine has been deliberate: it is neither my aim to save the principles and theses nor to discredit them. As a matter of fact, I think that this is hardly possible: for they are learned by examples and not by formulating general conditions, which makes our understanding of them somewhat undetermined and compatible with many precise renderings.

Nevertheless, I think that ($Ext^{\mathrm{T,Nec},\leftrightarrow}$) and (ThE.Nec) deserve to be called "principle of extensionality" and "thesis of extensionality". In the next section, they will be used as a model for how intensionality may enter into mathematics. For even if those principles and theses are questionable *in general* and even if suggestions for general explications of "synonymous" are viewed as failures — it is nonetheless quite possible that versions of the former will be true and methods for defining "synonymous" will succeed in the *restricted context of mathematical research.*

§2. Intensionality and (meta-) mathematics. In this section, I address the question

> Can intensionality play any role in the mathematical enterprise, and if so, which one?

in light of the general philosophical considerations presented in the previous section. I use the expression "mathematical enterprise" as a heading for the practice of the working mathematician, which is a blend of both mathematical and metamathematical, both formalized and informal reasoning. Thus, it does not only encompass (often formalized) mathematical theories and the languages wherein these are formulated, but also their (usually informal) metalanguages and metatheories.[25]

In the first subsection, last section's definitions of "synonymy" will be applied to and examples for *theses of intensionality* will be formulated for mathematical contexts. The second subsection contains a presentation of Feferman's important contribution to intensionality in metamathematics. His approach is compared with the one stemming from Section 1 in the concluding part of this section.

2.1. Application of the previous section to mathematics. Clearly, those theories which are *commonly* accepted as correct formalizations of "true" mathematics — like Peano Arithmetic (PA), Zermelo Fraenkel Set-theory (ZF),

[25]It is agreed that, from a systematic point of view, it is advisable to distinguish these levels from one another.

Second-order Number-theory (Z_2) and variants — prove principles of extensionality, like $(Ext^{=})$ and $(Ext^{\leftrightarrow})$. But it should be added that mathematical theories T have been developed for which this is not the case. Some of these theories are obtained from the more familiar ones through the restriction of principles of extensionality (see, e.g., [2], [14] and, in particular, [15]); others result by enriching quantificational languages L with nonclassical sentential operators (like "$\Box$") while keeping $(Ext^{=})$, say, to the original L (see, e.g., [2], [41]).[26] — Interesting as these alternatives to common mathematical practice may be, they will not be investigated in this paper.[27]

Yet, even if mathematical contexts are always extensional in the sense just mentioned, it does not follow that there is no place for intensionality in the mathematical enterprise. First, instances of $(Ext^{\equiv_1,\equiv_2})$ other than $(Ext^{=})$ or $(Ext^{\leftrightarrow})$ may be unprovable in the usual mathematical theories. And second, there may exist formulas which create *metamathematical* nonextensional contexts. I will deal with the latter idea first and address the first one as sort of a refinement of it afterwards.

For definiteness and simplicity, only formulas and theories formulated in L[PA], i.e. the first-order language of Peano Arithmetic in the vocabulary $\overline{0}$, $S, +, \cdot, \leq$, or in some metalanguage ML[PA] of L[PA], will be considered in what follows. ML[PA] should be so rich that syntactical conditions — like "a is a proof for b in T" — can be formulated in it; but at some places I will also assume that theories in which model-theoretic semantics for L[PA] can be developed are expressible in ML[PA].

In order to show that there are principles of extensionality formulated in ML[PA] which fail, it suffices to find a formula $C(x)$ from ML[PA] and sentences A, B such that A and B are coextensional, but C and $C[A/B]$ are not coextensional. Surely, formulas containing expressions standing for modalities and propositional attitudes suggest themselves as the first choice for such a C; but they simply do not belong to the common metalanguages of L[PA]. Instead, we find in them the metamathematical predicate "x is provable in T". Now, for sentences x, y of L[PA] it is certainly the case that

$$x = y \Longrightarrow (x \text{ is provable in } T \Leftrightarrow y \text{ is provable in } T).$$

Yet, at least in metamathematical research, what seems to be more interesting is (for formulas φ, ψ of L[PA])

$$(Ext^{\vdash,\leftrightarrow}) \qquad \mathcal{N} \models \varphi \longleftrightarrow \psi \Longrightarrow (T \vdash \varphi \Leftrightarrow T \vdash \psi).$$

This is similar to what I said as a motivation for $(Ext^{\text{Nec},\leftrightarrow})$. Moreover, $(Ext^{\vdash,\leftrightarrow})$ results from $(Ext^{\text{T,Nec},\leftrightarrow})$ simply be replacing "True" by "$\mathcal{N} \models$"

[26]In some texts, modal systems are even put forward as a sort of foundation-free foundation of mathematics; see [33] and [20].

[27]The main motivation for this decision is that they fail to be examples for that type of intensionality that is dealt with in [12], which is the one this paper is specifically about.

and "Nec" by "$T \vdash$".[28] And it is certainly natural to explain "φ is true" by "φ holds in $\mathcal{N}$" for sentences φ from L[PA].[29] Under the assumption of (SemP_2), "φ is coextensional with ψ" is thus defined as "$\mathcal{N} \models \varphi \leftrightarrow \psi$" (for sentences φ, ψ of L[PA]).

I therefore choose "x is provable in T" for the sought-after context $C(x)$ and formulate, in the style of (ThE.Nec), a first *thesis of intensionality* (for a theory T in L[PA]) as follows:

There are arithmetical sentences φ, ψ such that φ is coextensional with ψ, but "T proves φ" is not coextensional with "T proves ψ".

Note that whereas ($Ext^{\vdash,\leftrightarrow}$) is formulated in ML[PA], this thesis of intensionality is stated in a metalanguage of ML[PA].[30] Of course, theses of extensionality and intensionality are always metatheoretical statements: they are about expressions and, e.g., about the semantical values of these expressions. Principles of extensionality, on the other hand, are often formulated in languages which do not deal with linguistic entities; actually, this is so for most of the examples considered above. Thus, in some sense, there can be no mathematical theses of extensionality or intensionality, but only metamathematical, meta-metamathematical ones (and so on).[31]

Since it is simpler and more perspicious, I prefer to work with the reformulation of the above thesis I started with:

(ThI.1_T) *There are arithmetical sentences φ, ψ such that $\mathcal{N} \models \varphi \leftrightarrow \psi$ and $\neg(T \vdash \varphi \Leftrightarrow T \vdash \psi)$.*

Let's herewith come to a second metamathematical thesis of intensionality. It builds on the idea that (ThI/E) and its variant

If φ is synonymous with ψ, then $C(\varphi)$ is coextensional with $C(\psi)$

(for arbitrary sentences φ, ψ and contexts C of L) are plausible outside the domain of hyperintensional contexts. Since from the latter we (almost) obtain (ThE.1) for L by the addition of

φ is coextensional with $\psi \Longrightarrow \varphi$ is synonymous with ψ

---

[28] In addition, as sketched in Section 1, the phrase "x is necessary" may be explained as "x is provable in T" (for a suitable T; e.g., in the context dealt with here, PA).

[29] Let me add that, although there are widely shared beliefs about which mathematical sentences should be counted as true (or false) — some set-theoretic sentences which are independent from ZF being notable exceptions — I do not want to commit myself to the position that mathematical truths can be "intuited" apart from being provable in some acceptable theory (whatever *that* precisely is). Because of the ZF-definability of $\mathcal{N}$, such a "grasp" of the natural numbers (or: the standard model) can be avoided (in some sense).

[30] Moreover, it has, again, to be presupposed that L[PA] is included in ML[PA].

[31] Since a metamathematical thesis of extensionality is more or less equivalent to the statement that a mathematical principle of extensionality holds, there seems to be no harm in calling such a thesis a "mathematical thesis of extensionality".

(for φ, ψ of L), this principle may also be viewed as a way of expressing that all contexts of L are extensional. A second thesis of intensionality may thus be stated roughly as

There are arithmetical sentences φ, ψ such that φ is coextensional with ψ, but φ and ψ are not synonymous.

At this point, we have to deal with the two general schemata for explaining "is synonymous with" presented in Section 1 anew — and see how they fare with respect to mathematical formulas.

(i) Within the framework of possible worlds semantics, all mathematical truths are usually declared to be true in all possible worlds, whence all mathematical truths are true in the same possible worlds, which, by definition, means that they are synonymous.

(ii) Since biconditionals holding in $\mathcal{N}$ need not be provable in T (e.g., when PA is chosen for T), under the assumption of the provability approach coextensionality does not imply synonymy (with respect to T).

That coextensional sentences should be synonymous doesn't seem to be well-founded from an intuitive point of view — even when they are mathematical ones. Furthermore, such an implication should at least not follow so easily from *a definition*. This speaks in favour of the provability approach. Yet, (a) it has to be granted that the mere acceptance of the apparatus of possible worlds semantics does not carry any commitment to the claim that all arithmetical (let alone mathematical) truths have to be evaluated as true in each possible world; and (b) the dependency on some theory T which, in general, is left unspecified may be regarded as a weakness of the provability interpretation. Though probably true, these two remarks call for further comments.

With respect to (b), let me first note that in some cases the context of the investigation suggests a specific T — as, e.g., PA or subtheories of it, when one is doing number theory. What makes (a) unsatisfactory is that it does not provide any insight as to which possible worlds satisfy which arithmetical (or mathematical) truths. In fact, this sort of problem is well known from the cases in which possible worlds semantics is employed to distinguish between different forms of necessity — like logical necessity, metaphysical necessity or physical necessity. For it is common practice to do this by invoking different accessibility relations — relations which, in turn, are often specified through reference to theories T which hold in appropriate possible worlds.[32] But in this case, necessity and synonymy are eventually explicated by presupposing the provability interpretation — whence (b) turns out to present a problem for the possible worlds approach, too.

---

[32]For a "classical" example, see [4], p. 341: "The causally possible universes are all those logically possible universes in which these causal laws all hold". That is, we have $\langle W, R\rangle, w \models \Box\psi \Leftrightarrow \forall v \in W\ (wRv \Rightarrow \langle W, R\rangle,\ v \models \psi)$, where "$wRv$" gets explained as "$\langle D_v, \ldots\rangle \models T$".

For the mathematical languages dealt with here, I therefore opt for the provability interpretation of intensionality. This then, is the second *thesis of intensionality* for the metamathematics of arithmetic: for some arithmetical theory T

($\mathrm{ThI.2}_T$) *There are arithmetical sentences* φ, ψ *such that* $\mathcal{N} \models \varphi \leftrightarrow \psi$ *and* $T \nvdash \varphi \leftrightarrow \psi$.

In fact, the theses of intensionality put forward here are equivalent with each other and with

($\mathrm{ThI.Inc}_T$) *There are arithmetical sentences* φ *such that* $\mathcal{N} \models \varphi$ *and* $T \nvdash \varphi$.

For assume $\neg(\mathrm{ThI.1}_T)$ and let φ be a sentence such that $\mathcal{N} \models \varphi$, i.e., $\mathcal{N} \models \varphi \leftrightarrow \top$ for some tautology "$\top$"; then, by $\neg(\mathrm{ThI.1}_T)$, "$T \vdash \varphi \Leftrightarrow T \vdash \top$" follows, whence $T \vdash \varphi$. This contradicts ($\mathrm{ThI.Inc}_T$) (the missing implications are trivial).[33]

These are the (*meta-*) *metamathematical* theses of intensionality I wanted to present. Now, ($\mathrm{ThI.Inc}_T$) is reminiscent at the statement that T is incomplete. In fact, with the exception of $\mathrm{Th}(\mathcal{N})$, each consistent theory formulated in L[PA] is a witness for the theses of intensionality considered here. The upshot of this investigation is that in the metamathematics of arithmetic (or of, e.g., set theory), intensionality is ubiquitous. But this suggests that we may have dealt with rather uninteresting formulations of theses of intensionality. I will return to this challenge in a later chapter; let me close this part with the treatment of *mathematical* formulations of principles of extensionality and their failure.

Arithmetical principles of extensionality corresponding to the metamathematical ones just dealt with are easily obtained if the theories T addressed in the latter are arithmetically definable. Thus, let Σ be a set of formulas in L[PA], and let $\overline{\Sigma}$ be the set of formulas of L[PA] derivable from Σ in classical first-order logic (it is convenient to axiomatize it such that only *modus ponens* is a primitive rule of inference; see [12] for more details). If $S = \overline{\Sigma}$, Σ will be called an *axiom-set* for S.[34]

If τ is a representation of (the set of Gödel-numbers of) an *axiom-set* of the arithmetically definable theory T, the common arithmetizations of "proof in T", "provable in T" and "T is consistent" are:

$\mathrm{Proof}_\tau(x, y) :\leftrightarrow \mathrm{Seq}(x) \wedge y = x_{\mathrm{lh}(x)\dot{-}1} \wedge \forall v < \mathrm{lh}(x)\ (\mathrm{Log}Ax(x_v) \vee \tau(x_v) \vee \exists uw < v\ (x_w = x_u \dot{\rightarrow} x_v))$,

$\mathrm{Pr}_\tau(y) :\leftrightarrow \exists x\ \mathrm{Proof}_\tau(x, y)$,

33 Actually, this is just the reasoning mentioned in Example 2.

34 Note that S may fail to be recursively enumerable, i.e, it may not be *axiomatizable*; for lack of a better term, even in this case I will nonetheless use the term "axiom-set".

$\mathrm{Con}_\tau :\leftrightarrow \neg \mathrm{Pr}_\tau(\overline{\ulcorner \bot \urcorner})$.[35]

Here, "α is a representation of A" is defined as usual in metamathematical writings: for k-place α in L[PA] and $A \subseteq \omega^k$,

α is a representation of $A :\Leftrightarrow \forall n_1, \ldots, n_k \in \omega\ (\langle n_1, \ldots, n_k \rangle \in A \Leftrightarrow \mathcal{N} \models \alpha(\overline{n}_1, \ldots, \overline{n}_k))$.

Since "$\mathcal{N} \models \psi$" can be arithmetically simulated simply by writing "ψ" (for ψ in L[PA]), the, e.g., first metatheoretical principle of extensionality discussed above may be arithmetized by

$$(\varphi \leftrightarrow \psi) \longrightarrow \left(\mathrm{Pr}_\tau(\overline{\ulcorner \varphi \urcorner}) \leftrightarrow \mathrm{Pr}_\tau(\overline{\ulcorner \psi \urcorner})\right)$$

with φ, ψ in L[PA]. Given this schema, many formulations of theses of intensionality should be taken into account. For example, one has to distinguish between

(F-ThI.11$_T$-S) $\quad \exists \varphi\psi\ S \nvdash (\varphi \leftrightarrow \psi) \longrightarrow \left(\mathrm{Pr}_\tau(\overline{\ulcorner \varphi \urcorner}) \leftrightarrow \mathrm{Pr}_\tau(\overline{\ulcorner \psi \urcorner})\right)$

and

(F-ThI.12$_T$-S) $\quad \exists \varphi\psi\ S \vdash \neg((\varphi \leftrightarrow \psi) \rightarrow \left(\mathrm{Pr}_\tau(\overline{\ulcorner \varphi \urcorner}) \leftrightarrow \mathrm{Pr}_\tau(\overline{\ulcorner \psi \urcorner})\right))$,

where S is some theory formulated in L[PA].

Of course, (F-ThI.11$_T$-S) and (F-ThI.12$_T$-S) collapse into (ThI.1$_T$) if $S = \mathrm{Th}(\mathcal{N})$. Moreover, if S is (arithmetically) sound, (F-ThI.11$_T$-S.1) is true because it is a consequence of (ThI.1$_T$).[36] (F-ThI.12$_T$-S) is more interesting than (F-ThI.11$_T$-S) in that it depends more heavily on the relation between S and T than the latter principle. For example, let S, T be consistent r.e. extensions of $\mathrm{I}\Sigma_1$ with representations σ, τ in Σ^0_1. On the one hand, if $S \subseteq T$, then (F-ThI.12$_T$-S) is false. For $S \vdash \varphi \leftrightarrow \psi$ implies $T \vdash \varphi \leftrightarrow \psi$ and, thus, $T \vdash \mathrm{Pr}_\tau(\overline{\ulcorner \varphi \urcorner}) \leftrightarrow \mathrm{Pr}_\tau(\overline{\ulcorner \psi \urcorner})$ for arbitrary φ, ψ of L[PA]. But this is not compatible with (F-ThI.12$_T$-S) and the consistency of T. On the other hand, if $S = \mathrm{PA}$ and $T = \mathrm{I}\Sigma_1$, (F-ThI.12$_T$-S) is true. The reason is that $S \vdash \mathrm{Con}_{\mathrm{I}\Sigma_1}$, whence $S \vdash \mathrm{Con}_\tau \leftrightarrow \top$. Moreover, since $S \vdash \neg \mathrm{Pr}_{\mathrm{I}\Sigma_1}(\overline{\ulcorner \neg \mathrm{Pr}_{\mathrm{I}\Sigma_1}(\overline{\ulcorner \bot \urcorner}) \urcorner})$, we also have $S \vdash \neg(\mathrm{Pr}_\tau(\overline{\ulcorner \mathrm{Con}_\tau \urcorner}) \leftrightarrow \mathrm{Pr}_\tau(\overline{\ulcorner \top \urcorner}))$.

---

[35]It is assumed that

- for an expression t of L[PA], $\ulcorner t \urcorner$ is the Gödel-number and $\overline{\ulcorner t \urcorner}$ the Gödel-numeral of t;
- the usual "syntactic" metamathematical vocabulary — like "x is a sequence", "the length of (sequence) x", "x is the negation of y", "x is the conditional of y and z", "x is a formula", "x is a Π^0_n-formula" — is represented by Σ^0_0-formulas — like "$\mathrm{Seq}(x)$", "$\mathrm{lh}(x)$", "$x = \dot{\neg} y$", "$x = y \dot{\rightarrow} z$", "$\mathrm{Fml}(x)$", "$\Pi^0_n(x)$" — such that conditions characteristic for these predicates are provable in PA; see [19] and [22], and also the following subsection;
- Feferman's "dot-notation" is in use (see [12], [44]).

[36]If $S := \mathrm{PA} + \mathrm{Pr}_{\mathrm{pa}}(\overline{\ulcorner \bot \urcorner})$ and $T := \mathrm{PA}$, then $\forall \varphi\psi\ S \vdash \mathrm{Pr}_{\mathrm{pa}}(\overline{\ulcorner \varphi \urcorner}) \leftrightarrow \mathrm{Pr}_{\mathrm{pa}}(\overline{\ulcorner \psi \urcorner})$, whence (F-ThI.11$_T$-S) cannot be the case.

Let me leave it at this. The mathematical principles of extensionality obtained from the metamathematical ones considered here are as false as these (the proof of this is somewhat more involved than in case of the metamathematical principles, but also follows from well known results from the metamathematics of arithmetic). Noticing that intensionality reduces to incompleteness (or lack of soundness), a natural idea is to somehow restrict the range of admissable sentences A, B and theories T in order to get stronger, i.e. more *serious* cases of intensionality. I will deal with this more thoroughly in Section 3. Let me first present Feferman's conception of intensionality in metamathematics, however.

2.2. Feferman's conception. In this subsection, I will discuss some of the conceptual parts of Feferman's 1960 paper "Arithmetization of metamathematics". As far as I know, this paper is still the most detailed development of the topic of intensionality from the perspective of metamathematics.

When the word "intensional" first appears in [12], it is *the application of the method of arithmetization* which is characterized as being extensional or intensional (p. 35):

"... the application of the method [of arithmetization] can be classified as being *extensional* if essentially numerically correct definitions are needed, or *intensional* if the definitions must more fully *express* the notions involved, so that various of the general properties of these notions can be formally derived."[37]

Later, also *results* (p. 35) and *definitions* (p. 37) are called "intensional" (and "intensionally correct", in case of definitions (p. 36); see also [5]).

These classifications are hardly equivalent; moreover, they are at odds with the established philosophical use of "extensional" and "intensional" (as, e.g., sketched in Section 1).[38] From that point of view, the phrase "a definition is intensional", in particular, is hard to understand at all. In what follows, I will apply the predicates "is extensionally/intensionally correct"[39] solely to *arithmetizations*, i.e. arithmetical formulas.[40] This decision fits the passage quoted from [12], for arithmetizations are (the results of) the applications

[37] I will often refer to this passage, which contains the most explicit general explanation of "intensional" I have found in proof-theoretical texts.

[38] Feferman himself was quite aware that "intensional" as it is used in [12] may be a misnomer; in [13], p. 121, he writes: "In FEFERMAN [1962] and earlier publications, I spoke of giving a collection of axioms *intensionally* as opposed to giving it *extensionally*, i.e., merely as a set. To avoid confusion with the philosophical problem of *intensions* it seems now preferable to use other terminology such as (for a formula which defines a set), *formal presentation of a set*." "Formal presentation of a set" and "formal description of a set" are the phrases which Feferman preferred in subsequent writings.

[39] I interpret "is extensional/intensional" as a shorthand for "is extensionally/intensionally correct".

[40] Again, all formulas and theories dealt with will belong to L[PA]. — More generally, e.g., set-theoretical formulas could be taken into consideration.

of the method of arithmetization, and it avoids the first and third of the criticisms just put forward. With respect to the second criticism, let me remark that, on the one hand, there does not exist any general necessity to share the philosopher's *terminology*; moreover, Feferman's deviation from it does not show the irrelevance of his approach. On the other hand, it should be clear that we have more than a mere clash of terminologies here: the work done on intensionality in philosophy is (despite the difficulties in giving explications for the intensional vocabulary), especially when compared to the rather scattered remarks on this theme to be found in the metamathematical literature, so highly developed that reasons should be given for favouring a different conception of intensionality.

But maybe the difference between "the" philosophical and Feferman's treatment of intensionality is on a purely terminological level and, as such, merely superficial. Thus, let's have as closer look at how Feferman understands "extensionally correct" and "intensionally correct". Concerning the first, I take it that the formulation "[...] can be classified as being *extensional* if essentially numerically correct definitions are needed" (p. 35; see also [5], p. 113) suggests the following definition:[41] let $A \subseteq \omega^n$ and let α be a n-place formula of L[PA] (for some $n \in \omega$); then

$$\alpha \text{ is an extensional arithmetization of } A :\Longleftrightarrow \alpha \text{ is a representation of } A.$$

It is more important and much harder to give an explanation of "intensionally correct arithmetization", though. Certainly, an arithmetical formula α is never simply intensionally (or extensionally) correct, but only intensionally correct *relative to some A*.[42] For example, it may well be assumed that "$\mathrm{Pr}_{\mathrm{pa}}$" is an intensionally correct arithmetization of provability in PA; but it is certainly not an intensionally correct arithmetization of "ZF is consistent" or of being a formula in the language of PA. Moreover, it seems clear that for an arithmetization to be intensionally correct it has to be at least extensionally correct. But these are only minimal conditions on any possible explication of "α is an intensionally correct arithmetization of A".

Now, in the passage from [12] quoted above, we are told that for a definition (in the terminology adopted here: an arithmetization) to be intensional, it must

(Fef-1) "more fully *express* the notions involved" than a (merely) extensionally correct one, so that

(Fef-2) "various of the general properties of these notions can be formally derived."[43]

---

[41] In principle, I will accept it as adequate for the rest of this paper.

[42] See below for a discussion which entity such an A could be.

[43] In [5], only the second condition is repeated as characteristic (or necessary) for an *intensional application of the method of arithmetization*: "In the intensional approach to arithmetization of metamathematics, one gives formulas which define concepts such as "formula", "term",

The problem with this explanation is that we neither find a *general* and *precise* explication of "α expresses A" (e.g., a notion) in [12], nor are we informed what the general properties and notions are for which their arithmetizations should be derivable. Thus, it has to be admitted that *no definition* of "α is an intensionally correct arithmetization (or: definition) of A" has been presented in [12] (or, for that, in [5]). Moreover, (Fef-1) and (Fef-2) are not simply so clear that an explication of them would be superfluous. At the same time, this circumstance does not preclude having some understanding — even if it may be somewhat vague and speculative — of the predicate "intensionally correct" as it is used in [12]. In case of specific examples, in particular, one may have quite strong convictions as to which arithmetizations are intensionally correct or fail to be so. This seems to be so also for Feferman; for one of his most explicit and informative claims concerning intensionally correct arithmetizations appears in his discussion of the Rosser proof predicate "Prf'_τ" for T. There he writes ([12], p. 36):[44]

"Of course, $\mathrm{Prf}'_\tau(x, y)$ does not express the proof relation of T but rather the relation 'y is a proof in T of x and for all $z \leq y$, z is not a proof in T of the negation of x'."[45]

Let's call this "Feferman's claim". Quite clearly, Feferman does not argue for his claim by invoking formal theories S and formulas Ψ of L[S] containing "Prf_τ" such that Ψ is provable (in S) whereas $\Psi[\mathrm{Prf}'_\tau / \mathrm{Prf}_\tau]$ fails to be so. That is, condition (Fef-2) simply is not invoked by Feferman as a justification for his claim. His reason for asserting that "$\mathrm{Prf}'_\tau(x, y)$" expresses the Rosser-proof relation of T rather seems to be the fact that the *definiens* of the metatheoretical formula "x is a proof in T of y and for all $z \leq x$, z is not a proof in T of the negation of y" is *copied* by the *definiens* of "$\mathrm{Prf}'_\tau(x, y)$". *Structure preservation* is the central component of intensionally correct arithmetizations as explained in part (Fef-1):

if α expresses A, α should have (maybe: almost) the same structure as A.

Although it is at variance with some formulations of Feferman,[46] the interpretation I will put forward here — that in "α expresses A", A should be viewed as a metatheoretical formula — is in my opinion the most convincing reading or reconstruction of Feferman's reasoning.

For if it is relations (i.e., relations on ω) which are expressed (cf. the passage just quoted), it is hard to understand why Feferman's claim should

---

"substitution", "proof", "theorem", etc.; these definitions are said to be *intensional* provided the theory T can prove simple properties of these concepts" (p. 113). This section's investigations should make it plausible that each of (Fef-1) and (Fef-2) is important in its own right.

[44] As I will use "$\mathrm{Prf}'_\tau(x, y)$", it is defined as "$\mathrm{Prf}_\tau(x, y) \wedge \forall z \leq x \neg \mathrm{Prf}_\tau(z, \dot{\neg} y)$". Note that I have changed the order of the variables in the proof-predicate when compared to [12].

[45] The context suggests, moreover, that Feferman would also accept that "$\mathrm{Prf}_\tau(x, y)$" does express that x is a proof in T of y. Both assesments seem to be *opinio communis*.

[46] Actually, Feferman has been ambiguous on this point, and I have followed him in this.

be true. To see this, take a consistent (arithmetically definable) theory T (with an *axiom-set* which is represented by τ); then the relation $\{\langle a,b\rangle \mid$ a is a proof in T of b and for all $z \leq a$, z is not a proof in T of the negation of $b\}$ simply *is* the relation $\{\langle a,b\rangle \mid a$ is a proof in T of $b\}$. Now, since by assumption "Prf'_τ" expresses $\{\langle a,b\rangle \mid a$ is a proof in T of b and for all $z \leq a$, z is not a proof in T of the negation of $b\}$, it should also express $\{\langle a,b\rangle \mid$ a is a proof in T of $b\}$ — contrary to Feferman's claim.

Of course, there is a way out of this problem: simply deny that "α expresses A" is extensional at the second argument place; i.e., abandon the principle

$$\text{If } \alpha \text{ expresses } A \text{ and } A = B, \text{ then } \alpha \text{ expresses } B$$

if A, B are relations on ω. But there remains a difficulty: why does "Prf'_τ" express the Rosser-proof relation (for T) and not the usual one, where both are the same object?

And if what is expressed are *notions* (cf. the formulation of (Fef-1)), we have the problems with intensional entities already addressed in Section 1 — at least if we understand "notions" as, say, *relations-in-intension*: as *propositional functions*, i.e. functions from the set of possible worlds to certain sets of extensions, for example. As a matter of fact, Feferman provides neither an explanation of what "notion" could mean nor what the identity conditions for notions could be.[47] In addition, the above question can easily be transferred to the case considered here: why does "Prf'_τ" express the notion of being a Rosser-proof (in T) but not that of being a proof (in the usual sense)?[48]

These criticisms do not pertain to formulas when taken as the entities expressed: there are rather clear identity conditions for formulas; the principle of extensionality just mentioned is satisfied by them; and, as mentioned above, the two questions can be answered by recourse to structure preservation. Moreover, talk of structure preservation makes sense (and may be made precise) if α and A are both formulas of, say, quantificational languages. But it is much harder, if possible at all, to understand the assertion that an arithmetical formula α has the same structure as a relation-in-extension or a relation-in-intension. In particular, if the Rosser-proof-relation and the "usual" proof-relation (for T) had the same structure, it seems implausible that "Prf'_τ" would express the first but not the second one. But regarded as relations-in-extension, these relations do have the same structure (for they are the same relation); and considered as relations-in-intension, it is simply unclear whether they have the same structure or not.[49]

[47] This is no criticism of [12], of course, but rather a sign that Feferman refrains from postulating intensional entities.

[48] It is therefore of no help that notions are "preferable" to relations in that the above principle of extensionality may not be violated for them.

[49] One could project the structures of the metatheoretical formulations of "x is a Rosser-proof for y in T" and "x is a (usual) proof for y in T" onto the notions. This would probably amount

For notational clarity, I will write "F_A" for a formula from ML[PA] which has the relation A on ω as its extension. With this, I prefer the formulation "α expresses F_A" or even "α is synonymous with F_A" to "α expresses A", and correspondingly reformulate "α is an intensionally correct arithmetization of A" as "α is an intensionally correct arithmetization of F_A". If each intensionally correct arithmetization shall remain extensionally correct, a similar adjustment has to take place for extensionally correct arithmetizations: just replace an arithmetically definable relation A on ω by some metatheoretical formula F_A which has A as its extension (assuming that the metalanguage has, as it is usual, at least as much expressive power as L[PA]). There are, of course, different F_A's having A as an extension; but this creates no problem: for coextensional formulas from ML[PA] may at this point, where only extensional correctness matters, well replace each other.

Thus, what remains is the task of giving an explication of "α expresses F_A" for α of L[PA] and F_A from ML[PA]. The following explanation presupposes that ML[PA] is sort of a quantificational language itself (for clause (ii)). Furthermore, "basic" components could be, say, atomic subformulas. According to what I have said before, "α expresses F_A" should be defined approximately as

(i) α and F_A are coextensional;[50] and
(ii) the *definientia* (when formulated in primitive notation) of α and F_A have the same quantificational structure; and
(iii) the "basic" components c_α of α "correspond" to subformulas C_A of F_A.[51]

It seems plausible here that the correspondence of c_α with C_A should imply that c_α is synonymous with C_A rather than merely coextensional with it. But in this case, it has to be acknowlegded that this explication is not only quite vague; additionally, it has an air of circularity. And although it may be possible to circumvent it in favour of an inductive definition, the latter still seems to need some stipulations at the base level which are hard to characterize by using general conditions.

So much for condition (Fef-1). Let's turn now to condition (Fef-2) and see if it can come to help.

As I have noted, the obvious shortcoming of condition (Fef-2) is that we are not told which general properties of which notions have to be derivable in order for an arithmetization to be intensionally correct. It may be added that

to something like ***structured propositions*** (see [9]), and the relation of having the same structure would probably be close to Carnap's *intensional isomorphism* (see [8]). But under this approach, notions would be dependent on metatheoretical formulas, the latter being the primary objects.

[50] That is, there is a relation A on ω such that α represents A and A is the extension of F_A.

[51] I take it for granted that for each arithmetically definable relation B on ω which is the extension of some condition F_B from ML[PA], the phrase "β is an intensionally correct arithmetization of F_B" should be meaningful.

we do not even know in which theory the derivations have to take place; but let's not press this point and simply take PA for such a theory. Actually, I do not believe in the possibility of fixing a list of general properties or determining it by abstract conditions once and for all which correctly — as evaluated from an intuitive point of view — distinguishes intensional from merely extensional arithmetizations. And this seems to be so even if the collection of notions considered is severely restricted. In fact, I will illustrate this scepticism by dealing with just one "notion" — the metatheoretical provability-predicate for PA — for the rest of this subsection.[52]

As concerns the expressions "notions" and "general properties", I will take up the considerations on (Fef-1): talk of *notions A* should be replaced by the use of "metatheoretical formula F_A"; accordingly, under *general properties of notion A* which are derivable in PA one should understand formulas Φ of L[PA] which contain an arithmetization α of F_A as a subformula. Moreover, as I interpret (Fef-2), only such general properties of a notion should be "derivable" in PA which that notion actually has.

Thus, as explained through condition (Fef-2), a formula α of L[PA] can only be an intensionally correct arithmetization of F_A (from ML[PA]) if α is coextensional with F_A and for some — yet to be specified — set of formulas Φ from L[PA],

$$\mathcal{N} \models \Phi(\alpha) \Longrightarrow \mathrm{PA} \vdash \Phi(\alpha).$$

In addition, those Φ should represent the extensions (in $\mathcal{N}$) of those metatheoretical formulas F_Φ which are the proxies of the general properties mentioned in (Fef-2), and

$$\mathcal{N} \models \Phi(\alpha) \Longleftrightarrow F_\Phi(F_A)$$

should hold.

Since the notion I will deal with is PA-provability — more precisely: the usual (metatheoretical) PA-provability predicate, the (metatheoretical) Rosser PA-provability predicate or similar ones — a natural first choice for the "general properties" is provided for by the Löb-derivability conditions,[53] especially by the second (Löb 2) — the closure of the set of PA-theorems under *modus ponens* —

$$\text{(1)} \qquad \mathrm{PA} \vdash \varphi \wedge \mathrm{PA} \vdash \varphi \longrightarrow \psi \Longrightarrow \mathrm{PA} \vdash \psi$$

(for arbitrary $\varphi, \psi \in$ L[PA]) and the first (Löb 1)

$$\text{(2)} \qquad \mathrm{PA} \vdash \varphi \Longrightarrow \mathrm{PA} \vdash \mathrm{Pr}_{\mathrm{pa}}(\ulcorner \varphi \urcorner)$$

[52]Observations which are similar to what follows are made in [11] (cf. also [28]); but the philosophical conclusions drawn there are different from the ones put forward here.

[53]Of course, they do not exhaust the list of general conditions; see [12], p. 37/38, for a further criticism pertaining to them.

(for arbitrary φ in L[PA]). If an arithmetical formula α is to be an intensionally correct arithmetization of provability in PA, it should thus be provable in PA that

$$\forall x, y\ (\alpha(x) \wedge \alpha(x \dot{\rightarrow} y) \rightarrow \alpha(y)) \tag{3}$$

and that

$$\forall x\ (\alpha(x) \rightarrow \alpha(\overline{\ulcorner \mathrm{Pr}_{\mathrm{pa}}(\dot{x}) \urcorner})). \tag{4}$$

For this, I have taken it for granted that the sentential operators and quantifiers of ML[PA] are translated into the same sentential operators and quantifiers of L[PA].[54]

Now, when we replace "$\alpha(x)$" by "$\mathrm{Pr}_{\mathrm{pa}}(x)$" in (3) and (4), we indeed obtain theorems of PA — which supports the assessment that "$\mathrm{Pr}_{\mathrm{pa}}$" is an intensionally correct arithmetization of the metatheoretical (usual) PA-provability predicate. In addition, since

$$\forall x, y\ \left(\mathrm{Pr}'_{\mathrm{pa}}(x) \wedge \mathrm{Pr}'_{\mathrm{pa}}(x \dot{\rightarrow} y) \rightarrow \mathrm{Pr}'_{\mathrm{pa}}(y)\right) \tag{5}$$

is not (for all arithmetizations of Rosser-provability; see [18]) provable in PA, we have reason for asserting that "$\mathrm{Pr}'_{\mathrm{pa}}$" is not an intensionally correct arithmetization of that predicate.

This much is as expected and intuitively convincing. But now, let's consider "$\mathrm{Pr}'_{\mathrm{pa}}$" in a slightly different role: both intuitively and in the light of (Fef-1), "$\mathrm{Pr}'_{\mathrm{pa}}$" is just as intensionally correct an arithmetization of the metatheoretical *Rosser-provability-in-*PA *predicate* (for which I write "PA $\vdash^R$") as "$\mathrm{Pr}_{\mathrm{pa}}$" is of the PA-provability predicate. Now we have for arbitrary φ, ψ in L[PA]

$$\mathrm{PA} \vdash^R \varphi \wedge \mathrm{PA} \vdash^R \varphi \longrightarrow \psi \Longrightarrow \mathrm{PA} \vdash^R \psi; \tag{6}$$

thus, if "$\mathrm{Pr}'_{\mathrm{pa}}$" were an intensionally correct arithmetization of the metatheoretical Rosser-provability-in-PA predicate as explained by (Fef-2), (5) should be provable in PA — which, as just noted, is not the case (in general). Thus, it seems we have a "tension" (if not contradiction) between conditions (Fef-1) and (Fef-2) and between our intuitions and (Fef-2) *with respect to* "$\mathrm{Pr}'_{\mathrm{pa}}$".

In addition, it should be noted that there are general properties of the (usual) metatheoretical PA-provability predicate, too, whose arithmetization through "$\mathrm{Pr}_{\mathrm{pa}}$" is *not* provable in PA. For example, consider the converse of Löb 1, i.e.

$$\mathrm{PA} \vdash \mathrm{Pr}_{\mathrm{pa}}(\overline{\ulcorner \varphi \urcorner}) \Longrightarrow \mathrm{PA} \vdash \varphi \tag{7}$$

---

[54]In case of the quantifiers, there is an alternative: arithmetize, e.g., (1) as

$$\alpha(\overline{\ulcorner \varphi \urcorner}) \wedge \alpha(\overline{\ulcorner \varphi \rightarrow \psi \urcorner}) \longrightarrow \alpha(\overline{\ulcorner \psi \urcorner})$$

for arbitrary φ, ψ of L[PA]. In general, this seems to be too weak, however.

(for all φ in L[PA]). If "$\mathrm{Pr}_{\mathrm{pa}}$" were an intensionally correct arithmetization of the metatheoretical PA-provability predicate, we should have

$$\text{(8)} \qquad \mathrm{PA} \vdash \overline{\mathrm{Pr}_{\mathrm{pa}}\left(\ulcorner \mathrm{Pr}_{\mathrm{pa}}(\ulcorner\varphi\urcorner)\urcorner\right) \longrightarrow \mathrm{Pr}_{\mathrm{pa}}(\ulcorner\varphi\urcorner)}$$

for arbitrary φ of L[PA] — which is, of course, not the case.

For a second example, take a variant of the metatheoretical claim for the consistency of PA: for all φ in L[PA],

$$\text{(9)} \qquad \mathrm{PA} \vdash \varphi \Longrightarrow \mathrm{PA} \nvdash \neg\varphi.$$

Again, its formalization with "$\mathrm{Pr}_{\mathrm{pa}}$",

$$\text{(10)} \qquad \forall x \left(\mathrm{Pr}_{\mathrm{pa}}(x) \rightarrow \neg\, \mathrm{Pr}_{\mathrm{pa}}(\dot\neg x)\right)$$

fails to be provable in PA.[55]

Because of examples such as these, (Fef-2) should even make us doubt whether "$\mathrm{Pr}_{\mathrm{pa}}$" is an intensionally correct arithmetization of the PA-provability predicate. Yet, the strength of these examples may itself be put into question: for they rest on the assumptions that PA is consistent or sound (to some extent). When (formalizations of) these assumptions are included as further premises in the formalizations of (7) and (9), formulas result which *are* provable in PA — or so the story goes.

The first part of this attempted defense is certainly true. But a claim of the sort

The PA-*provability predicate satisfies some general condition, but a formalization of this very assertion is not provable in* PA

already implies that PA is consistent. Thus, there is no way around the assumption of the consistency of PA. And if, instead of PA, a T is taken which fails to be sound, a statement of the form

$$\text{(11)} \qquad T \vdash \forall xy \left(\mathrm{Pr}_\tau(x) \rightarrow \mathrm{Pr}_\tau(y)\right)$$

may well be trivial — as it is in the case of $T = \mathrm{PA} + \mathrm{Pr}_\tau(\ulcorner\bot\urcorner)$. Furthermore, the statements (7) and (9) themselves, and not only their weakened forms obtained by the addition of metatheoretical correctness claims as premises, *are* metatheorems. This is the case at least if ZF is taken as the theory which is admitted for proving metatheorems (which is the usual decision; if ZF were replaced by, say, PA in this role, examples using finitely axiomatizable subtheories of PA instead of PA would do a similar job).

A further method for defending the judgement that "$\mathrm{Pr}_{\mathrm{pa}}$" is an intensionally correct arithmetization of the usual PA-provability predicate consists in

[55] By the way: what does "general properties" mean if not a metatheoretical formula, but a sentence is to be arithmetized in an intensionally correct way? Does it mean that, if this sentence obtains, its arithmetization has to be provable? Well, take then as a (true) metatheoretical sentence "PA is consistent". If we arithmetize in the standard way we obtain "$\mathrm{Con}_{\mathrm{pa}}$", which is not provable in PA. On the other hand, the Rosser variant "$\mathrm{Con}'_{\mathrm{pa}}$" is provable in PA.

pointing out that its competitors are not superior. Thus, it could be claimed that neither

(12) $$\forall x \left(\mathrm{Pr}'_{\mathrm{pa}} \left(\overline{\ulcorner \mathrm{Pr}_{\mathrm{pa}}(\dot{x}) \urcorner} \right) \to \mathrm{Pr}'_{\mathrm{pa}}(x) \right)$$

nor

(13) $$\forall x \left(\mathrm{Pr}'_{\mathrm{pa}}(x) \to \neg \mathrm{Pr}'_{\mathrm{pa}}(\dot{\neg} x) \right)$$

is provable in PA. Though true in case of (12), a small change in the definition of "$\mathrm{Pr}'_{\mathrm{pa}}$" results in a variant "$\mathrm{Pr}^R_{\mathrm{pa}}$" for which (13) is provable in PA: $\mathrm{Prf}^R_\tau(x,y) :\leftrightarrow \mathrm{Prf}_\tau(x,y) \wedge \forall z \leq x \, (\neg \mathrm{Prf}_\tau(z, \dot{\neg} y) \wedge \neg \exists u \leq y \, (y = \dot{\neg} u \wedge \mathrm{Prf}_\tau(z,u)))$. Moreover, this line of reasoning does, if successful, not really show that "$\mathrm{Pr}_{\mathrm{pa}}$" is intensionally correct; rather, its consequence might be that there are no intensionally correct arithmetizations of the predicate "provable in PA".

As a final example: what is an intensionally correct arithmetization of

(14) $$\mathrm{PA} \vdash \varphi \iff \mathrm{PA} \vdash^R \varphi$$

(for all φ in L[PA])? Intuitively, it is

(15) $$\forall x \left(\mathrm{Pr}_{\mathrm{pa}}(x) \leftrightarrow \mathrm{Pr}'_{\mathrm{pa}}(x) \right)$$

— which is not provable in PA. The fact that "$\forall x \ (\mathrm{Pr}_{\mathrm{pa}}(x) \leftrightarrow \mathrm{Pr}_{\mathrm{pa}}(x))$" and "$\forall x \ (\mathrm{Pr}'_{\mathrm{pa}}(x) \leftrightarrow \mathrm{Pr}'_{\mathrm{pa}}(x))$" *are* provable in PA constitutes no reason for regarding these as intensionally correct, however.

This example leads to a suggestion which I will only mention: it may be hoped that by demanding intensionally correct arithmetizations of *many* metatheoretical formulas, nonintended arithmetizations are excluded. I do not know if this idea works. For one, this is hard to decide given the vagueness of what the general properties mentioned in (Fef-2) are. Moreover, there may be some form of give-and-take here: the effects of nonintended arithmetizations at one place may be balanced by effects of nonintended arithmetizations at other places to such an extent that they become "invisible".[56]

Let me stop the investigation of (Fef-2) here. It is clear that one could go on inventing examples like these, and that I have only touched the surface of what an explication of (Fef-2) could be.

How to explain "α is an intensionally correct arithmetization of F_A", then? Of course, an arithmetization can only be intensional if both (Fef-1) and (Fef-2) hold for it. Thus, let the assumptions placed on ML[PA] and on the

[56]For example, what if the metamathematical predicates employed in order to arithmetize "x ist provable in PA" as usual — like "concatenation", "length" and "formula" — are arithmetized in uncommon, maybe "intensionally incorrect" ways? Does this make "$\mathrm{Pr}_{\mathrm{pa}}$" intensionally incorrect, too?

Φ's mentioned above be fulfilled, α be a k-place formula of L[PA] and F_A be a k-place formula of ML[PA]; then

α is an intensionally correct arithmetization of F_A $:\Leftrightarrow$ (i) α expresses F_A and (ii) for (specific, suitable) arithmetical formulas Φ, PA $\vdash \Phi(\alpha) \Leftrightarrow \mathcal{N} \models \Phi(\alpha)$.

As a test case for this "definition", consider the unusual representation of the usual *axiom-set* of PA discovered and investigated by Feferman himself in [12]:[57]

$$\mathrm{pa}^*(x) :\longleftrightarrow \mathrm{pa}(x) \wedge \forall z \leq x \ \mathrm{Con}_{\mathrm{pa}\lceil z} .$$

"$\mathrm{Pr}_{\mathrm{pa}^*}$" is a representation of the theory PA. Furthermore, it has a good chance for being an intensionally correct arithmetization of its metatheoretical analogue "x is Feferman-provable in PA": condition (Fef-1) simply presents no problem, and it turns out that — since "$\mathrm{Pr}_{\mathrm{pa}^*}$" is just a usual provability predicate and only the representation of an *axiom-set* has been chosen strangely — many general properties of "x is Feferman-provable in PA" are transferred into PA-theorems for "$\mathrm{Pr}_{\mathrm{pa}^*}$". Granted, these are not the same as in the case of "$\mathrm{Pr}_{\mathrm{pa}}$"; but whether this fact makes the latter intensionally correct or the former is not answered by the definition put forward here.[58]

Neither is it clear to me that this result is intended, nor does it seem more plausible than a similar result would be for Rosser-provability, i.e., that "'$\mathrm{Pr}^R_{\mathrm{pa}}$' were an intensionally correct arithmetization of 'PA $\vdash^R$'", which is probably not the case, however. What I would certainly question is that "$\mathrm{Pr}_{\mathrm{pa}^*}$" is an intensionally correct arithmetization of PA; but, as a matter of fact, this way of talking has not been explained given the way I interpret Feferman's approach.

2.3. Comparison of Sections 2.1 and 2.2. According to Section 2.1, intensionality in the mathematical enterprise amounts to some theory T (in L[PA]) and, e.g., the existence of arithmetical sentences φ, ψ such that $\mathcal{N} \models \varphi \leftrightarrow \psi$, $T \vdash \varphi$ and $T \nvdash \psi$. Alternatively, it is the claim that there are coextensional sentences which fail to be synonymous (with respect to T) which is taken to be of prime relevance. This is not so for Feferman's conception of intensionality — or at least not explicitly so: in [12], no explanation or discussion of "A is synonymous with B" is provided and no principles of extensionality or theses of extensionality — or their negations — are presented. Rather, Feferman is

[57] $\mathrm{pa}\lceil z(x) :\leftrightarrow \mathrm{pa}(x) \wedge x \leq z$.

[58] It is remarkable, however, that for the variants of (Löb 1) formulated with "$\mathrm{Pr}'_{\mathrm{pa}}$", "$\mathrm{Pr}^R_{\mathrm{pa}}$" or "$\mathrm{Pr}'_{\mathrm{pa}^*}$" instead of "$\mathrm{Pr}_{\mathrm{pa}}$", their arithmetizations employing "$\mathrm{Pr}_{\mathrm{pa}}$", i.e.

$$\forall x \left(\left(\mathrm{Pr}_{\mathrm{pa}}(x) \rightarrow \mathrm{Pr}_{\mathrm{pa}}(\overline{\ulcorner \mathrm{Pr}'_{\mathrm{pa}}(\dot{x}) \urcorner}) \right) \right),$$

$$\forall x \left(\left(\mathrm{Pr}_{\mathrm{pa}}(x) \rightarrow \mathrm{Pr}_{\mathrm{pa}}(\overline{\ulcorner \mathrm{Pr}^R_{\mathrm{pa}}(\dot{x}) \urcorner}) \right) \right),$$

and

$$\forall x \left(\left(\mathrm{Pr}_{\mathrm{pa}}(x) \rightarrow \mathrm{Pr}_{\mathrm{pa}}(\overline{\ulcorner \mathrm{Pr}_{\mathrm{pa}^*}(\dot{x}) \urcorner}) \right) \right),$$

are provable in PA; see [45].

interested in distinguishing intensionally correct arithmetizations from merely extensionally correct ones.

Certainly, Feferman holds that there are extensionally correct arithmetizations of, e.g., "PA is consistent" which are not intensionally correct — "$\neg \mathrm{Pr}'_{\mathrm{pa}}(\ulcorner\bot\urcorner)$" being an example. Now, here we have an arithmetical provability predicate which does not even have the same structure as the usual one. Thus, let's consider a stronger claim than the one just addressed: there are extensionally correct arithmetizations of "PA is consistent", both of which are formulated employing only *the usual* arithmetized provability predicate for PA, such that only one of them is intensionally correct. I think that this claim — let's call it "Feferman's Intensionality Thesis" — is, although not formulated in [12],[59] nevertheless a natural extension of what is presented there:

(ThI.Fef) *There are arithmetical formulas* $\mathrm{pa}_1(x)$ *and* $\mathrm{pa}_2(x)$ *of* L[PA] *such that*

(i) $\mathrm{pa}_1(x)$ *and* $\mathrm{pa}_2(x)$ *are representations of* axiom-sets *of* PA;[60]

(ii) $\forall\Phi\ (\mathrm{PA} \vdash \Phi(\mathrm{Con}_{\mathrm{pa}_1}) \Leftrightarrow \mathcal{N} \models \Phi(\mathrm{Con}_{\mathrm{pa}_1}))$;

(iii) $\neg\forall\Phi\ (\mathrm{PA} \vdash \Phi(\mathrm{Con}_{\mathrm{pa}_2}) \Leftrightarrow \mathcal{N} \models \Phi(\mathrm{Con}_{\mathrm{pa}_2}))$.[61]

Following last subsection's explanation of "α is an intensionally correct arithmetization of F_A", Φ is assumed to range over all the "general properties" which are relevant for specifying what an intensionally correct arithmetization of "PA is consistent" is.[62]

In order to compare now ($\mathrm{ThI.1_{PA}}$) and ($\mathrm{ThI.2_{PA}}$) with (ThI.Fef), the former theses have to be formulated by employing "$\mathrm{Con}_{\mathrm{pa}_1}$" and "$\mathrm{Con}_{\mathrm{pa}_2}$", with "$\mathrm{pa}_1$" and "$\mathrm{pa}_2$" chosen as above. One obtains

There are arithmetical formulas $\mathrm{pa}_1(x)$ *and* $\mathrm{pa}_2(x)$ *representing* axiom-sets *of* PA *such that*

(I) $\mathcal{N} \models \mathrm{Con}_{\mathrm{pa}_1} \leftrightarrow \mathrm{Con}_{\mathrm{pa}_2}$; PA *is consistent* $\Leftrightarrow \mathcal{N} \models \mathrm{Con}_{\mathrm{pa}_1} \Leftrightarrow \mathcal{N} \models \mathrm{Con}_{\mathrm{pa}_2}$;

(II) $\mathrm{PA} \nvdash \mathrm{Con}_{\mathrm{pa}_1} \leftrightarrow \mathrm{Con}_{\mathrm{pa}_2}$ (*or* $\neg(\mathrm{PA} \vdash \mathrm{Con}_{\mathrm{pa}_1} \Leftrightarrow \mathrm{PA} \vdash \mathrm{Con}_{\mathrm{pa}_2})$).

Whereas Feferman's thesis and these instances of ($\mathrm{ThI.1_{PA}}$) and ($\mathrm{ThI.2_{PA}}$) are more or less the same in their "extensional" parts ((I) follows from (i)), they seem to be quite different in their "intensional" parts. But this is not

[59] Actually, Feferman deals with variations on the usual provability predicates similar to the ones discussed here in [12] (in chapter 7).

[60] This implies that for all φ in L[PA], $\mathcal{N} \models \mathrm{Pr}_{\mathrm{pa}_1}(\ulcorner\varphi\urcorner) \Leftrightarrow \mathrm{PA} \vdash \varphi \Leftrightarrow \mathcal{N} \models \mathrm{Pr}_{\mathrm{pa}_2}(\ulcorner\varphi\urcorner)$, and that $\mathcal{N} \models \mathrm{Con}_{\mathrm{pa}_1} \Leftrightarrow$ PA is consistent $\Leftrightarrow \mathcal{N} \models \mathrm{Con}_{\mathrm{pa}_2}$.

[61] The first condition and footnote 59 should express that "$\mathrm{Con}_{\mathrm{pa}_1}$" and "$\mathrm{Con}_{\mathrm{pa}_2}$" are extensionally correct arithmetization of the consistency of PA; the second is to state that "$\mathrm{Con}_{\mathrm{pa}_1}$" is also intensionally correct, whereas the third is supposed to convey that "$\mathrm{Con}_{\mathrm{pa}_2}$" fails to be so.

[62] Of course, we do not know which conditions exactly are admissable here. But certainly, Φ's such that $\Phi(A) \equiv A$ or $\Phi(A) \equiv (A \leftrightarrow \mathrm{Con}_{\mathrm{pa}_2})$ should not belong to them.

really so, if the relevant general conditions Φ employed here are *extensional*: that is, if from $\mathrm{PA} \vdash \mathrm{Con}_{\mathrm{pa}_1} \leftrightarrow \mathrm{Con}_{\mathrm{pa}_2}$, it follows that

$$\mathrm{PA} \vdash \Phi(\mathrm{Con}_{\mathrm{pa}_1}) \longleftrightarrow \Phi(\mathrm{Con}_{\mathrm{pa}_2}).$$

For assume (ii) from (ThI.Fef) and ¬(II); then $\mathrm{PA} \vdash \mathrm{Con}_{\mathrm{pa}_1} \leftrightarrow \mathrm{Con}_{\mathrm{pa}_2}$, and by extensionality and because of

$$\mathcal{N} \models \Phi(\mathrm{Con}_{\mathrm{pa}_1}) \longleftrightarrow \Phi(\mathrm{Con}_{\mathrm{pa}_2}),$$

(ii) implies

$$\forall\Phi\left(\mathrm{PA} \vdash \Phi(\mathrm{Con}_{\mathrm{pa}_2}) \Leftrightarrow \mathcal{N} \models \Phi(\mathrm{Con}_{\mathrm{pa}_2})\right).$$

But this contradicts (iii) from (ThI.Fef). That is, Feferman's Intensionality thesis implies the intensionality thesis ($\mathrm{ThI.2_{PA}}$) with "$\mathrm{Con}_{\mathrm{pa}_1}$" and "$\mathrm{Con}_{\mathrm{pa}_2}$" as witnesses. Thus, although Feferman has not explicitly proposed theses of intensionality in [12], it seems that (ThI.Fef) can be regarded as one.

§3. Serious intensionality.

3.1. Theses of serious intensionality. With Gödel's incompleteness theorems in the background, it is no wonder that there are arithmetical sentences A and B such that $\mathcal{N} \models A \leftrightarrow B$, but, say, $\mathrm{PA} \nvdash A \leftrightarrow B$. Thus, taking the usual consistency assertion "$\mathrm{Con}_{\mathrm{pa}}$" and "$\top$" as examples, one has

$$\mathcal{N} \models \mathrm{Con}_{\mathrm{pa}} \longleftrightarrow \top \quad \text{and} \quad \mathrm{PA} \nvdash \mathrm{Con}_{\mathrm{pa}} \longleftrightarrow \top.$$

Yet, from an intuitive perspective, nobody would regard "$\mathrm{Con}_{\mathrm{pa}}$" and "$\top$" as being synonymous. That is why, at the end of Section 2.1, I suggested that it would result in particularly "serious" forms of intensionality if *we had arithmetical sentences A and B such that $\mathcal{N} \models A \leftrightarrow B$, $T \nvdash A \leftrightarrow B$ and as many further conditions Φ on A, B and T were satisfied as would seem reasonable.*

Let me point out that (ThI.Fef), though being an example for such a strengthening of ($\mathrm{ThI.2_{PA}}$), is not motivated by the goal of finding examples for serious intensionality. With (ThI.Fef), one wants to distinguish between intensionally correct arithmetizations and merely extensionally correct ones; in case of serious intensionality, the aim is to find formulas α and α' both of which are intensionally correct arithmetizations of F_A — but still fail to be synonyms (in the sense of Section 2.1).[63] And for this, (ThI.Fef) is simply not the right thesis: for example, it may well be that "$\mathrm{pa}^*(x)$" is not excluded from being a formula "$\mathrm{pa}_1(x)$" as considered in (ThI.Fef) (cf. Section 2.2.); but not both "$\mathrm{Pr}_{\mathrm{pa}^*}$" and "$\mathrm{Pr}_{\mathrm{pa}}$" should be taken to express "x is provable in PA" (as understood in the usual sense).

[63] Surely, it is quite implausible that an intensionally correct and merely extensionally correct arithmetization (of F_A) should be synonymous with each other.

Now, in order to obtain more serious forms of intensionality, the most promising move is simply to pile up everything which has been taken into consideration as a possible restriction of the admissable arithmetizations of "x is provable in T". To start with, only the usual arithmetized provability-predicates, i.e., "$\mathrm{Pr}_\tau(x)$", should be allowed (this is already so in (ThI.Fef)). Moreover, these provability predicates should satisfy the Löb-derivability conditions, with proofs for them given in, say, PA or weaker theories. If "$\mathrm{Con}_{\mathrm{pa}'}$" is the consistency assertion induced by such a provability predicate "$\mathrm{Pr}_{\mathrm{pa}'}(x)$" — "$\mathrm{pa}'(x)$" being a representation of an *axiom-set* of PA — it is guarenteed in this case that, e.g., PA $\nvdash \mathrm{Con}_{\mathrm{pa}'}$. Finally — thinking at "pa*", which is of complexity Δ^0_2 — what comes to mind as a further condition on the representations "pa′" of "axiomsets" of PA is that they should be Σ^0_0 or Σ^0_1, but not more complex.[64]

Given these assumptions as generalized to arbitrary arithmetically definable theories, let me now state two versions of the *Serious Intensionality Thesis* for a theory T (in L[PA]):

(SerIT_T.1) *There are arithmetical formulas* $\tau(x)$ *and* $\tau'(x)$ *representing* axiom-sets *of* T *such that*

- *The complexity of* τ *and* τ' *— from the point of view of the arithmetical hierarchy — is less then or equal to the complexity of* T;[65]
- *the Löb-derivability conditions hold for* "$\mathrm{Pr}_\tau(x)$" *and* "$\mathrm{Pr}_{\tau'}(x)$";
- $T \nvdash \mathrm{Con}_\tau \leftrightarrow \mathrm{Con}_{\tau'}$.

I will also deal with a weakened form of this thesis:

(SerIT_T.2) *There are arithmetical formulas* $\tau(x)$ *and* $\tau'(x)$ *representing* axiom-sets *of* T *such that*

- *The complexity of* τ *and* τ' *— from the point of view of the arithmetical hierarchy — is less then or equal to the complexity of* T;
- *the Löb-derivability conditions hold for* "$\mathrm{Pr}_\tau(x)$" *and* "$\mathrm{Pr}_{\tau'}(x)$";
- $T \nvdash \forall x\,(\mathrm{Pr}_\tau(x) \leftrightarrow \mathrm{Pr}_{\tau'}(x))$.

Let me emphasize that the conditions formulated in (SerIT_T.1) and (SerIT_T.2) are *not intended* as the full list of general properties which could provide a generally adequate explication of "α is an intensionally correct arithmetization of 'x is provable in T'" (or "T is consistent", resp.). But the whole topic "serious intensionality" is certainly a contribution to Feferman's approach to intensionality, and it is motivated by the goal to formulate such general conditions as demanded by condition (Fef-2).

---

[64]In particular, Feferman's nonstandard provability predicate will not be admitted into consideration as an intensionality-generating formula, for PA $\vdash \mathrm{Con}_{\mathrm{pa}^*}$ and because it is too complex. — See [29] for more on these conditions.

[65]That is, if T is Σ^0_k or Δ^0_{k+1}, resp., then τ and τ' should be Σ^0_k or Δ^0_{k+1}, resp. ($k \geq 1$).

Now it may be that one manages to establish one of these intensionality theses, but succeeds only

- for "uninteresting" theories T, or
- by using representations for *axiom-sets* of T which are not "natural".

Whatever "interesting theories" may be — PA and ZF are certainly examples, whereas PA + the sentence with Gödel number k or arithmetically definable completions of PA do not always belong to them. Furthermore, I have discussed definitions of

$$\alpha \text{ is a natural representation of } A$$

(for a k-place arithmetical formula α and $A \subseteq \omega^k$) in some detail in [29] (see also [31]). Let me here refer to these papers and only mention that, from the perspective of the proposals from [29], the representations of *axiom-sets* considered in the following constructions are natural.[66]

Thus, I take it as an important additional goal to find an interesting theory T and *natural* representations of *axiom-sets* of T witnessing $(\mathrm{SerIT}_T.1)$ and $(\mathrm{SerIT}_T.2)$.

3.2. Examples for $(\mathrm{SerIT}_T.1)$ and $(\mathrm{SerIT}_T.2)$. In this subsection, I will present two methods for obtaining examples of $(\mathrm{SerIT}_T.1)$ and $(\mathrm{SerIT}_T.2)$.

Since the dependency on the choice of the representation τ of an *axiom-set* of theory T is the main topic of this section, let me be more explicit in this than before. In particular, I will deal with Ax(PA), the usually chosen *axiom-set* of PA, and $\mathrm{Tr}_{\Sigma^0_k}$, the set of true Σ^0_k-sentences ($k \geq 1$);[67] Ax(PA) is taken to be (naturally) represented by the Σ^0_0-formula "pa(x)", and "$\mathrm{Tr}_{\Sigma^0_k}(x)$" is used for the Σ^0_k-formula representing $\mathrm{Tr}_{\Sigma^0_k}$ ($k \geq 1$; see [22] for details).

In addition, the schema of *local ω-consistency* will be considered: for an arithmetical formula τ representing an *axiom-set* of T,

$$\omega\text{-}\mathrm{Con}[\tau] := \big\{\, \mathrm{Pr}_\tau(\overline{\ulcorner \exists x\, \psi(x) \urcorner}) \rightarrow \exists x\, \neg\, \mathrm{Pr}_\tau(\overline{\ulcorner \neg \psi(\dot{x}) \urcorner}) \mid \psi \text{ in L[PA] with exactly one free variable } x \big\}.$$

If τ is Σ^0_0, this set contains only Σ^0_2-sentences. I also write "ω-Con[τ]" for its (natural) representation; i.e. (somewhat sloppily; "$\mathrm{Fml}_1(z)$" arithmetizes "z is a formula in L[PA] with one free variable")

$$\omega\text{-}\mathrm{Con}[\tau](x) :\longleftrightarrow \exists z \leq x\, \big(\mathrm{Fml}_1(z) \wedge x = \overline{\ulcorner \mathrm{Pr}_\tau(\overline{\ulcorner \exists u \dot{z} \urcorner}) \dot{\rightarrow} \exists u\, \neg\, \mathrm{Pr}_\tau(\overline{\ulcorner \dot{\neg} z[\dot{u}] \urcorner}) \urcorner}\big).$$

[66] One remark on terminology: in [31], I did not distinguish between "α expresses A" and "α is a natural representation of A", understanding both in the same sense as "α is a natural representation of A" in [29], where I used only this phrase. Since in this paper I interpret "α expresses F_A" in Feferman's sense, it is to be sharply distinguished from "α is a natural representation of A".

[67] See [19] and [22] for the theories and the background on recursion theory and the arithmetical hierarchy.

Furthermore, RFN[σ] ($k \geq 1$) is the uniform reflection principle for S (σ, to be more precise), i.e. the set of all formulas "$\forall x\ (\mathrm{Pr}_\sigma(\overline{\ulcorner \psi(\dot{x}) \urcorner}) \rightarrow \psi(x))$" for arbitrary formulas ψ in L[PA]; $\mathrm{RFN}_{\Sigma^0_k}[\sigma]$ is the restriction of RFN[σ] to the Σ^0_k-formulas.

EXAMPLE 1. Take for T the theory $\mathrm{PA} + \mathrm{Tr}_{\Sigma^0_2}$. — One (natural) *axiom-set* of T is $\mathrm{Ax(PA)} \cup \mathrm{Tr}_{\Sigma^0_2}$. A natural representation of this set is τ, which is defined by

$$\tau(x) :\longleftrightarrow \mathrm{pa}(x) \vee \mathrm{Tr}_{\Sigma^0_2}(x).$$

Since $\omega\text{-}\mathrm{Con[pa]} \subseteq \mathrm{Tr}_{\Sigma^0_2}$, it follows that $\mathrm{T} = \mathrm{PA} + \mathrm{Tr}_{\Sigma^0_2} + \omega\text{-}\mathrm{Con[pa]}$. Therefore, the set $\mathrm{Ax(PA)} \cup \mathrm{Tr}_{\Sigma^0_2} \cup \omega\text{-}\mathrm{Con[pa]}$ is also an *axiom-set* for T; it is naturally represented by $\tau\tau$, where

$$\tau\tau(x) :\longleftrightarrow \mathrm{pa}(x) \vee \mathrm{Tr}_{\Sigma^0_2}(x) \vee \omega\text{-}\mathrm{Con[pa]}(x).$$

All assumptions stated in (SerIT_T.2) are satisfied, with the possible exception of the last one — to which I now turn. Assume that for each formula ψ of L[PA] (with free variable x)

$$\mathrm{T} \vdash \forall x\ \big(\mathrm{Pr}_\tau(\overline{\ulcorner \psi(\dot{x}) \urcorner}) \leftrightarrow \mathrm{Pr}_{\tau\tau}(\overline{\ulcorner \psi(\dot{x}) \urcorner})\big).$$

Then

$$\left(\mathrm{T} + \mathrm{RFN}_{\Sigma^0_2}[\tau]\right) \vdash \mathrm{RFN}_{\Sigma^0_2}[\tau\tau].$$

A generalization of the next lemma is proved in [30] (see also [32]):

LEMMA 1. *If* $\mathrm{PA} \subseteq \mathrm{T}$, *then* $\mathrm{T} + \mathrm{RFN}_{\Sigma^0_k}[\mathrm{pa}] \vdash \mathrm{RFN}_{\Sigma^0_k}[\mathrm{pa} + \mathrm{Tr}_{\Sigma^0_k}]$ ($k \geq 1$).

Thus, we obtain

$$\left(\mathrm{T} + \mathrm{RFN}_{\Sigma^0_2}[\mathrm{pa}]\right) \vdash \mathrm{RFN}_{\Sigma^0_2}\left[\mathrm{pa} + \mathrm{Tr}_{\Sigma^0_2}\right].$$

Moreover, it is trivial that

$$\left(\mathrm{T} + \mathrm{RFN}_{\Sigma^0_2}[\tau\tau]\right) \vdash \mathrm{RFN}_{\Sigma^0_2}\left[\mathrm{pa} + \omega\text{-}\mathrm{Con[pa]}\right].$$

Taken together, these lines imply

$$\left(\mathrm{T} + \mathrm{RFN}_{\Sigma^0_2}[\mathrm{pa}]\right) \vdash \mathrm{RFN}_{\Sigma^0_2}\left[\mathrm{pa} + \omega\text{-}\mathrm{Con[pa]}\right].$$

In [43], the following lemma is stated (see also [42] for a sketch of a proof[68]):

LEMMA 2. $\mathrm{PA} + \mathrm{RFN}_{\Sigma^0_2}[\mathrm{pa} + \omega\text{-}\mathrm{Con[pa]}] \vdash \mathrm{RFN}_{\Sigma^0_2}[\mathrm{pa} + \mathrm{RFN[pa]}]$.

---

[68] A detailed, though quite different proof is presented in [30].

With this lemma,

$$\left(\mathrm{T}+\mathrm{RFN}_{\Sigma^0_2}[\mathrm{pa}]\right) \vdash \mathrm{RFN}_{\Sigma^0_2}\left[\mathrm{pa}+\mathrm{RFN}[\mathrm{pa}]\right]$$

follows from the above. A further application of Lemma 1 finally implies

$$\left(\mathrm{T}+\mathrm{RFN}_{\Sigma^0_2}[\mathrm{pa}]\right) \vdash \mathrm{RFN}_{\Sigma^0_2}\left[\mathrm{pa}+\mathrm{RFN}[\mathrm{pa}]+\mathrm{Tr}_{\Sigma^0_2}\right].$$

But this contradicts a generalization of Gödel's second incompleteness theorem to T.[69]

What I take to be a weakness of this example is that $\mathrm{PA}+\mathrm{Tr}_{\Sigma^0_2}$ is not axiomatizable.[70] Usually, nonaxiomatizable theories are regarded as "monsters", maybe of some interest as an object of investigation, but not as its means, and therefore not as valuable as axiomatizable theories. Thus, the task is to find an axiomatizable example. Here it is:

EXAMPLE 2. Take the theory PA. — Now consider for each $n \in \omega$ the sentence

$$\neg\,\mathrm{Proof}_{\mathrm{zf}}\left(\overline{n}, \ulcorner\bot\urcorner\right),$$

where "$\mathrm{zf}(x)$" is a Σ^0_0-formula naturally representing one of the usual axiomatizations of ZF (see, e.g, [24]). Assuming that ZF is consistent, each of these sentences is a true Σ^0_0-sentence, whence provable in PA. That is,

$$\overline{\mathrm{PA}} = \overline{\mathrm{Ax}(\mathrm{PA}) \cup \{\neg\,\mathrm{Proof}_{\mathrm{zf}}(\overline{n}, \ulcorner\bot\urcorner) \mid n \in \omega\}}.$$

A natural representation of the (metamathematical predicate corresponding to the) *axiom-set* $\mathrm{Ax}(\mathrm{PA}) \cup \{\neg\,\mathrm{Proof}_{\mathrm{zf}}(\overline{n}, \ulcorner\bot\urcorner) \mid n \in \omega\}$ is "$\mathrm{pa}^{\mathrm{zf}}$", which is defined by

$$\mathrm{pa}^{\mathrm{zf}}(x) :\longleftrightarrow \mathrm{pa}(x) \vee \exists y \leq x \left(x = \ulcorner\neg\,\mathrm{Proof}_{\mathrm{zf}}(\dot{y}, \ulcorner\bot\urcorner)\urcorner\right).$$

Thus, "pa" and "$\mathrm{pa}^{\mathrm{zf}}$" are both representations of *axiom-sets* of PA; both are Σ^0_0-formulas, whence of lower complexity than PA itself; and by general metamathematical results, both "$\mathrm{Pr}_{\mathrm{pa}}$" and "$\mathrm{Pr}_{\mathrm{pa}^{\mathrm{zf}}}$" satisfy all of Löb's derivability conditions (provably in PA).

Assume then

$$(*) \qquad \mathrm{PA} \vdash \mathrm{Con}_{\mathrm{pa}} \longrightarrow \mathrm{Con}_{\mathrm{pa}^{\mathrm{zf}}}\,.$$

---

[69] Actually, the Löb-derivability conditions hold for $\mathrm{PA}+\mathrm{Tr}_{\Sigma^0_k}$ with representation "$\mathrm{pa}(x) \vee \mathrm{Tr}_{\Sigma^0_k}(x)$" ($k \geq 1$).

[70] What I like with respect to it is that the result is surprising; it uses nontrivial metamathematics which is useful for other results, too; and the ingredients involved — $\mathrm{Tr}_{\Sigma^0_2}$ and local ω-consistency — have been of some interest in metamathematics quite independently of it.

By the definition of $\mathrm{pa}^{\mathrm{zf}}$,

$$\mathrm{PA} \vdash \forall uv \left(u = \overline{\ulcorner \neg \mathrm{Proof}_{\mathrm{zf}}(\dot{v}, \overline{\ulcorner \bot \urcorner}) \urcorner} \rightarrow \mathrm{Pr}_{\mathrm{pa}^{\mathrm{zf}}}(u)\right), \text{ i.e.}$$

$$\mathrm{PA} \vdash \forall v \ \mathrm{Pr}_{\mathrm{pa}^{\mathrm{zf}}} \left(\overline{\ulcorner \neg \mathrm{Proof}_{\mathrm{zf}}(\dot{v}, \overline{\ulcorner \bot \urcorner}) \urcorner}\right).$$

Now "$\neg \mathrm{Proof}_{\mathrm{zf}}(v, \overline{\ulcorner \bot \urcorner})$" is Σ^0_0 and $\mathrm{RFN}_{\Pi^0_1}[\mathrm{pa}^{\mathrm{zf}}]$ is over PA equivalent to "$\mathrm{Con}_{\mathrm{pa}^{\mathrm{zf}}}$". Therefore

$$\mathrm{PA} + \mathrm{Con}_{\mathrm{pa}^{\mathrm{zf}}} \vdash \forall v \ \neg \mathrm{Proof}_{\mathrm{zf}} \left(v, \overline{\ulcorner \bot \urcorner}\right).$$

Assuming $(*)$, this implies

$$\mathrm{PA} + \mathrm{Con}_{\mathrm{pa}} \vdash \mathrm{Con}_{\mathrm{zf}}.$$

But this is not the case.[71]

Of course, there are numerous modifications and extensions of this example:

(i) Similar results providing as many theories T witnessing (SerIT_T.1) as one would like to have can be built along the same lines. For example, one can show by the same argument as given for PA:

THEOREM. *If T is an r.e. extension of* PA[72] *there exists* (*natural*) *representations of axiom-sets of T such that* (SerIT_T.1) *holds for these representations.*

(ii) If one is not willing to presuppose the consistency of ZF, it is enough to assume the consistency of, e.g., $\mathrm{PA} + \mathrm{Con}_{\mathrm{pa} + \mathrm{Con}_{\mathrm{pa}}}$ instead; and the consistency of this theory can be established by transfinite induction up to ε_0.

(iii) One may be dissatisfied by the occurence of an arithmetization of a metamathematical predicate in Example 2. This can be circumvented, however, by employing results on the independence of combinatorial principles from PA. Just assume that ψ is a sentence in L[PA] such that

- ψ is Π^0_1, i.e. $\psi \equiv \forall x \varphi$;
- ψ is not provable in PA, but for each n, $\mathrm{PA} \vdash \varphi(\overline{n})$;
- $\mathrm{PA} + \psi \vdash \mathrm{RFN}_{\Sigma^0_1}[\mathrm{pa}]$;

ψ may be taken as a sentence expressing some combinatorial principle, like the variant of the Ramsey theorem from Paris-Harrington; see, e.g., [22].

By simply substituting $\varphi(s)$ for "$\neg \mathrm{Proof}_{\mathrm{zf}}(s, \overline{\ulcorner \bot \urcorner})$" in the proof just given, one shows that, in fact, $\mathrm{PA} = \overline{\mathrm{PA} \cup \{\varphi(\overline{n}) \mid n \in \omega\}}$, but that this is not provable in PA.

---

[71] In fact, $\mathrm{PA} \vdash (\mathrm{Con}_\tau \rightarrow \mathrm{Con}_\sigma) \leftrightarrow \forall x \ \mathrm{Pr}_\tau(\overline{\ulcorner \neg \mathrm{Proof}_\sigma(\dot{x}, \overline{\ulcorner \bot \urcorner}) \urcorner})$ if σ and τ are Σ^0_0-formulas representing extensions of Q.

[72] Actually, T may be a much weaker theory (like $\mathrm{I}\Delta_0$).

§4. Concluding remarks. Let me close with some brief remarks on the application of the concepts of intensionally correct arithmetizations and serious intensionality to philosophy, in particular to Quinean themes.

First, as has been pointed out by Quine, it is particularly difficult to make sense of *interlingual synonymy.* But "α is an intensionally correct arithmetization of F_A" may provide an *explicans* for "α is synonymous with F_A" for α of L[PA] (more generally, mathematical languages) and F_A from ML[PA]. This is so in particular when L[PA] and ML[PA] are disjoint languages. For in this case, the definitions of "α is synonymous with F_A" supplied by the possible worlds and the provability approaches are simply not applicable.[73]

Second, let's assume that α and α' (of L[PA]) are both intensionally correct arithmetizations of F_A (from ML[PA]). Do they have to be synonymous with one another? Well, if we assume the definition[74]

α is synonymous with α' $:\Leftrightarrow$ there is a formula F_A from ML[PA] such that α and α' are intensionally correct arithmetizations of F_A,

the answer is trivially "yes". But synonymy between formulas of L[PA] has already been explained in a different way:

α is synonymous with α' (with respect to T) $\Leftrightarrow T \vdash \alpha \leftrightarrow \alpha'$.

The examples given for (SerIT_T.1) and (SerIT_T.2) establish that — even if quite a few conditions on T and the formulas α, α' admitted are added — synonymy in the second (here accepted) sense is by no means a consequence of synonymy in the first sense.

Third, an intensionally correct arithmetization is similar to an interpretation (see [12], [19]). An interpretation from theory S in theory T is a function $\mathcal{I}$ of L[S] to L[T], preserving the quantificational structure of the formulas of L[S] such that for all $\psi \in S$, $\mathcal{I}(\psi) \in T$. In our case, F_A is mapped to an arithmetization α. Structure preservation is taken care of in (Fef-1); and S is replaced by the vaguely given, "open" set of informally stated "general properties" addressed in (Fef-2).

Fourth, in view of the third observation, serious intensionality may be considered as a version of Quine's thesis of the *indeterminacy of translation* (see [38, 39]) — which is one of his arguments against the meaningfulness of "x is synonymous with y". Quine's reasoning has the following structure: there could be cases where there are functions f and g from some language L to

[73]If L[PA] is a sublanguage of ML[PA], "$\alpha \leftrightarrow F_A$" is well-formed; but if ML[PA] fails to be a formal language, neither the evaluation of this biconditional in possible worlds models nor its provability will be defined. Replacing an informal ML[PA] by a formal counterpart FML[PA] and explaining synonymy between formulas A of L[PA] and their counterparts A_F of FML[PA] is of no help here: it begs the question why A and F_A should be synonymous with each other.

[74]Under the assumption of higher expressive power of ML[PA], synonymy thus defined is an equivalence relation.

L' such that for some expression A in L, both $f(A)$ and $g(A)$ are acceptable as translations of A, but fail to be synonymous. Then there is "no fact of the matter" whether f or g is "the right" translation, and there is "no fact of the matter" whether the expressions occuring are synonymous or not; "x is synonymous with y" is not well-defined.

Now, each function $\mathcal{A}$ mapping "x is provable in PA" (or: "PA is consistent") to one of its arithmetizations as it is demanded in (SerIT$_T$.1) (or (SerIT$_T$.2)) may be regarded as such a translation. And if "synonymous with" is defined in the second way considered above (under *second*), $\mathcal{A}(\varphi)$ need not be synonymous with $\mathcal{A}'(\varphi)$ (for φ from ML[PA]). Nonetheless, synonymy is well-defined for formulas from L[PA], and each $\mathcal{A}$ is precisely determined, too. Thus, it seems that Quine's indeterminacy thesis does not really support his scepticism concerning the understandability or meaningfulness of "synonymous".[75]

Let me conclude: the vagueness of the expressions employed in the *definiens* of "intensionally correct arithmetization" notwithstanding, and although I have mainly presented problems for them — I think that (Fef-1) and (Fef-2) are intuitively quite strong and convincing conditions, and that they should also be fruitful for work outside the domain of metamathematics.

REFERENCES

[1] J. Barwise and J. Perry, ***Situations and attitudes***, MIT Press, Cambridge, Mass., 1983.

[2] M. Beeson, ***Foundations of constructive mathematics***, Springer, New York, 1985.

[3] G. Boolos, ***The logic of provability***, Cambridge UP, Cambridge, 1993.

[4] A. W. Burks, ***Chance, cause, reason***, The University of Chicago Press, Chicago, London, 1963.

[5] S. Buss, *First-order proof theory of arithmetic*, ***Handbook of proof theory*** (S. Buss, editor), Elsevier, Amsterdam, 1998, pp. 79–147.

[6] R. Carnap, ***Der logische Aufbau der Welt***, Meiner, Hamburg, 1928, Quotations are from ***The logical structure of the world***, University of California Press, Berkeley, 1967.

[7] ———, ***Logische Syntax der Sprache***, Springer, Wien, 1934, Quotations are from ***The logical syntax of language***, Routledge and Kegan Paul, 1937.

[8] ———, ***Meaning and necessity***, University of Chicago Press, Chicago, 1947.

[9] M. Cresswell, ***Structured meanings***, MIT Press, Cambridge, Mass., 1985.

[10] D. Davidson, ***Inquiries into truth and interpretation***, Clarendon Press, Oxford, 1984.

[11] M. Detlefsen, ***Hilbert's program***, Reidel, Dordrecht, 1986.

[12] S. Feferman, *Arithmetization of metamathematics in a general setting*, ***Fundamenta Mathematicae***, vol. XLIX (1960), pp. 35–92.

[13] ———, *Autonomous transfinite progressions and the extent of predicative mathematics*, ***Logic, methodology and philosophy of science III*** (B. van Rootselaar and J. F. Staal, editors), North-Holland, Amsterdam, 1968, pp. 121–135.

[75] These sketchy remarks are, of course, no proper treatment of this complex subject. Nonetheless, let me add another one: Quine's famous "Gavagai"-example is supposed to provide acceptable translations f("gavagai") and g("gavagai") (of the term or predicate "gavagai") which are not even coreferential. Why does Quine's reasoning against the meaningfulness of "synonymous" not show the meaninglessnes of "refers to", but rather shows the relativity of reference?

[14] ———, *Constructive theories of functions and classes*, ***Logic colloquium '78*** (M. Boffa et al., editors), Springer, Berlin, 1979, pp. 159–224.

[15] ———, *Intensionality in mathematics*, ***The Journal of Symbolic Logic***, vol. 14 (1985), pp. 41–55.

[16] G. FREGE, *Über Sinn und Bedeutung*, ***Zeitschrift für Philosophie und philosophische Kritik***, NF, vol. 100 (1892), pp. 25–50, Reprinted in G. Patzig (ed.): *G. Frege: Funktion, Begriff, Bedeutung, Fünf logische Studien*, Vandenhoeck and Ruprecht, Göttingen 1962, 40–65.

[17] K. GÖDEL, *Eine Interpretation des intuitionistischen Aussagenkalküls*, ***Ergebnisse eines mathematischen Kolloquiums***, vol. 4 (1933), pp. 39–40.

[18] D. GUASPARI and R. M. SOLOVAY, *Rosser sentences*, ***Annals of Mathematical Logic***, vol. 16 (1979), pp. 81–99.

[19] P. HÁJEK and P. PUDLÁK, ***Metamathematics of first-order arithmetic***, Springer, Berlin, 1993.

[20] G. HELLMAN, ***Mathematics without numbers***, Clarendon, Oxford, 1989.

[21] G. HUGHES and M. CRESSWELL, ***A companion to modal logic***, Methuen, London/New York, 1984.

[22] R. KAYE, ***Models of peano arithmetic***, Clarendon Press, Oxford, 1991.

[23] S. KRIPKE, *Semantical considerations on modal logic*, ***Acta Philosophica Fennica***, vol. 16 (1963), pp. 83–94.

[24] K. KUNEN, ***Set theory***, Elsevier, Amsterdam, 1980.

[25] C. I. LEWIS and C. H. LANGFORD, ***Symbolic logic***, Century, New York, 1932.

[26] R. B. MARCUS, *Extensionality*, ***Mind***, vol. 69 (1960), pp. 55–62.

[27] Y. MOSCHOVAKIS, *Sense and denotation as algorithm an value*, ***Logic colloquium '90*** (J. Oikkonen and J. Väänänen, editors), Springer, Berlin, 1993.

[28] A. MOSTOWSKI, ***Thirty years of foundational studies***, Blackwell, Oxford, 1966.

[29] K. G. NIEBERGALL, *"Natural" representations and extensions of Gödel's second theorem*, forthcoming.

[30] ———, ***Zur Metamathematik nichtaxiomatisierbarer Theorien***, CIS, München, 1996.

[31] ———, *On the limits of Gödel's second incompleteness theorem*, ***Argument und Analyse. proceedings of GAP4*** (C. U. Moulines and K. G. Niebergall, editors), mentis, Paderborn, 2002, pp. 109–136.

[32] K. G. NIEBERGALL and M. SCHIRN, *Hilbert's programme and Gödel's theorems*, ***Dialectica***, vol. 56 (2002), pp. 347–370.

[33] H. PUTNAM, *Mathematics without foundations*, ***Journal of Philosophy***, vol. 64 (1967), pp. 5–22.

[34] W. V. O. QUINE, *Notes on existence and necessity*, ***Journal of Philosophy***, vol. 40 (1943), pp. 113–127.

[35] ———, ***Mathematical logic***, revised ed., Harvard University Press, Cambridge, 1951.

[36] ———, *The problem of meaning in linguistics*, ***From a logical point of view*** (W. V. O. Quine, editor), Harvard University Press, Cambridge, Mass., 1953, pp. 47–64.

[37] ———, *Reference and modality*, ***From a logical point of view*** (W. V. O. Quine, editor), Harvard University Press, Cambridge, Mass., 1953, pp. 139–159.

[38] ———, ***Word and object***, M.I.T. Press, Cambridge, Mass., 1960.

[39] ———, *Ontological relativity*, ***Ontological relativity and other essays***, Columbia University Press, New York, 1969.

[40] B. RUSSELL, ***An inquiry into meaning and truth***, Unwin Paperbacks, London, 1950.

[41] S. Shapiro (editor), ***Intensional mathematics***, North-Holland, Amsterdam, 1985.

[42] C. SMORYNSKI, *Consistency and related metamathematical properties*, Technical Report 75–02, Mathematisch Instituut, Amsterdam, 1975.

[43] ———, *The incompleteness theorems*, ***Handbook of mathematical logic*** (J. Barwise, editor), North-Holland, Amsterdam, 1977.

[44] ———, ***Self-reference and modal logic***, Springer, Berlin, 1985.

[45] A. VISSER, *Peano's smart children: a provability logical study of systems with built-in consistency*, ***Notre Dame Journal of Formal Logic***, vol. 30 (1989), pp. 161–196.

SEMINAR FÜR PHILOSOPHIE, LOGIK UND WISSENSCHAFTSTHEORIE
PHILOSOPHIE-DEPARTMENT
LUDWIG-MAXIMILIANS-UNIVERSITÄT MÜNCHEN
LUDWIGSTR. 31, D-80539 MÜNCHEN, GERMANY
E-mail: kgn@lrz.uni-muenchen.de

REPRESENTATION THEOREM FOR MODELS OF DYNAMIC INTENSIONAL LOGIC

ANDREJA PRIJATELJ

Abstract. In this paper the representation theorem for the class of models of dynamic intensional logic is established. It relies on a particular universal construction resulting from a new axiomatization of the class of models considered.

§1. **Introduction.** The motivation for the present paper has arisen from a talk given by M. Stokhof on Dynamic Montague Grammar. In the early nineties this has become a hot topic dealing with a dynamic interpretation of natural language, yielding among others an adequate interpretation of anaphoric relations between quantificational expressions and pronouns ([5], [6]). However, we have been not exploring the subject as a whole. We have primarily been interested in the logical setting used to accomplish their linguistic task. The logical setting itself consists of:

(I) a formal language, called dynamic intensional logic (**DIL**), including a distinguished non-empty set of discourse markers;
(II) a class of **DIL** models;
(III) an interpretation of **DIL** terms with respect to a given **DIL** model, state and assignment of values to variables.

(I), (II) and (III) represent a dynamic extension of the respective static counterparts of intensional logic **IL** as formulated by Gallin [4] (see appendix A). This introductory part is basically aimed at pointing out the additional dynamic components (see [5]). Nevertheless, a full definition of the class of **IL** models will be given since it is also a building stone for our equivalent axiomatization of the class of **DIL** models in Section 2. It will also be seen that the latter enables a natural universal construction of a **DIL** model that finally results in the representation theorem for the class of **DIL** models.

To start with (I): the additional primitive symbols and syntactic rules involved with new syntactic categories, added to those of the syntactic system

Intensionality
Edited by Reinhard Kahle
Lecture Notes in Logic, 22

of **IL** are given below:

(i) $\{, \}, /,$

(ii) (a) If $x \in \mathcal{DM}$, then $x \in \mathrm{Con}_{(\mathtt{s},\mathtt{e})}$,

(b) If $x \in \mathcal{DM}$, $A \in Tm_{\mathtt{e}}$ and $B \in Tm_\tau$, then $\{A/x\}B \in Tm_\tau$,

where $\mathcal{DM}$ denotes the set of discourse markers of type (s, e) and $\{A/x\}$ a state switcher acting on any term of an arbirtary type τ.

To continue with (II): an axiomatic description of **DIL** models based on the static **IL** models will be laid out at length later on in this paper. For this purpose, let us first present a definition of simple one-sorted types, originating from the theory of types introduced by Church [3].

DEFINITION 1.1. Let $\mathtt{e}$, $\mathtt{t}$, $\mathtt{s}$ be any distinct objects, none of which is an ordered pair. The set $\mathcal{T}$ of simple one-sorted types is the smallest set satisfying:

i. $\mathtt{e}, \mathtt{t} \in \mathcal{T}$,

ii. if $\tau, \tau' \in \mathcal{T}$, then $(\tau, \tau') \in \mathcal{T}$,

iii. if $\tau \in \mathcal{T}$, then $(\mathtt{s}, \tau) \in \mathcal{T}$.

Further, we shall call $\mathcal{T}_i = \{(\mathtt{s}, \tau);\ \tau \in \mathcal{T}\} \subset \mathcal{T}$ the set of intensional types and $\mathcal{T}_n = \mathcal{T} \setminus \mathcal{T}_i$ the set of non-intensional types.

Remark: Observe, that $\mathtt{s}$ itself is not a type.

DEFINITION 1.2. Let $\mathcal{D}$ and $\mathcal{S}$ be any non-empty sets.

1. The standard frame based on $\mathcal{D}$ and $\mathcal{S}$ is represented recursively by the indexed family $(\mathcal{D}_\tau)_{\tau \in \mathcal{T}}$ of sets satisfying:
   i. $\mathcal{D}_{\mathtt{e}} = \mathcal{D}$,
   ii. $\mathcal{D}_{\mathtt{t}} = \{0, 1\}$,
   iii. $\mathcal{D}_{(\tau,\tau')} = \mathcal{D}_{\tau'}^{\mathcal{D}_\tau}$,
   iv. $\mathcal{D}_{(\mathtt{s},\tau)} = \mathcal{D}_\tau^{\mathcal{S}}$.
2. A standard **IL** model based on $\mathcal{D}$ and $\mathcal{S}$ is an ordered pair $\langle (\mathcal{D}_\tau)_{\tau \in \mathcal{T}}, F \rangle$ satisfying:
   i. $(\mathcal{D}_\tau)_{\tau \in \mathcal{T}}$ is a standard frame based on $\mathcal{D}$ and $\mathcal{S}$,
   ii.
   $$F : \bigcup_{\tau \in \mathcal{T}} \mathrm{Con}_\tau \longrightarrow \bigcup_{\tau \in \mathcal{T}} \mathcal{D}_\tau^{\mathcal{S}}$$
   is an interpretation function such that $F(c_\tau) \in \mathcal{D}_\tau^{\mathcal{S}}$ for any $c_\tau \in \mathrm{Con}_\tau$, where $\tau \in \mathcal{T}$.

The basic idea used for the definition of a **DIL** model given below is as follows. Among the standard class of **IL** models one distinguishes a particular subclass satisfying some additional semantic postulates. By these postulates certain restrictions are imposed on the interpretation function F, and on the equality relation in the set of states S, yielding the intended dynamic interpretation of expressions within the scope of a state switcher.

We now present the definition of a **DIL** model, as given in [5].

DEFINITION 1.3. A standard **IL** model $\langle \mathcal{S}, \mathcal{D}, F \rangle$ is a **DIL** model if the following three postulates are satisfied:

1. **Rigidity postulate:** For all $\tau \in \mathcal{T}$, $c \in \mathrm{Con}_{(s,\tau)}$ and $s, s' \in \mathcal{S}$
$$F(c)(s) = F(c)(s').$$
We write $V(c)$ for $F(c)(s)$.
2. **Distinctness postulate:** If $s = s'(F)$ for $s, s' \in \mathcal{S}$, then $s = s'$, where $s = s'(F)$ is an abbreviation for: $V(c)(s) = V(c)(s')$ for any $c \in \bigcup_{\tau \in \mathcal{T}} \mathrm{Con}_{(s,\tau)}$ and $F(c)(s) = F(c)(s')$ for all constants of any non-intensional type.
3. **Update postulate:** For all $s \in \mathcal{S}$, $x \in \mathcal{DM}$, $d \in \mathcal{D}$, there is an $s' \in \mathcal{S}$ such that:
(i) $s' \in V(x)^{-1}(d)$,
(ii) $s = s'(F_x)$,
where $s = s'(F_x)$ is an abbreviation for: $V(c)(s) = V(c)(s')$ for any $c \in \bigcup_{\tau \in \mathcal{T}} \mathrm{Con}_{(s,\tau)} - \{x\}$ and $F(c)(s) = F(c)(s')$ for all constants of any non-intensional type.

Remarks: The surjectivity of the function $V(x)$ follows by (i), while (ii) means that s and s' can not be distinguished by F restricted to $\bigcup_{\tau \in \mathcal{T}} \mathrm{Con}_\tau - \{x\}$.

Let us finally give a sketch of (III), i.e. the intended dynamic interpretation. The usual notation for an extension of any **DIL** term will be used here: say $V_{s,a}(B)$, where $B \in Tm_\tau$, with the parameters s and a referring to a given state and an assignment of values to variables respectively, while omitting the parameter M that specifies a **DIL** model. The additional semantic rules corresponding to their syntactic counterparts given in (I) are the following:

(a) for $x \in \mathcal{DM}$ $V_{s,a}(x) := F(x)(s)$, where F is an interpretation function of a **DIL** model, in particular, $F(x) \in \mathcal{D}^{\mathcal{S}}_{(s,e)}$;
(b) $V_{s,a}(\{A/x\}B) := V_{s',a}(B)$, where s' is the uniquely determined state such that the following holds: $V_{s,a}(x)(s') = V_{s,a}(A)$ and $V(c_{(s,\tau)})(s) = V(c_{(s,\tau)})(s')$ for all constants of an intensional type except perhaps for the discourse marker x itself, as well as $F(c_\tau)(s) = F(c_\tau)(s')$ for all constants of any non-intensional type.

Remarks: clause (a) above tells us that the extension of a discourse marker x at parameters s and a is determined by the interpretation function F (of a given **DIL** model). Since $\mathcal{DM}$ is just a particular non-empty set of constants of type (s,e), this matches the standard interpretation of intensional logic constants. On the other hand, clause (b) above is the real key to the intended dynamic interpretation. In fact, it reveals the semantic role imposed on a particular state switcher. To be precise, given a **DIL** model, the extension of $\{A/x\}B$ with parameters s and a is just the extension of the term B where solely the initial state parameter s is switched to the uniquely determined state s' subject to the following constraints: the extension of the discourse marker x applied to s' is exactly the individual determined by the extension of the term A at the initial parameters s and a, and moreover, roughly speaking, within a given **DIL** model, the two states s and s' can not be distinguished by F except perhaps on the indicated discourse marker x. The latter fact is often abbreviated by $s = s'(F_x)$. It has to be emphasized that existence and uniqueness of the switched state s' considered are granted by the Distinctness and the Update postulates specifying a class of **DIL** models given above. Despite the pleasant fact that the two postulates mentioned are the proper carriers of the intended dynamic interpretation they have nevertheless presented the main difficulty in giving a construction of a **DIL** model. And as far as we know, this has been quite an intriguing open problem for long, thus calling for a detailed elaboration in what follows. At this point, we can offer the reader just the basic idea of our approach. In order that a given "static" **IL** model be equipped with the proper dynamic features we shall, on top of it introduce a so-called Update function $U : S \times DM \times D \to S$ resulting in an equivalent axiomatization of the class of given **DIL** models. The relevance of our axiomatization can be justified by further results obtained in this paper, namely: a universal construction of a **DIL** model yielding a representation theorem for the corresponding class of models.

§2. **DIL models.** In this section, we shall give an alternative axiomatization of the class of **DIL** models suitable for the purposes discussed above.

DEFINITION 2.1. A model of dynamic intensional logic, **DIL** model, based on $\mathcal{D}$ and $\mathcal{S}$ is an ordered triple $\langle (\mathcal{D}_\tau)_{\tau\in\mathcal{T}}, F, U\rangle$ where $\langle (\mathcal{D}_\tau)_{\tau\in\mathcal{T}}, F\rangle$ is a standard **IL** model based on $\mathcal{D}$ and $\mathcal{S}$ and $U : \mathcal{S} \times \mathcal{DM} \times \mathcal{D} \to \mathcal{S}$ is the Update function, satisfying the following conditions:

1. for all $\tau \in \mathcal{T}_i$, $c_\tau \in \mathrm{Con}_\tau$ and $s, s' \in \mathcal{S}$
$$F(c_\tau)(s) = F(c_\tau)(s').$$
We write $V(c_\tau)$ for $F(c_\tau)(s)$.
2. for $s \in \mathcal{S}$, $d, d' \in \mathcal{D}$, $x \in \mathcal{DM}$
   i. $U(s, x, V(x)(s)) = s$,
   ii. $U(U(s, x, d), x, d') = U(s, x, d')$,
   iii. $U(U(s, x, d), x', d') = U(U(s, x', d'), x, d)$ for all $x' \in \mathcal{DM} \setminus \{x\}$.
3. for $s \in \mathcal{S}$, $x \in \mathcal{DM}$, $d \in \mathcal{D}$
   i. $V(x)(U(s, x, d)) = d$,
   ii. $V(c_\tau)(U(s, x, d)) = V(c_\tau)(s)$ for all $c_\tau \in \mathrm{Con}_\tau \setminus \{x\}$ with $\tau \in \mathcal{T}_i$,
   iii. $F(c_\tau)(U(s, x, d)) = F(c_\tau)(s)$ for all $c_\tau \in \mathrm{Con}_\tau$ with $\tau \in \mathcal{T}_n$.

It is easy to observe the following useful

FACT 2.2. Within any **DIL** model based on $\mathcal{D}$ and $\mathcal{S}$ the following holds: $U(U(s, x, d), x, V(x)(s)) = s$ for all $s \in \mathcal{S}$, $x \in \mathcal{DM}$ and $d \in \mathcal{D}$.

PROOF. Straightforward by Definition 2.1, (2.ii) and (2.i). ⊣

Since the representation theorem for the class of **DIL** models is our final goal we shall next define the notion of an isomorphism between **DIL** models.

DEFINITION 2.3. An isomorphism based on $\Phi_{\mathbf{e}}$, $\Phi_{\mathbf{t}}$, $\Phi_{\mathbf{s}}$ of a **DIL** model $\langle(\mathcal{D}_\tau)_{\tau\in\mathcal{T}}, F, U\rangle$ based on $\mathcal{D}$ and $\mathcal{S}$, onto a **DIL** model $\langle(\mathcal{D}'_\tau)_{\tau\in\mathcal{T}}, F', U'\rangle$ based on $\mathcal{D}'$ and $\mathcal{S}'$ is a family of bijective mappings $(\Phi_\alpha)_{\alpha\in\mathcal{T}\cup\{\mathbf{s}\}}$ where $\Phi_{\mathbf{s}} : \mathcal{S} \to \mathcal{S}'$ and $\Phi_\tau : \mathcal{D}_\tau \to \mathcal{D}'_\tau$ for all $\tau \in \mathcal{T}$ satisfying:

1. for $\alpha \in \mathcal{T} \cup \{\mathbf{s}\}$ and $\tau \in \mathcal{T}$

$$\Phi_{(\alpha,\tau)}(f) = \Phi_\tau \circ f \circ \Phi_\alpha^{-1} \quad \text{where } f \in \mathcal{D}_{(\alpha,\tau)},$$

   i.e. the following diagram commutes for every $f \in \mathcal{D}_{(\alpha,\tau)}$:

$$\begin{array}{ccc} \mathcal{D}_\alpha & \xrightarrow{\Phi_\alpha} & \mathcal{D}'_\alpha \\ {\scriptstyle f}\big\downarrow & & \big\downarrow{\scriptstyle \Phi_{(\alpha,\tau)}(f)} \\ \mathcal{D}_\tau & \xrightarrow{\Phi_\tau} & \mathcal{D}'_\tau \end{array}$$

2. for $c_\tau \in \mathrm{Con}_\tau$ with $\tau \in \mathcal{T}$

$$\Phi_{(\mathbf{s},\tau)}(F(c_\tau)) = F'(c_\tau),$$

3. for $s \in \mathcal{S}$, $x \in \mathcal{DM}$, $d \in \mathcal{D}$

$$\Phi_{\mathbf{s}}(U(s, x, d)) = U'(\Phi_{\mathbf{s}}(s), x, \Phi_{\mathbf{e}}(d)).$$

Remark: Given bijective mappings $\Phi_{\mathbf{e}}$, $\Phi_{\mathbf{t}}$, $\Phi_{\mathbf{s}}$, observe that, the family $(\Phi_\alpha)_{\alpha\in\mathcal{T}\cup\{\mathbf{s}\}}$ of bijections between the two standard frames can be obtained recursively by (1) above.

The time has come to have a closer look at the set of states S of an arbitrary **DIL** model. For this purpose, several equivalence relations are to be introduced in S, as follows.

DEFINITION 2.4. Let **M** be a **DIL** model based on $\mathcal{D}$ and $\mathcal{S}$. We say that

1. $s, s' \in \mathcal{S}$ are weakly-accessible, $s \sim_M s'$, if and only if

$$V(x)(s) = V(x)(s')$$

   for all but finitely many $x \in \mathcal{DM}$.
2. $s, s' \in \mathcal{S}$ are accessible, $s \approx_M s'$, if and only if for some natural number n there exist $x_1, x_2, \ldots, x_n \in \mathcal{DM}$ and $d_1, d_2, \ldots, d_n \in \mathcal{D}$ such that

$$s' = U(\ldots U(U(s, x_1, d_1), x_2, d_2) \ldots, x_n, d_n);$$

3. $s, s' \in \mathcal{S}$ are similar, $s \equiv_M s'$, if and only if

$$V(x)(s) = V(x)(s')$$

   for all $x \in \mathcal{DM}$.

Remark: Indeed, all the above given relations are equivalence relations. Symmetry of the accessibility relation can be verified by using Fact 2.2, the rest is trivial. From now on, we will omit the subscript **M** in the denotation of the above given relations whenever no ambiguity can occur.

Moreover, we will use $q_\sim : \mathcal{S} \to \mathcal{S}/\sim$, $q_\approx : \mathcal{S} \to \mathcal{S}/\approx$, and $q_\equiv : \mathcal{S} \to \mathcal{S}/\equiv$ to stand for the respective natural mappings.

A useful observation concerning the properties of accessible states is this:

LEMMA 2.5. *Given a* **DIL** *model* **M** *based on* $\mathcal{D}$ *and* $\mathcal{S}$, *the following holds: if* $s, s' \in \mathcal{S}$ *are accessible, then for some natural number* m *there are some distinct* $x_1, x_2, \ldots, x_m \in \mathcal{DM}$ *and some* $d_1, d_2, \ldots, d_m \in \mathcal{D}$ *such that*

$$s' = U\big(\ldots U\big(U\big(s, x_1, d_1\big), x_2, d_2\big) \ldots, x_m, d_m\big).$$

Moreover,

i. $V(x_i)(s') = d_i$ *for* $i = 1, \ldots, m$,
ii. $V(y)(s') = V(y)(s)$ *for* $y \in \mathcal{DM} \setminus \{x_1, x_2, \ldots, x_m\}$.

PROOF. Since $s \approx s'$ we have $s' = U(\ldots U(U(s, x_1', d_1'), x_2', d_2') \ldots, x_n', d_n')$ for some $x_1', x_2', \ldots, x_n' \in \mathcal{DM}$ and $d_1', d_2', \ldots, d_n' \in \mathcal{D}$. First, 2.1, (2.ii) and (2.iii) are to be applied to s' a number of times, possibly zero, in order to get

$$s' = U\big(\ldots U\big(U\big(s, x_{i_1}', d_{i_1}'\big), x_{i_2}', d_{i_2}'\big) \ldots, x_{i_m}', d_{i_m}'\big),$$

where $x_{i_1}', x_{i_2}', \ldots, x_{i_m}'$ are precisely all the distinct discourse markers among the given $x_1', x_2', \ldots, x_n'$. This justifies the first statement above. The rest of the lemma can easily be proved by using 2.1, (3.i) and (3.ii) a number of times. ⊣

The final observation reveals a fundamental interplay between the equivalence relations introduced above.

FACT 2.6. Given a **DIL** model **M** based on $\mathcal{D}$ and $\mathcal{S}$, any $s, s' \in \mathcal{S}$:

i. if $s \approx s'$, then $s \sim s'$,
ii. if $s \equiv s'$, then $s \sim s'$,
iii. if $s \approx s'$ and $s \equiv s'$, then $s = s'$.

PROOF. (i) by Lemma 2.5, (ii),
(ii) trivial,
(iii) straightforward by definitions, using essentially Lemma 2.5, (i) and 2.1 (2.i) a number of times. ⊣

§3. Universal construction. In what follows, a universal construction of a **DIL** model will be laid out.

Given an arbitrary non-empty set $\mathcal{D}^*$ we will first introduce an equivalence relation in $\mathcal{D}^{*\mathcal{DM}}$, as follows.

Let $p, p' \in \mathcal{D}^{*\mathcal{DM}}$ be pre-accessible, then $p \asymp_{\mathcal{D}^*} p'$ if and only if $p(x) = p'(x)$ for all but finitely many $x \in \mathcal{DM}$. However, from now on, we shall omit the subscript $\mathcal{D}^*$ in the notation for the relation just introduced whenever no ambiguity can occur. Moreover, let $q_{\asymp} : \mathcal{D}^{*\mathcal{DM}} \to \mathcal{D}^{*\mathcal{DM}}/\asymp$ denote the corresponding natural mapping.

Next, we define the set of states

$$\mathcal{S}^* = \bigcup_{\mathcal{P} \in \mathcal{D}^{*\mathcal{DM}}/\asymp} \mathcal{P} \times \mathcal{E}_{\mathcal{P}},$$

where $(\mathcal{E}_{\mathcal{P}})_{\mathcal{P} \in \mathcal{D}^{*\mathcal{DM}}/\asymp}$ is any family of pairwise disjoint sets whose union $\mathcal{E} = \bigcup_{\mathcal{P} \in \mathcal{D}^{*\mathcal{DM}}/\asymp} \mathcal{E}_{\mathcal{P}}$ is non-empty.

The standard frame based on $\mathcal{D}^*$ and $\mathcal{S}^*$, namely $(\mathcal{D}^*_\tau)_{\tau \in \mathcal{T}}$, is obtained by Definition 1.2, (1).

Further we introduce a function $U^* : \mathcal{S}^* \times \mathcal{DM} \times \mathcal{D}^* \to \mathcal{S}^*$ in the following way:

for any $s \in \mathcal{S}^*$, where $s = (p, e) \in \mathcal{P} \times \mathcal{E}_{\mathcal{P}}$ for some $\mathcal{P} \in \mathcal{D}^{*\mathcal{DM}}/\asymp$, and for all $x \in \mathcal{DM}$ and all $d \in \mathcal{D}$

$$U^*((p, e), x, d) = (p', e),$$

where $p' \in \mathcal{P}$ is such that for all $y \in \mathcal{DM}$

$$p'(y) = \begin{cases} d; & y = x, \\ p(y); & y \neq x. \end{cases}$$

Note that U^* is well-defined since $p' \asymp p$ by definition. Next we define a function $V^*(x) : \mathcal{S}^* \to \mathcal{D}^*$ for each $x \in \mathcal{DM}$ as follows: for any $s \in \mathcal{S}^*$, where $s = (p, e) \in \mathcal{P} \times \mathcal{E}_{\mathcal{P}}$ for some $\mathcal{P} \in \mathcal{D}^{*\mathcal{DM}}/\asymp$,

$$V^*(x)(p, e) = p(x).$$

Finally, we shall define a function

$$F^* : \bigcup_{\tau \in \mathcal{T}} \mathrm{Con}_\tau \longrightarrow \bigcup_{\tau \in \mathcal{T}} \mathcal{D}_\tau^{*\mathcal{S}^*}.$$

For this definition we have to consider constants of intensional types and constants of non-intensional types separately.

Given an arbitrary function

$$f : \bigcup_{\tau' \in \mathcal{T}} \mathrm{Con}_{(\mathbf{s},\tau')} \backslash \mathcal{DM} \longrightarrow \bigcup_{\tau' \in \mathcal{T}} \mathcal{D}_{\tau'}^{*\,\mathcal{E}}$$

such that for every $c_{(\mathbf{s},\tau')} \in \mathrm{Con}_{(\mathbf{s},\tau')} \backslash \mathcal{DM}$ with $\tau' \in \mathcal{T}$

$$f\left(c_{(\mathbf{s},\tau')}\right) \in \mathcal{D}_{\tau'}^{*\,\mathcal{E}}$$

we define $V^*(c_{(\mathbf{s},\tau')}) : \mathcal{S}^* \to \mathcal{D}^*_{\tau'}$ as follows: for any $s \in \mathcal{S}^*$, where $s = (p, e) \in \mathcal{P} \times \mathcal{E}_{\mathcal{P}}$ for some $\mathcal{P} \in \mathcal{D}^{*\mathcal{DM}}/\asymp$

$$V^*\left(c_{(\mathbf{s},\tau')}\right)(p, e) = f\left(c_{(\mathbf{s},\tau')}\right)(e).$$

Thus, $V^*(c_\tau)$ has been defined for all constants of all intensional types including the set $\mathcal{DM}$. Moreover, let for all $c_\tau \in \mathrm{Con}_\tau$, where $\tau \in \mathcal{T}_i$,

$$F^*(c_\tau)(s) = V^*(c_\tau)$$

for any $s \in \mathcal{S}^*$.

Further, given an arbitrary function

$$g : \bigcup_{\tau \in \mathcal{T}_n} \mathrm{Con}_\tau \longrightarrow \bigcup_{\tau \in \mathcal{T}_n} \mathcal{D}_\tau^{*\mathcal{E}}$$

such that for every $c_\tau \in \mathrm{Con}_\tau$ where $\tau \in \mathcal{T}_n$

$$g(c_\tau) \in \mathcal{D}_\tau^{*\mathcal{E}}$$

we define $F^*(c_\tau) : \mathcal{S}^* \to \mathcal{D}_\tau^*$ as follows:

for any $s \in \mathcal{S}^*$, where $s = (p, e) \in \mathcal{P} \times \mathcal{E}_\mathcal{P}$ for some $\mathcal{P} \in \mathcal{D}^{*\mathcal{DM}}/\asymp$

$$F^*(c_\tau)(p, e) = g(c_\tau)(e).$$

The reader is now invited to verify that $\langle (\mathcal{D}_\tau^*)_{\tau \in \mathcal{T}}, F^*, U^* \rangle$ is indeed a **DIL** model based on $\mathcal{D}^*$ and $\mathcal{S}^*$.

Any **DIL** model obtained by the universal construction introduced above will be called a $*$-model. Observe that every $*$-model is generated by the parameters $\mathcal{D}^*$, $(\mathcal{E}_\mathcal{P})_{\mathcal{P} \in \mathcal{D}^{*\mathcal{DM}}/\asymp}$, f and g.

§4. **Representation theorem for the class of DIL models.** In this final section we shall justify the relevance of the universal construction obtained. In fact, we are going to prove that every **DIL** model is isomorphic to a $*$-model. To this end, we proceed as follows. Given a **DIL** model M based on D and S, we introduce a bijective function mapping each state s of S to an ordered pair with the first component an element of $\mathcal{D}^{\mathcal{DM}}$ and the second component the class of accessible states with respect to s. Roughly speaking, the first component is to be uniquely determined by the interpretation of discourse markers within a given **DIL** model. Afterwards, it will be seen that the codomain of the mapping considered in fact satisfies the description of the set $\mathcal{S}^*$ in the universal construction above.

DEFINITION 4.1. Given a **DIL** model **M** based on $\mathcal{D}$ and $\mathcal{S}$ we define the following functions:

- $\Psi_M : \mathcal{S} \to \mathcal{D}^{\mathcal{DM}}$ such that for any $s \in \mathcal{S}$,

  $$\Psi_M(s)(x) = V(x)(s)$$

  for all $x \in \mathcal{DM}$.
- $\Phi_M = (\Psi_M, q_\approx) : \mathcal{S} \to \bigcup_{\mathcal{P} \in \mathcal{D}^{\mathcal{DM}}/\asymp} \mathcal{P} \times \Psi_M^{-1}(P)/\approx$ such that for all $s \in \mathcal{S}$,

  $$\Phi_M(s) = (\Psi_M(s), q_\approx(s)).$$

Remark: To verify that Φ_M is well-defined, observe, firstly, that given any $s \in \mathcal{S}$

$$\Psi_M(s) \in q_{\asymp}(\Psi_M(s))$$

and, secondly, by Fact 2.6, (i)

$$q_{\approx}(s) \in q_{\sim}(s)/\approx .$$

Now we can assert

$$\Psi_M^{-1}(q_{\asymp}(\Psi_M(s))) = q_{\sim}(s)$$

by the definitions.

LEMMA 4.2. *Given a* **DIL** *model* **M** *based on* $\mathcal{D}$ *and* $\mathcal{S}$ *the function* Φ_M *is bijective.*

PROOF. (a) Injectivity of Φ_M is shown as follows: suppose $\Phi_M(s) = \Phi_M(s')$ for some $s, s' \in \mathcal{S}$. By definition this gives $\Psi_M(s) = \Psi_M(s')$ and $q_{\approx}(s) = q_{\approx}(s')$, yielding $s \equiv s'$ and $s \approx s'$. And hence, $s = s'$ by Fact 2.6, (iii).

(b) Surjectivity of Φ_M can be proved in the following way: let (p, e) be an arbitrary element of $\mathcal{P} \times \Psi_M^{-1}(P)/\approx$ for some $P \in \mathcal{D}^{\mathcal{DM}}/\asymp$. Then there exists $s \in \mathcal{S}$ such that $q_{\approx}(s) = e$ and $\Psi_M(s) \in \mathcal{P}$. Clearly, $p \asymp \Psi_M(s)$, which means that $p(x) = \Psi_M(s)(x)$ for all but finitely $x \in \mathcal{DM}$. If $p = \Psi_M(s)$, we are done.

Otherwise, consider $s' = U(\dots U(U(s, x_1, p(x_1)), x_2, p(x_2)), \dots, x_n, p(x_n))$, where $x_1, x_2, \dots, x_n$ are exactly those elements of $\mathcal{DM}$ at which p and $\Psi_M(s)$ are not equal. By definition, $s \approx s'$ and thus $q_{\approx}(s') = e$. Moreover, $\Psi_M(s') = p$ essentially by Lemma 2.5, (i) and (ii). And hence, $\Phi_M(s') = (p, e)$, which completes the proof. ⊣

Comments: in particular, given a finite set of discourse markers $\mathcal{DM}$, for any **DIL** model based on D and S the following holds. The only equivalence class of pre-accessible elements in $\mathcal{D}^{\mathcal{DM}}$ is the whole set itself, while $\Psi_M^{-1}(\mathcal{D}^{\mathcal{DM}})/\approx$ is just the quotient set of states identified by the accessibility relation in S. This indeed shows that any $s \in \mathcal{S}$ is uniquely determined by the interpretation of discourse markers and all other constants within a **DIL** model given. In short, for arbitrary $s, s' \in \mathcal{S}$, if $s = s(F)$, then $s = s'$. On the other hand, if DM is an infinite set the whole reasoning remains valid. Only this time the codomain of the function Φ_M is the union of similar "finite" structures with respect to the pre-accessibility relation in $\mathcal{D}^{\mathcal{DM}}$.

At last, we are ready to prove our central result.

THEOREM 4.3 (Representation Theorem). *Any* **DIL** *model is isomorphic to a* $*$*-model.*

PROOF. Given a **DIL** model $\boldsymbol{M} = \langle(\mathcal{D}_\tau)_{\tau\in\mathcal{T}}, F, U\rangle$ based on $\mathcal{D}$ and $\mathcal{S}$ we are going to construct an isomorphic $*$-model $\boldsymbol{M}^* = \langle(\mathcal{D}^*{}_\tau)_{\tau\in\mathcal{T}}, F^*, U^*\rangle$ and the corresponding isomorphism $(\Phi_\alpha)_{\alpha \in \mathcal{T} \cup \{\mathbf{s}\}}$ based on $\Phi_\mathbf{e}$, $\Phi_\mathbf{t}$, $\Phi_\mathbf{s}$ simultaneously.

To begin with the universal construction put $\mathcal{D}^* = \mathcal{D}$ and $\mathcal{E}_\mathcal{P} = \Psi_M^{-1}(P)/\approx$ for all $\mathcal{P} \in \mathcal{D}^{\mathcal{DM}}/\asymp$, with Ψ_M being given by 4.1. Clearly,

$$(\Psi_M^{-1}(P)/\approx)_{\mathcal{P} \in \mathcal{D}^{\mathcal{DM}}/\asymp}$$

is a family of pairwise disjoint sets, whose union, $\mathcal{E}$, is non-empty and hence, following the universal construction

$$\mathcal{S}^* = \bigcup_{\mathcal{P} \in \mathcal{D}^{\mathcal{DM}}/\asymp} \mathcal{P} \times \Psi_M^{-1}(P)/\approx .$$

Moreover, the standard frame $(\mathcal{D}^*{}_\tau)_{\tau \in \mathcal{T}}$ based on $\mathcal{D}^*$ and $\mathcal{S}^*$ can now be obtained by 1.2, (1).

Next, let, both, $\Phi_\mathbf{e} : \mathcal{D} \to \mathcal{D}^*$ and $\Phi_\mathbf{t} : \{0, 1\} \to \{0, 1\}$ be identity functions. Further, put $\Phi_\mathbf{s} = \Phi_M : \mathcal{S} \to \mathcal{S}^*$, where Φ_M is given by 4.1 and bijective by 4.2. Now, the whole family of bijective mappings $(\Phi_\alpha)_{\alpha \in \mathcal{T} \cup \{\mathbf{s}\}}$, where $\Phi_\tau : \mathcal{D}_\tau \to \mathcal{D}^*{}_\tau$ for every $\tau \in \mathcal{T}$, can be obtained by 2.3, (1); witness remark of Definition 2.3.

It is now straightforward to spell out the validity of 2.3, (3) for our case. Namely, using relevant definitions and once Lemma 2.5 once in a straightforward way we can confirm

$$\Phi_\mathbf{s}(U(s, x, d)) = U^*(\Phi_\mathbf{s}(s), x, \Phi_\mathbf{e}(d))$$

for all $s \in \mathcal{S}$, $x \in \mathcal{DM}$, $d \in \mathcal{D}$, where U^* is given by the universal construction.

Moreover, put

$$F^*(c_\tau) = \Phi_{(\mathbf{s},\tau)}(F(c_\tau))$$

for all $c_\tau \in \mathrm{Con}_\tau$ with $\tau \in \mathcal{T}$, as imposed by 2.3, (2). It remains to be seen that this definition is, in fact, compatible with the definition of F^* given by the universal construction. This means that the following has to be justified.

Firstly, $V^*(c_\tau)$ given for every $c_\tau \in \mathrm{Con}_\tau$ with $\tau \in \mathcal{T}_i$ by

$$V^*(c_\tau) = F^*(c_\tau)(s),$$

with $s \in \mathcal{S}^*$, is well-defined. This follows from the fact that for any $s', s'' \in \mathcal{S}^*$ we have $F^*(c_\tau)(s') = F^*(c_\tau)(s'')$ as the relevant definitions reduce to 2.1 (1).

In particular, for all $x \in \mathcal{DM}$

$$V^*(x)(p', e') = p'(x)$$

holds for any $(p', e') \in \mathcal{P} \times \mathcal{E}_{\mathcal{P}}$ with $\mathcal{P} \in \mathcal{D}^{\mathcal{DM}}/\asymp$. Namely, using the relevant definitions we get

$$\begin{aligned}(F^*(x)(p,e))(p',e') &= (\Phi_{(\mathbf{s},(\mathbf{s},\mathbf{e}))}(F(x))(p,e))(p',e') \\ &= (\Phi_{(\mathbf{s},\mathbf{e})}(F(x)(\Phi_{\mathbf{s}}^{-1}(p,e))))(p',e') \\ &= (\Phi_{(\mathbf{s},\mathbf{e})}(V(x)))(p',e') \\ &= \Phi_{\mathbf{e}}(V(x)(\Phi_{\mathbf{s}}^{-1}(p',e'))) \\ &= V(x)(s') = \Psi_M(s')(x) = p'(x).\end{aligned}$$

The last equality follows from the fact that $\Phi_{\mathbf{s}}^{-1}(p', e') = s'$. Thus, $\Phi_{\mathbf{s}}(s') = (\Psi_M(s'), q_{\approx}(s')) = (p', e')$ and hence, $\Psi_M(s')(x) = p'(x)$.

Secondly, we will point out that the generator

$$f : \bigcup_{\tau' \in \mathcal{T}} \mathrm{Con}_{(\mathbf{s},\tau')} \setminus \mathcal{DM} \longrightarrow \bigcup_{\tau' \in \mathcal{T}} \mathcal{D}_{\tau'}^{*\,\mathcal{E}}$$

of M^* given by

$$f(c_{(\mathbf{s},\tau')})(e) = V^*(c_{(\mathbf{s},\tau')})(p,e)$$

for all constants of intensional types except discourse markers and for all $e \in \mathcal{E}_{\mathcal{P}}$ and $p \in \mathcal{P} \in \mathcal{D}^{\mathcal{DM}}/\asymp$, is well-defined. That is to say, we have to justify

$$V^*(c_{(\mathbf{s},\tau')})(p',e) = V^*(c_{(\mathbf{s},\tau')})(p'',e)$$

for arbitrary $p', p'' \in \mathcal{P}$ with $\mathcal{P} \in \mathcal{D}^{\mathcal{DM}}/\asymp$. Again, using the relevant definitions in a straightforward way the above equality can be reduced to

$$\Phi_{\tau'}(V(c_{(\mathbf{s},\tau')})(s'))\Phi_{\tau'}(V(c_{(\mathbf{s},\tau')})(s'')),$$

where $s' = \Phi_{\mathbf{s}}^{-1}(p', e)$ and $s'' = \Phi_{\mathbf{s}}^{-1}(p'', e)$. Since, clearly, $q_{\approx}(s') = q_{\approx}(s'') = e$ means $s' \approx s''$, we have

$$s'' = U(\ldots U(U(s', x_1, d_1), x_2, d_2) \ldots, x_n, d_n)$$

by 2.4 (2). Thus, using 2.1 (3.ii) n times successively we get

$$V(c_{(\mathbf{s},\tau')})(s') = V(c_{(\mathbf{s},\tau')})(s'')$$

which yields the equality in question.

Analogously, the reader may show that the generator

$$g : \bigcup_{\tau \in \mathcal{T}_n} \mathrm{Con}_{\tau} \longrightarrow \bigcup_{\tau \in \mathcal{T}_n} \mathcal{D}_{\tau}^{*\mathcal{E}}$$

of M^* given by

$$g(c_\tau)(e) = F^*(c_\tau)(p,e)$$

for all constants of non-intensional types and for all $e \in \mathcal{E}$ and $p \in \mathcal{P} \in \mathcal{D}^{\mathcal{DM}}/\asymp$, is well-defined.

This completes the proof. ⊣

Appendix A. Intensional logic–IL.

A.1. Syntax. Primitive symbols: For each $\tau \in \mathcal{T}$ we have a denumerable list of *variables*

$$x_\tau^0, x_\tau^1, x_\tau^2, \ldots$$

and non-logical *constants*

$$c_\tau^0, c_\tau^1, c_\tau^2, \ldots$$

of type τ, together with the improper symbols

$$\doteq, \lambda, \hat{}, \check{}, (,).$$

Terms: We characterize recursively the set Tm_τ of *terms* of **IL** of type τ, as follows:

1. Every variable of type τ belongs to Tm_τ.
2. Every constant of type τ belongs to Tm_τ.
3. $A \in Tm_{(\tau,\tau')}$, $B \in Tm_\tau$ imply $A(B) \in Tm_{\tau'}$.
4. $A \in Tm_{\tau'}$, x variable of type τ imply $\lambda x A \in Tm_{(\tau,\tau')}$.
5. $A, B \in Tm_\tau$ imply $(A \doteq B) \in Tm_{\mathfrak{t}}$.
6. $A \in Tm_\tau$ implies $\hat{}A \in Tm_{(\mathbf{s},\tau)}$.
7. $A \in Tm_{(\mathbf{s},\tau)}$ implies $\check{}A \in Tm_\tau$.

A.2. Semantics. A standard model $M = \langle (\mathcal{D}_\tau)_{\tau\in\mathcal{T}}, F\rangle$ of **IL** being already given in this paper we proceed as follows in order that a term A_τ for any $\tau \in \mathcal{T}$ is given a proper interpretation.

Let $As(M)$ denote the set of all *assignments* over M, i.e., all functions a on the set of variables of **IL** such that $a(x_\tau) \in \mathcal{D}_\tau$ for every variable x_τ of type τ.

If $a \in As(M)$, x_τ is a variable of type τ, and $d \in \mathcal{D}_\tau$ then $a(x_\tau/d)$ denotes the assignment a' whose value $a'(y)$ for a variable y is equal to d if $y = x_\tau$ and $a(y)$ otherwise.

We define the extension $V_{s,a}^M(A_\tau)$ in M of the term $A_\tau$with respect to the state $s \in \mathcal{S}$ and the assignment a, by the following recursion on the term A_τ of **IL** (we suppress the superscript "M"):

1. $V_{s,a}(x_\tau) = a(x_\tau)$,
2. $V_{s,a}(c_\tau) = F(c_\tau)(s)$,
3. $V_{s,a}(A_{(\tau,\tau')}(B_\tau)) = V_{s,a}(A_{(\tau,\tau')})[V_{s,a}(B_\tau)]$,
4. $V_{s,a}(\lambda x_\tau A_{\tau'})$ = the function G on $\mathcal{D}_\tau$ whose value at $d \in \mathcal{D}_\tau$ is equal to $V_{s,a'}(A_{\tau'})$, where $a' = a(x_\tau/d)$,
5. $V_{s,a}(A_\tau \doteq B_\tau) = 1$ if $V_{s,a}(A_\tau)V_{s,a}(B_\tau)$, and 0 otherwise,
6. $V_{s,a}(\hat{}A_\tau)$ = the function G on $\mathcal{S}$ whose value at $s' \in \mathcal{S}$ is equal to $V_{s',a}(A_\tau)$,
7. $V_{s,a}(\check{}A_{(s,\tau)}) = V_{s,a}(A_{(s,\tau)})(s)$.

REFERENCES

[1] H. Barendregt, ***The lambda calculus: its syntax and semantics***, North-Holland, Amsterdam, 1984.

[2] J. van Benthem, ***A manual of intensional logic***, CSLI Lecture Notes, CSLI, Stanford, 1988.

[3] A. Church, *A formulation of the simple theory of types*, ***The Journal of Symbolic Logic***, vol. 5 (1940), pp. 56–68.

[4] D. Gallin, ***Intensional and higher-order modal logic***, North-Holland, Amsterdam, 1975.

[5] J. Groenendijk and M. Stokhof, *Dynamic montague grammar*, ***Papers from the second symposium on logic and language*** (L. Kálmán and L. Pólos, editors), Akadémiai Kiadó, Budapest, 1990, pp. 3–48.

[6] ———, *Dynamic predicate logic*, ***Linguistics and Philosophy***, vol. 14 (1991), pp. 39–100.

[7] R. Montague, ***Formal philosophy***, Yale University Press, New Haven, 1974, Edited by R. Thomason.

[8] A. Prijatelj, ***Existence theorems for models of dynamic intensional logic***, IMFM Preprint Series, no. 28, University of Ljubljana, 1990.

[9] A. S. Troelstra and D. van Dalen, ***Constructivism in mathematics. An introduction***, vol. I and II, North Holland, Amsterdam, 1988.

DEPARTMENT OF MATHEMATICS
UNIVERSITY OF LJUBLJANA
SLOVENIA

Obituary notice by Prof. Anne Troelstra, Amsterdam.

March 31, 2002, dr. Andreja Prijatelj died at Ljubljana, 48 years old, after an illness of 6 months. Andreja, daughter of the mathematician prof. Niko Prijatelj, was born on October 6, 1953. In 1978 she graduated in technical mathematics at the University of Ljubljana. She studied pedagogy and andragogy, and in 1984 she did post-graduate studies in Mathematical Logic at the Department of Mathematics of the University of Belgrado.

From 1985 till 1991 she held a position as teaching assistant at the Faculty of Electro-Engineering in Ljubljana. For the academic year 1990-1991 she obtained a scholarship of the Slovenian Research Association, to study at the Institute of Logic, Language and Computation (ILLC) of the University of Amsterdam. From 1991 till 1994 she held a PhD-assistant position at the University of Amsterdam. She obtained her doctorate at the University of Amsterdam on January 6, 1995, with a thesis entitled “Investigating Bounded Contraction”. Her thesis work was supervised by Johan van Benthem and Anne Troelstra. She then taught logic for the students in artificial intelligence at the ILLC until March 1995. After that she spent a 6-month scholarship of the Slovenian Research Foundation at the University of Cambridge (UK), Imperial College in London and Stanford.

In 1996 she was appointed as a full-time researcher at the Department of Mathematics in Ljubljana, and in 1997 she became Lecturer at the same department. These tasks she fulfilled with great energy and enthusiasm. Most of her work in mathematical logic is connected with resource-conscious logics such as linear logic and the Lambek calculus.

Andreja always remembered her years in Amsterdam as an especially happy period in her life. Enjoying contacts with other people, warm-hearted, and sincerely interested in the well-being of others, she is remembered with affection by all who knew her.

INTENSION, INTENTION

KRISTER SEGERBERG

Abstract. Intensionality has generally been of more concern to logicians than intentionality. But also the latter merits their interest. This paper, a contribution to the logic of action, involves both concepts—the former implicitly, the latter explicitly.

§1. Informal background. Many logicians and philosophers, the present writer included, are familiar with the term "intension" (with an *s*) but rather less familiar with the term "intention" (with a *t*). Outside philosophy and logic the situation is reverse: "intention" is used by all, "intension" by few, if any. A miniature history of the development of the two terms was given by E. J. Lemmon [6]:

> The medieval term *intentio* was originally employed as a translation of the Arabic term *ma'na*, a form in the soul identified with a meaning or a notion, and meant throughout medieval epistemology a natural sign in the soul. Later the *Port Royal Logic* distinguished between the *comprehension* and *extension* of a general term in something of the way in which Mill later distinguished connotation and denotation: whilst the extension is the set of things to which the term applies, its comprehension is the set of attributes which it implies. Sir William Hamilton replaced "comprehension" by "intension", faultily spelling the word with an "s" by analogy with "extension". Since then, the term "intentionality" has gone one way, via Brentano to Chisholm, and the word "intensionality" another via Carnap to Quine.

This elegant quotation is offered for what it is worth. It is of some help in explaining philosophers' terminology, but unfortunately it leaves unexplained the connexion with everyday usage in the context of action. And it is the latter that is of concern in this paper.

We study a system in which one agent ("the agent") operates in some environment ("the world"). The world may or may not be dynamic, in the sense that things can happen even if the agent does not do anything; when it is, it is convenient to postulate yet another agent, called "Nature". Both the

Intensionality
Edited by Reinhard Kahle
Lecture Notes in Logic, 22

agent and Nature, if it exists, may bring change about, but only the agent is capable of action in a strict sense of the word, and it is only the agent who has intentions. Nature is an agent by courtesy only.

The concept of history is handy for describing the situation of the agent at a certain point. Except at the beginning of things, there is a past history of what has taken place so far. Except at the end of things, there is a number of possible future histories. The past has already been realized and is now fixed. The future is yet undetermined, but however many future histories are now possible, in the end exactly one will be realized. It would of course be useful for the agent to be able to select the unique future history that will be realized, but that happy possibility is reserved for omnipotent agents. On the other hand, unless the agent is completely powerless, he is able to influence the future: by his action he can trim the number of possible future histories. For if he performs a certain action, then he has made certain that those histories in which this very action was not made on that very occasion are no longer possible. But, in general, future developments may, but need not, be in accord with the agent's intentions. This may be represented by taking the agent in effect to divide the set of possible futures into two mutually exclusive subsets: those that we may label "intended" and the rest. From the agent's point of view, a future is intended if and only if it would realize all the intentions of the agent if it were allowed to run its course.

So what are histories? Records of what happens! A complete history is a complete record, a sequence in the proper order of all events and actions that take place, from the beginning to the end. In this paper we will only discuss actions of one kind, namely, "real" actions — actions that are related to possible change in the world (*res*). In a future paper we hope to discuss, in addition to real actions, "mental" actions — actions that the agent performs in "making up his mind" or "changing his mind" (*mens*).

It should be not only admitted but emphasized that the modelling presented here is simplistic: there will be no mention of beliefs, desires, goals, and the like. But the purpose of the present paper is not to develop a complete theory of action. Rather, we wish to devise a modelling that allowes a rudimentary representation of intention in action.

§2. **Formal modelling.** Let U be a given nonempty set (*universe*). Elements of U are called *world states* or simply *points*. A *path* in U is a nonempty sequence of points; for simplicity, in this paper we assume paths to be finite[1]. A path $(u_0, \dots, u_n)$, where $n \geq 0$, is said to be of length n and is sometimes written just $u_0 \dots u_n$. If p is a path, we write $p(*)$ for the first and $p(\#)$ for the last element of p. If $p = u_0 \dots u_m$ and $q = v_0 \dots v_n$ are paths of length m and n, respectively, and if w is a point such that $p(\#) = u_m = w = v_0 = q(*)$,

[1] This is clearly not a necessary restriction. Even discreteness is not necessary — cf. [9].

then $u_0 \ldots u_{m-1} w v_1 \ldots v_{n-1}$ is a path of length $m + n$ and is denoted by pq; note that pq is well-defined only if $p(\#) = q(*)$. We consider that, for paths p, q and r, $(pq)r = p(qr)$. An *event type* or simply *event* is a set of paths. If a is an event and p is a path such that $p \in a$, then we say that p is a *possible realization* of a or, somewhat carelessly, that p *realizes* a. We assume that the set of events is closed under the operations of union, intersection, (direct) composition and indirect composition. That is to say, if a and b are events then $a \cup b$ (the *union* of a and b), $a \cap b$ (the *intersection* of a and b), $a \cdot b$ (the *composition* of a and b, in that order) and $a \cdot\cdot b$ (the *indirect composition* of a and b, in that order) are also events, where

$$a \cdot b = \{pq : p \in a \,\&\, q \in b \,\&\, p(\#) = q(*)\}$$
$$a \cdot\cdot b = \{pqr : p \in a \,\&\, r \in b \,\&\, p(\#) = q(*) \,\&\, q(\#) = r(*)\}.$$

Note that both compositions are associative. An *individual event* is a pair (a, p) where a is an event type and p is a path that realizes a.

In a more general treatment there would be a set of agents, perhaps with a structure of its own. For simplicity, in this paper we recognize only one agent, who will be denoted by 1. For convenience we also treat Nature, denoted by 0, as an agent, but as an agent in a degenerate sense. An *(individual) action* is a triple (i, a, p), where i is an agent and (a, p) is an individual event. An action (i, a, p) where $i = 0$ is said to be an *action by Nature* or *natural event* (cf. the expression "act of God"!).

As stated in the introduction, histories are identified as (possible) chronicles of actions. Formally, a *history* is a sequence

$$h = (i_0, a_0, p_0) \cdots (i_{n-1}, a_{n-1}, p_{n-1})$$

where $n \geq 0$ and, for each $k < n$, (i_k, a_k, p_k) is an action, not necessarily atomic; if $n = 0$ we say that h is the *empty* history and denote it by the symbol $\emptyset$. If $n > 0$ then the *beginning* of h is $h(*) = p_0(*)$ and the *end* of h is $h(\#) = p_{n-1}(\#)$. It is clear that if h and g are histories, then the *concatenation* hg of h and g is also a history; note that, if f is also a history, then $h(gf) = (hg)f$. The empty history has no beginning and no end, but we still consider that it can be concatenated with other histories and that $h\emptyset = \emptyset h = h$. We define a relation $\approx$ between histories as the smallest equivalence relation that satisfies the following two conditions:

(i) if $i = i_0 = \cdots = i_{n-1}$ and $a = a_0 \cdot \ldots \cdot a_{n-1}$ and $p = p_0 \ldots p_{n-1}$ then $(i_0, a_0, p_0) \cdots (i_{n-1}, a_{n-1}, p_{n-1}) \approx (i, a, p)$,
(ii) if $g \approx g'$ then $hgf \approx hg'f$.

If $h \approx h'$, we say that h and h' are (*historically*) *equivalent*. Evidently, every history is equivalent to a unique history in which every action is atomic.

A history h is *complete* with respect to a set S of histories if h is nonempty and $h \in S$ and, for all histories g, if $hg \in S$ or $gh \in S$ then $g = \emptyset$. A set H of histories is *complete-in-itself* if all elements of H are complete with respect to H. Let H be a set of histories complete-in-itself. A proper initial of an element of H is said to be a (*possible*) *past* (in H), while the complement is said to be a (*possible*) *future* (in H). An *articulated history* is a pair (h, g) such that $hg \in H$. Notice that there is a kind of perspective inherent in an articulated history: it is as if the complete history hg were viewed from a present "now" situated between the past h and the future g. (When h and g are nonempty, the "now" is at $h(\#) = g(*)$. If either h or g is empty, then hg is viewed from the very beginning or from the very end, respectively.) If h is a possible past in H we write

$$\mathrm{fut}_H(h) = \{g : hg \in H\}$$

for the set of possible futures of h in H.

The final ingredient is the notion of an *intention function* I that selects a subset $I(\mathrm{fut}_H(h))$ of $\mathrm{fut}_H(h)$; $I(\mathrm{fut}_H(h))$, which we will also write $\mathrm{int}_H(h)$, is the set of *intended* continuations of h.[2] (Intuitively, the agent plans to make, or is committed to making, the actual future be one of those he intends.)

As a summary of these considerations, let us say that an *action structure* is a set (U, E, H, I) where U is a universe, E is a set of events, H is a set of histories complete-in-itself and I is an intention function. Some conventions: we will use (sometimes with super- or subscripts) g, h for histories, a, b for events, p, q for paths, i for elements of $\{0, 1\}$ and n for natural numbers.

§3. **Formal semantics.** It will be useful to introduce by steps a formal language suited to action structures. Two basic syntactic categories are recognized, *formulæ* and *terms*. We assume denumerable and nonoverlapping supplies of primitive formulæ ((*real*) *propositional letters*) and primitive terms ((*real*) *event letters*). An *action model* is an action structure together with a valuation that assigns *meanings* or *intensions* to all primitive formulæ and terms: to every primitive formula ϕ a subset $\|\phi\|$ of the universe and to every primitive term ϱ an event type $\|\varrho\|$. An object language suitable for this kind of model would contain a truth-functionally complete set of Boolean connectives with the usual conditions: $\|\varphi \wedge \psi\| = \|\varphi\| \cap \|\psi\|$, $\|\varphi \vee \psi\| = \|\varphi\| \cup \|\psi\|$, $\|\neg\varphi\| = U \setminus \|\varphi\|$ (where U is the universe), etc. Similarly, such a language would contain the (regular) sum operator $(+)$, the product operator $(\bullet)$ and

[2] In a slightly more general treatment we would make I a function that selects, from each subset S of $\mathrm{fut}_H(h)$, a subset $IS \subseteq S$: IS would be the set of futures of h that are *intended relative to* S. The set $I(\mathrm{fut}_H(h))$ would still be the set of intended futures of h. With the help of this definition of I we would be able to handle *conditional intention*.

two concatenation operators (; and ;;) with the following readings:

$$\begin{array}{rl} \alpha + \beta & \alpha \text{ or } \beta, \\ \alpha \bullet \beta & \alpha \text{ and (simultaneously) } \beta, \\ \alpha ; \beta & \alpha \text{ and then (immediately) } \beta, \\ \alpha ;; \beta & \alpha \text{ and then (prehaps later) } \beta. \end{array}$$

The corresponding conditions are $\|\alpha + \beta\| = \|\alpha\| \cup \|\beta\|$ and $\|\alpha \bullet \beta\| = \|\alpha\| \cap \|\beta\|$ and $\|\alpha ; \beta\| = \|\alpha\| \cdot \|\beta\|$ and $\|\alpha ;; \beta\| = \|\alpha\| \cdot\cdot \|\beta\|$.

We will give the truth-conditions of formulæ with respect to articulated histories in an action model, and we will write $(h, g) \models \varphi$ to express that a formula φ is true (in the given action model) with respect to the articulated history (h, g). We stipulate that, for any propositional letter ϕ,

$$(h, g) \models \phi \text{ iff either } h(\#) \neq \emptyset \text{ and } h(\#) \in \|\phi\| \text{ or}$$
$$g(*) \neq \emptyset \text{ and } g(*) \in \|\phi\|.$$

(By definition, h and g are not both empty. If both h and g are nonempty, then of course $h(\#) = g(*)$.)

A basic tenet in the semantic tradition in which this paper is written is that the intension of any formula is a proposition and the intension of every term is an event. Thus the propositional letters, as treated above, may be said to express propositions about the world (real propositions). In general, any purely Boolean formula — formula in which every propositional operator is a Boolean connective — expresses a real proposition.

§4. Propositional operators. We introduce operators [F], [P], ⟨F⟩, ⟨P⟩, [AFTER α] and [BEFORE α] (for terms α), operating on formulæ to form formulæ:

[F] φ	it will always be the case that φ,
[P] φ	it has always been the case that φ,
⟨F⟩ φ	it will some time be the case that φ,
⟨P⟩ φ	it has some time been the case that φ,
[AFTER α] φ	after α, it will be the case that φ,
[BEFORE α] φ	before α, it was the case that φ,
(UNTIL φ) θ	until it is the case that φ, it will be the case that θ,
(SINCE φ) θ	ever since it was the case that φ, it has been the case that θ.

Here, α is a term (standing for an event) and φ is a formula (standing for a proposition). A fuller rendering of $[\text{AFTER } \alpha]\varphi$ is, "if α is just about to happen, then just after α has been completed it will be the case that φ". We give the truth-conditions for the forward-looking (future-oriented) operators:

$$(h,g) \models [\text{F}]\varphi \text{ iff } \forall g', g''(g \approx g'g'' \Rightarrow (hg', g'') \models \varphi),$$

$$(h,g) \models [\text{AFTER } \alpha]\varphi \text{ iff } \forall g', g'', n, i_0, \ldots, i_{n-1}, a_0, \ldots, a_{n-1}, p_0, \ldots, p_{n-1}$$
$$((g \approx g'g'' \,\&\, g' = (i_0, a_0, p_0) \cdots (i_{n-1}, a_{n-1}, p_{n-1})$$
$$\&\, a_0 \cdot \ldots \cdot a_{n-1} = \|\alpha\|) \Rightarrow (hg', g'') \models \varphi),$$

$$(h,g) \models (\text{UNTIL } \varphi)\,\theta \text{ iff } \forall g', g''(((1)\,\&(2)) \Rightarrow (3)),$$

where (1)–(3) are the following conditions:

(1) $h \approx g'g'' \,\&(hg', g'') \models \varphi$,
(2) $\forall g_0, g_1((g \approx g_0 g_1 \,\&(hg_0, g_1) \models \varphi) \Rightarrow \exists f(g_0 \approx g'f))$,
(3) $\forall g_0, g_1((g \approx g_0 g_1 \,\&\, \exists f(g' \approx g_0 f)) \Rightarrow (hg_0, g_1) \models \theta)$.

The conditions for the backward-looking (past-oriented) operators are analogous. The operators $\langle\text{F}\rangle$ and $\langle\text{P}\rangle$, duals of $[\text{F}]$ and $[\text{P}]$, are readily defined: $\langle\text{F}\rangle\varphi =_{df} \neg[\text{F}]\neg\varphi$ and $\langle\text{P}\rangle\varphi =_{df} \neg[\text{P}]\neg\varphi$. Two further operators are $[\text{H}]$ and $[\text{I}]$ with the truth conditions:

$$(h,g) \models [\text{H}]\varphi \text{ iff } \forall g'(g' \in \text{fut}_H(h) \Rightarrow (h, g') \models \varphi),$$
$$(h,g) \models [\text{I}]\varphi \text{ iff } \forall g'(g' \in \text{int}_H(h) \Rightarrow (h, g') \models \varphi).$$

Four comments. First, dual operators $\langle\text{H}\rangle$ and $\langle\text{I}\rangle$ are again readily definable: $\langle\text{H}\rangle\varphi =_{df} \neg[\text{H}]\neg\varphi$ and $\langle\text{I}\rangle\varphi =_{df} \neg[\text{I}]\neg\varphi$. Second, the reading of $[\text{H}]$ is natural:

$[\text{H}]\varphi$ it is necessary that φ.

However, even though in the following section we shall be able to give a natural interpretation of the operation [I], it seems to have no really natural counterpart in ordinary language; we will return to this question below. Third, the necessity expressed by [H] is not meant to be logical necessity but rather a notion of "historical" necessity (Scott, Chellas) or unavoidability (R. H. Thomason). Fourth, our "after"-operator must not be confused with the "after"-operator of dynamic logic: the nearest counterpart in our theory of Vaughan Pratt's operator $[\alpha]$ is $[\text{H}]\,[\text{AFTER } \alpha]$.

§5. **Term operators.** We introduce the term operators **does**, **reals**, **occs** and **done**, **realled**, **occed**, operating on terms to form formulæ:

does α	the agent is just about to α (to do α, to begin doing α),
reals α	the agent is just about to α (to realize α, to begin realizing α),
occs α	α is just about to occur;
done α	the agent has just α-ed (done α, finished doing α),
realled α	the agent has just α-ed (realized α, finished realizing α),
occed α	α has just occurred.

For the forward-looking operators, the formal truth-conditions with respect to a given action structure are:

$$(h,g) \models \mathbf{does}\,\alpha \text{ iff } \exists g', g'', a, p(g \approx g'g'' \,\&\, g' = (1,a,p) \,\&\, a = \|\alpha\|),$$

$$(h,g) \models \mathbf{reals}\,\alpha \text{ iff } \exists g', g'', a, p(g \approx g'g'' \,\&\, g' = (1,a,p) \,\&\, p \in \|\alpha\|),$$

$$(h,g) \models \mathbf{occs}\,\alpha \text{ iff } \exists g', g'', n, i_0, \ldots, i_{n-1}, a_0, \ldots, a_{n-1}, p_0, \ldots, p_{n-1} \\ (g = g'g'' \,\&\, g' = (i_0, a_0, p_0) \cdots (i_{n-1}, a_{n-1}, p_{n-1}) \,\& \\ p_0 \ldots p_{n-1} \in \|\alpha\|).$$

The truth-conditions for the backward-looking operators — **done**, **realled** and **occed** — are analogous.

The informal readings above are given for heuristic purposes: the "real meaning" of the operators is given by the truth-conditions. In the informal description of actions it is sometimes preferable to use a verb phrase, sometimes a noun phrase[3]. The two formal expressions **does** α and **reals** α were given one common reading ("the agent is just about to α") in order to draw attention to an ambiguity often found in casual descriptions of actions. But there is no claim that informal descriptions of actions in terms of the given readings are automatically translatable into our formal language.

All the operators introduced so far form formulæ that may be said to express *real* propositions — propositions about the world. Our last two operators **int**° and **int** are different:

[3]Compare "The play opens Wednesday" with "The opening of the play is on Wednesday", and "They met in Stockholm" with "Their meeting took place in Stockholm". Pairs such as these may not have exactly the same meaning, in the informal sense of the term, but their truth-conditions seem to be identical.

int° α the agent intends (in a narrow sense) to α,
the agent intends (in a narrow sense) to do α,

int α the agent intends to α,
the agent intends to do α.

Formulæ of type **int**° α or **int** α do not express a real proposition but rather *intentional* or *mental* ones — not about the world but about the agent's mind. Truth-conditions:

$$(h,g) \models \mathbf{int}^\circ\ \alpha \text{ iff } \forall g'(g' \in \mathrm{int}_H(h) \Rightarrow \exists g_0, g_1, g_2, a, p(g' \approx g_0 g_1 g_2\ \& \\ g_1 = (1, a, p)\ \&\ a = \|\alpha\|\ \&\ p \in a)),$$

$$(h,g) \models \mathbf{int}\,\alpha \text{ iff } \forall f, f', g', g''(((1)\ \&(2)) \Rightarrow (3)),$$

where (1)–(3) are as follows:

(1) $f \in \mathrm{cont}_H(h)\ \&\ f \approx g'g''\ \&\ f' \in \mathrm{cont}_H(fg')$,
(2) $\sim\exists g_0, g_1, g_2, a, p(g' \approx g_0 g_1 g_2\ \&\ g_1 = (1, a, p)\ \&\ a = \|\alpha\|\ \&\ p \in a)$,
(3) $\exists g_0, g_1, g_2, f'', a, p(g'f' \approx g_0 g_1 g_2\ \&\ g'f'' \approx g_0 g_1\ \&\ g_1 = (1, a, p)\ \&$
$a = \|\alpha\|\ \&\ p \in a)$.

These truth-conditions may look forbidding, but they are actually quite natural. The former: **int**° α is true with respect to (h, g) if the agent does α in every intended continuation of h. The latter: **int** α is true with respect to (h, g) if, if at the end of every (not necessarily intended) continuation g' of h (if there is one), the agent has not yet performed α, then he will eventually have performed α in every intended continuation of hg'.

Our object language is now complete. Because of the tentative nature of the thoughts put forward in this paper, no effort has been made to try to axiomatize the set of valid formulæ, that is, the set of formulæ true with respect to all articulated histories under all valuations in all action structures. The temporal operators are of course wellknown, and the operators [H] and [I] are normal in the sense of modal logic; with no conditions on action structures the latter are both K45-operators. The validity of the following schemata may be noted:

$$[\text{AFTER } \alpha\, ;\beta]\,\varphi \equiv [\text{AFTER } \alpha]\,[\text{AFTER } \beta]\,\varphi,$$

$$[\text{AFTER } \alpha\, ;;\beta]\,\varphi \equiv [\text{AFTER } \alpha]\,\langle \text{F} \rangle\,[\text{AFTER } \beta]\,\varphi;$$

$$[\text{AFTER } \alpha]\,\mathbf{done}\,\alpha;$$

$$\mathbf{does}(\alpha\, ;\beta) \equiv (\mathbf{does}\,\alpha \wedge [\text{AFTER } \alpha]\,\mathbf{does}\,\beta),$$

$$\mathbf{does}(\alpha\, ;;\beta) \equiv (\mathbf{does}\,\alpha \wedge [\text{AFTER } \alpha]\,\langle \text{F} \rangle\,\mathbf{does}\,\beta),$$

$$\mathbf{does}(\alpha\, ;\beta) \supset \mathbf{does}(\alpha\, ;;\beta);$$

$$\mathbf{reals}(\alpha + \beta) \equiv (\mathbf{reals}\,\alpha \vee \mathbf{reals}\,\beta),$$
$$\mathbf{reals}(\alpha \bullet \beta) \equiv (\mathbf{reals}\,\alpha \wedge \mathbf{reals}\,\beta);$$
$$\mathbf{does}\,\alpha \supset \mathbf{reals}\,\alpha,$$
$$\mathbf{reals}\,\alpha \supset \mathbf{occs}\,\alpha;$$
$$\mathbf{int}^{\circ}\,\alpha \equiv [\textsc{i}]\,\langle\textsc{f}\rangle\,\mathbf{does}\,\alpha.$$

§6. **Philosophical comments.** The naïve theory presented here does not go very far. It is published nevertheless because, in spite of its naïvité, it suggests that the techniques of modal logic can throw a certain light on some topics that have been much discussed by philosophers, some of them for centuries. We end the paper with some remarks in support of this claim.

1. *Action versus intentional action.* Philosophers have long been interested in the question of the relationship between action and intentional action. A simple-minded answer would be that an intentional action is an action that is performed by an agent who intends to perform it. To coin a slogan in the form of an equation,

$$\text{intentional action} = \text{intention} + \text{action}.$$

This slogan is of course not very informative except against a background analysis of intention and action. But with the modelling presented in this paper, there are several ways of making it precise. Consider the following propositions, where α is some verb phrase (or some noun phrase):

(1) The agent intends to α (intends to do α).
(2) The agent αs (or does α).
(3) The agent intentionally αs (or intentionally does α).

Proposition (1) can be rendered by either $\mathbf{int}^{\circ}\,\alpha$ or $\mathbf{int}\,\alpha$, but also by either $[\textsc{i}]\,\langle\textsc{f}\rangle\,\mathbf{does}\,\alpha$ or $(\textsc{until}\ \mathbf{done}\,\alpha)\,[\textsc{i}]\,\langle\textsc{f}\rangle\,\mathbf{done}\,\alpha$ — notice that the schemata

$$\mathbf{int}^{\circ}\,\alpha \equiv [\textsc{i}]\,\langle\textsc{f}\rangle\,\mathbf{done}\,\alpha,$$
$$\mathbf{int}\,\alpha \equiv (\textsc{until}\ \mathbf{done}\,\alpha)\,[\textsc{i}]\,\langle\textsc{f}\rangle\,\mathbf{done}\,\alpha$$

are generally valid. Evidently, there are two different readings here; call the former narrow, the latter wide. The wide reading, which seems to agree with the intuitions of Bratman [2], is clearly the more interesting one. (Readers will now be able to appreciate the difficulty, mentioned at the end of Section 4, of finding a colloquial reading of the operator [I]: there seems to be no direct counterpart in natural language. To render $[\textsc{i}]\,\varphi$ as "The agent intends that φ" would in effect be to adopt the narrow, less interesting reading.)

The formalization of proposition (2) is of course not trivial; even in our schematic formalism there are several possibilities. In this connexion it is well to caution that the heuristic readings given above are approximations at best and sometimes, if taken literally, misleading. Thus it might occasionally be acceptable to cast (2) as **does** α, but usually something more is required. Think of a routine or program as associated with the agent's action: he performs the action by running this routine or program, thereby setting off various events of which some are intended, others not, some foreseen, others not. A serious difficulty here is to say more precisely what can be meant by the phrase "setting off an event": do we have in mind logical necessity, or causal necessity, or something weaker? Without solving this difficulty, we note the following nontrivial, possible rendering of (2):

$$\exists\xi(\textbf{does}\ \xi \wedge [\text{H}]\,[\text{AFTER}\ \xi]\,\langle\text{F}\rangle\ \textbf{realled}\ \alpha),$$

where $\exists$ is a quantifier over event types. (Our present theory does not provide for such quantification, but it would not be difficult to define an extension that does.)

With this analysis of doing and intending, our slogan above yields several possibilities for rendering (3), one of which is

$$\textbf{int}\ \alpha \wedge \exists\xi(\textbf{does}\ \xi \wedge [\text{I}]\ \textbf{does}\ \xi \wedge [\text{H}]\,[\text{AFTER}\ \xi]\,\langle\text{F}\rangle\ \textbf{realled}\ \alpha).$$

This kind of reading is (perhaps) reminiscent of Davidson's dictum that an action is intentional if it is intentional under some description [4].

Needless to say, the preceding remarks are not meant as a final analysis of these matters; for one thing, we have not discussed causality, and our modelling lacks concepts such as beliefs and probability. The purpose of the discussion in this subsection is merely to suggest that the present kind of formalism may usefully contribute to the analysis of action.

2. *Actions versus events.* The relationship between actions and events has been the subject of considerable controversy. According to one influential view, the class of actions forms a subclass of the class of events: an action is an event with a special causal history (Davidson [4]). According to another influential view, actions are not events, even though for every action there is a corresponding event (von Wright [9]). Our theory agrees with the latter view at least to the extent that, for each individual action $(1, a, p)$, there is the corresponding individual event (a, p).

It should be noted that what Davidson and von Wright discuss is individual human action, while the theory presented here is meant to apply to all kinds of action. Furthermore, the important topic of causality, fundamental to the theories of Davidson and von Wright, has not been touched upon in this paper.

3. *Weakness of will.* Whether "weakness of the will" is possible has been much debated by philosophers. For better or worse, in our theory it is: the sets

$$\{[\text{I}]\ \textbf{does}\ \varrho, \neg\ \textbf{does}\ \varrho\},$$
$$\{\textbf{int}\ \varrho, [\text{F}]\neg\ \textbf{does}\ \varrho\},$$

where ϱ is an event letter, are both consistent (since they are satisfiable in our modelling). That is to say, the agent may intend to do something now and not do it, and he may intend to do something (at some time) and never do it. Someone not wishing to accept this will have to formulate conditions outlawing the models in which these sets are satisfiable.

4. *Unintended consequences of actions.* Suppose that, in every history in a set H of histories, any realization of an event a is invariably followed by a realization of b. Is it possible to intend to bring about a and yet not intend to bring about b? The answer, with respect to the formalism given above, is negative, as witnessed by the validity of the schema

$$(\dagger) \qquad [\textsc{h}]\,[\textsc{f}](\mathbf{occed}\,\alpha \supset [\textsc{after}\ \alpha]\,\langle\textsc{f}\rangle\,\mathbf{occed}\,\beta) \supset (\mathbf{int}\,\alpha \supset \mathbf{int}\,\beta).$$

Moral philosophers will regard this as a defect of our formalism. An easy way to repair it — not entirely *ad hoc* — would be to introduce yet another primitive concept, which is of importance in its own right. Informally, we have understood the set H in an action structure (U, E, H, I) as the set of "historically possible" histories that are relevant (for a certain unstated purpose or from a certain unstated perspective). Let us now introduce the new primitive K which is to be a set of histories that includes H and which, informally, is thought of as the set of all "logically possible" histories. In general, H is a proper subset of K. We also introduce a new modal operator $\Box$ with the reading "it is logically necessary that" and with the truth-condition

$$(h, g) \models \Box\varphi \text{ iff } \forall g'(g' \in \mathrm{fut}_K(h) \Rightarrow (h, g') \models \varphi).$$

The intention function I is redefined by the condition $\mathrm{int}_K(h) = I(\mathrm{fut}_K(h))$, and the truth-conditions for [I] and **int** are modified in the obvious manner, namely, by replacing the subscript H by K in the defining conditions.

Schema ($\dagger$) is no longer valid in the new setting where validity is revised to be with respect to all action structures (U, E, K, H, I). That is to say, even if an occurrence of a is followed by an occurrence of b in all historically possible futures of h (that is, elements of $\mathrm{fut}_H(h)$), it is still possible to intend a without intending b, provided that at least one logically possible future of h (that is, element of $\mathrm{fut}_K(h)$) contains an occurrence of a without a subsequent occurrence of b. Notice that the schema

$$(\ddagger) \qquad \Box\,[\textsc{f}](\mathbf{occs}\,\alpha \supset [\textsc{after}\ \alpha]\,\langle\textsc{f}\rangle\,\mathbf{occs}\,\beta) \supset (\mathbf{int}\,\alpha \supset \mathbf{int}\,\beta)$$

is valid in the revised sense of validity. Thus even in the modified formalism it is not possible to intend a without intending b if b's occurring is a logical consequence of a's occurring. Perhaps philosophers will find this acceptable.

5. *Mental acts.* The possibility of mental acts (or actions, in our jargon) has been disputed by some philosophers. But it seems harmless to think

that there could be actions that alter the agent's intentions without altering anything in the world. Agents change their mind from time to time: they form new intentions just as they give up old ones. Of course, there is a sense in which changes of mind happen even in our modelling. For example, if it is false (with respect to a certain articulated history (h, g)) that **int** α and true that [AFTER γ] **int** α, then α will be added to the agent's stock of intentions as soon as γ has occurred (if it occurs). But all that this means is that the agent already now has a certain conditional intention, namely, to do α if γ occurs: if history moves in such a way that γ occurs, then the intention to do α is triggered without any effort on the part of the agent. In a general theory, the agent should be able to change his mind "on the spot", as it were.

One interesting example of an action whose analysis requires mental actions is disjunctive action. We have already met with the regular sum operator, which is interpreted as set-theoretical union. As a logical "sum" of actions, it has limited application: there is a need also for another kind of disjunction, which we might term "intentional sum" and which cannot be interpreted in our modelling. The need for such a concept is felt in cases where one tries to analyse the position of an agent who has decided to do a-or-b with a view either to do a or to do b but has yet to decide which. In other words, the agent now has the intention to do a-or-b, and to carry it out he must choose one of a and b over the other, and then carry out the chosen action. It is natural to regard "choosing a over b" as a mental action.

One way to go might be to introduce intentional states along with world states: we might try to develop a theory in which a fundamental rôle is played by total states (s, u), where s is an intentional state and u is a world state. Action and event types would be sets of finite paths of total states. Purely real actions would be those that leave the intentional state unchanged, purely mental actions those that leave the world state unchanged; in addition, there would be "mixed" actions. But there is a difficulty here! Intentional states should reflect the intentions held by the agent. Moreover, it is tempting to identify intentions with intentions to do an action. And actions are defined in terms of total states, which include intentional states. So here is a circle. The trick would be to find a way to work out this sketch without the circle turning vicious. It has not been attempted in this paper and has to await another occasion.

§7. Historical remarks. The action logic presented here is inspired by Georg Henrik von Wright's work, especially [9] and [10]; cf. [8]. Technically it is in the Prior/Scott/Chellas tradition but augmented by Vaughan Pratt's ideas about dynamic logic. For a somewhet different, yet related, approach, see the work of Nuel Belnap and his collaborators [1, 5]. Our representation of intention is influenced by Michael Bratman's analysis [2]. For another modelling of intention that goes far beyond that of the present paper, see the work of Philip Cohen and Hector Levesque, for example, [3][4].

REFERENCES

[1] Nuel Belnap, Michael Perloff, and Ming Xu, ***Facing the future: agents and choices in our indeterministic world***, Oxford University Press, New York, 2001.

[2] Michael E. Bratman, ***Intention, plans, and practical reason***, Harvard University Press, Cambridge, MA, 1987.

[3] Philip R. Cohen and Hector J. Levesque, *Intention is choice with commitment*, ***Artificial Intelligence***, vol. 42 (1990), pp. 213–261.

[4] Donald Davidson, ***Essays on actions and events***, Clarendon Press, Oxford, 1980.

[5] John F. Horty, ***Agency and deontic logic***, Clarendon Press, Oxford, 2001.

[6] E. J. Lemmon, ***An introduction to modal logic***, American Philosophical Quarterly monograph series, vol. 11, Oxford, 1977, Written in collaboration with Dana Scott and published posthumously.

[7] Krister Segerberg, *Getting started: beginnings in the logic of action*, ***Studia Logica***, vol. 51 (1992), pp. 347–378.

[8] ———, *Results, consequences, intentions*, ***Actions, norms, values: discussion with Georg Henrik von Wright*** (Georg Meggle, editor), Walter de Gruyter, Berlin, New York, 1999, pp. 147–157.

[9] ———, *Outline of a logic of action*, ***Advances in modal logic. vol. 3*** (Frank Wolter et al., editors), vol. 3, World Scientific Publishing Co., Singapore, 2002, pp. 365–387.

[10] Georg Henrik von Wright, ***Norm and action: a logical enquiry***, Routledge & Kegan Paul, London, 1963.

[11] ———, ***The varieties of goodness***, Routledge & Kegan Paul, London, 1963.

FILOSOFISKA INSTITUTIONEN
UPPSALA UNIVERSITET
DROTTNINGGATAN 4
753 10 UPPSALA, SWEDEN
E-mail: Krister.Segerberg@filosofi.uu.se

[4]The paper was finished while the author was a Fellow during 2001-2 at S.C.A.S.S.S. (The Swedish Collegium for Advanced Study in the Social Sciences). He is indebted to an anonymous referee for a number of corrections and good suggestions.

MODALITY, MOOD, AND DESCRIPTIONS

KAI FREDERICK WEHMEIER

Abstract. In this paper, I argue that Kripke's modal argument against description theories of proper names is fallacious. More precisely, it is shown that the argument crucially depends on a failure to distinguish between the subjunctive and indicative moods of English. I propose a formal system in which such a distinction is implemented in a straightforward way, and show that Kripke's argument is invalid when formalized in this setting. The formal language introduced is compared to modal predicate logic with and without an actuality operator and to a standard first-order language with explicit quantification over possible worlds.

§1. Introduction. By means of what semantic features is a proper name tied to its bearer? This is a puzzling question indeed: proper names — like "Aristotle" or "Paris" — are syntactically simple, and it therefore does not seem possible to reduce their meanings, by means of a principle of compositionality, to the meanings of more basic, and hence perhaps more tractable, linguistic elements.

It used to be widely held that the syntactical simplicity of proper names is misleading, that names are just abbreviations, in some sense or other, for certain other expressions, whose referential mechanism is clearer. The most prominent such theory, sometimes attributed to Frege and Russell, holds that the expressions for which proper names go proxy are definite descriptions, like "the man who taught Alexander the Great" or "the capital of France". While not without problems of its own, this description theory of names has a great intuitive appeal, and yields a beautifully unified account of the phenomenon of singular reference.

Thirty years ago, however, an influential argument against the description theory was put forward by Saul Kripke. Appealing to the referential behaviour of names and descriptions, respectively, in contexts of metaphysical possibility and necessity, it has come to be known as the "modal argument". According to Kripke, names are "rigid designators", referring to the same individual no matter what actual or counterfactual situation we may be speaking about, whereas definite descriptions usually designate, if at all, non-rigidly, shifting their referents from possible world to possible world. The argument runs roughly as follows:

Intensionality
Edited by Reinhard Kahle
Lecture Notes in Logic, 22

Consider the proper name "Aristotle" and some candidate for a description it might abbreviate — "the teacher of Alexander", say. If the name abbreviates the description, the two expressions should be synonymous, and hence one should be substitutable for the other in all (but quotational) contexts *salva veritate*. Now

(1) Under certain counterfactual circumstances, Aristotle would not have taught Alexander

is true: After all, Philip II of Macedonia might have decided to educate his son all by himself, in which case Philip, but not Aristotle, would have taught Alexander. However, the result of replacing "Aristotle" by "the teacher of Alexander" in (1),

(2) Under certain counterfactual circumstances, the teacher of Alexander would not have taught Alexander,

is false[1]. The reason for this divergence in truth value is, according to Kripke, that "Aristotle" rigidly designates Aristotle even in discourse about counterfactual situations in which he would not have done some of the things he did do, whereas "the teacher of Alexander" refers, with respect to any counterfactual situation, to the individual (if any) who would, in that situation, have taught Alexander. In any case, the assumption of synonymy is refuted by the difference in truth-value between (1) and (2). And the example can easily be generalised: most properties we might use to identify an individual descriptively are possessed by that individual only contingently, so that an argument analogous to the one given above can be constructed.

In this paper, I shall argue that the modal argument is fallacious. The observation on which my criticism is based is this: If we read "the teacher of Alexander" in (2) as "the man who would have taught Alexander", the sentence is indeed false. But if we read it as "the man who taught Alexander" instead, it is true, so that the latter description appears to be substitutable for "Aristotle" *salva veritate*. The modal argument thus seems to depend on an ambiguity concerning verb mood.[2]

Concerning any such claims of ambiguity, Kripke (1980, pp. 59–60, footnote 22) has remarked:

> Some philosophers have thought that descriptions, in English, are ambiguous, that sometimes they non-rigidly designate, in each

---

[1] Or rather, as Kripke points out, it has a reading which makes it false, namely the one according the description narrow scope with respect to the modal operator: when "x teaches Alexander" is symbolised as "$\mathsf{T}x$" and (1) is rendered as "$\Diamond\neg\mathsf{T}a$", the reading under which (2) is false is "$\Diamond[(\iota x)(\mathsf{T}x)]\neg\mathsf{T}(\iota x)(\mathsf{T}x)$", which, according to Russell's theory of descriptions, expands to "$\Diamond(\exists! x\ \mathsf{T}x \wedge \exists x\ (\mathsf{T}x \wedge \neg\mathsf{T}x))$".

[2] For a critical discussion of two other popular strategies against the modal argument, viz. the so-called "wide scope theory" and the "actualised descriptions" approach, see Soames (1998).

> world, the object (if any) satisfying the description, while sometimes they *rigidly* designate the object actually satisfying the description. (...) I find any such alleged ambiguities dubious. I know of no clear evidence for them which cannot be handled either by Russell's notion of scope or by the considerations alluded to in footnote 3, p. 25.

The main task in justifying my objection to the modal argument is, then, to show that the distinction of moods cannot be explained away by appeal to Russell's notion of scope for definite descriptions.[3] To this end, it will be necessary to examine in some detail the main tool of logical analysis for ordinary modal discourse, viz., standard modal predicate logic. The results of this investigation seem to be relevant beyond the question of the modal argument's soundness, for they indicate that a modal logic adequately reflecting ordinary discourse should be extensional.

The paper is organised as follows: I explain my objection in more detail in §2. The (hypothetical) Kripkean reply, relying on Russell's notion of scope, is outlined in §3: according to it, verb mood simply indicates whether the respective occurrences of a predicate lie within or without the scope of a modal operator. This reply is found to be insufficient in §4, where it is shown that there are innocuous pieces of ordinary modal discourse whose formalisation in modal logic is precluded by such an explanation of mood. After discussing, in §6, another objection (outlined in §5) based on the consideration of a special kind of modal context, I develop, in §7, a formal system S5* of modal predicate logic that allows for the distinction of moods, and is thereby capable of formalising the aforementioned statements inexpressible in ordinary quantified S5. Within this framework, the informal objection to the modal argument presented in §2 is analysed formally in §8. Some remarkable properties of S5*, like extensionality and absence of non-rigidity, are discussed in §9. §10 explores the relation of S5* and ordinary S5 (with and without actuality operator) to the two-sorted predicate calculus with explicit quantification over possible worlds. Finally, I discuss some consequences of my analysis for the philosophy of modality and the philosophy of language in §11.

§2. Descriptions in the mood. Let us try to repeat the argument of §1, this time by using "the man who taught Alexander" instead of the more elliptical phrase "the teacher of Alexander". Sentence (2) now corresponds to

(3) Under certain counterfactual circumstances, the man who would have taught Alexander would not have taught Alexander.

---

[3] Kripke's footnote 3 refers to the distinction between speaker's reference and semantic reference, as used in his discussion of Donnellan-type ambiguities between referential and attributive uses of descriptions; it does not appear to be relevant in the present case.

Clearly, (3), like (2), is false, and hence the non-synonymy of "Aristotle" and "the man who taught Alexander" follows, just as before.

There is a problem, however: The description "the man who taught Alexander" does not occur in (3). What we have in fact shown is that "the man who *would have taught* Alexander" is not synonymous with "Aristotle". Let us try again:

(4) Under certain counterfactual circumstances, the man who taught Alexander would not have taught Alexander.

This time, we got the substitution right. But — sentence (4) is not false. Clearly, it might have been the case that the man who *taught* Alexander (viz., Aristotle) *would not have taught* him.

The situation seems to be this: If we substitute the expression whose non-synonymy with the name we are trying to establish — the indicative description "the man who taught Alexander" —, no change in truth-value is forthcoming. Such a truth-value change occurs only if we use the corresponding subjunctive[4] expression "the man who would have taught Alexander" instead — but no-one would claim that "Aristotle" should be synonymous with it in the first place.[5]

---

[4]Strictly speaking, "would have taught" is actually not a subjunctive, but a "conditional mood" form. A word of explanation for the apparently idiosyncratic use of the term "subjunctive" throughout this paper is thus in order. Before I justify the terminology with respect to the English language, let me mention that, unlike English and most modern Romance languages, both Latin and German employ the subjunctive (more precisely, the *coniunctivus irrealis* and the *Konjunktiv* II, respectively) in both the protasis and the apodosis of counterfactual conditionals (witness "Si *tacuisses*, philosophus *mansisses*" and "Wenn du *geschwiegen hättest*, *wärest* du ein Philosoph *geblieben*", respectively), and hence also in modal constructions such as those used above. With regard to these languages, my terminology would therefore be perfectly adequate. As far as I am aware, English, French and the other Romance languages also used to work this way, and a separate "conditional mood" verb form only occurred in later stages of their development. Even in contemporary English, where the subjunctive proper seems to be almost extinct, remnants of the earlier practice can still be observed, as e.g., in the phrase "Were that true, there were no more to say" (Fowler 1983, 596; Fowler notes that the conditional form "would be" is the modern equivalent for the subjunctive "were" of the apodosis). From a systematic, semantic point of view it seems clear that the conditional and the subjunctive are to be grouped together, for they both refer not to how things stand in the real world, but to counterfactual, hypothetical situations. It does, therefore, not seem too much of an injustice to official grammatical classifications to apply the term "subjunctive" also to these conditional forms, especially in the light of Fowler's (quoting the Oxford English dictionary) explication that *subjunctive* "is taken to mean the use of a verb-form different from that of the indicative mood in order to 'denote an action or a state of affairs as conceived (and not as a fact), and [expressing] ... a contingent, hypothetical, or prospective event'" (Fowler 1983, 595; the parentheses and square brackets are Fowler's, the ellipsis is mine).

[5]The subjunctive phrase is, after all, not even a designator. For who is the man who would have taught Alexander? The question does not make sense, unless a counterfactual situation has been specified with respect to which the subjunctive is to be interpreted. That is, the subjunctive phrase is at best an incomplete singular term, much like "his father" when no value has been assigned to the variable "he". Just as one would naturally ask, in reaction to an abrupt utterance

Why did the original argument go through so smoothly? Only because "the teacher of Alexander" is homonymous between the indicative "the man who taught Alexander" — this *is* a designator and hence a plausible synonymy candidate for "Aristotle" — and the subjunctive "the man who would have taught Alexander", as which it must be read in order to make the result of the substitution come out false. In other words: the modal argument as presented in §1 rests on an equivocation. Any defence of the modal argument will have to explain the indicative-subjunctive distinction away.

§3. Objection I: Verb mood and modal scope. Is the analysis of the preceding section not much too simplistic? Does it not rest on a quite elementary logical error? After all, the Kripkean can tell the following story in defence against the objection from grammatical mood:

"Consider modal predicate logic. Sentence (1), for example, can there be expressed as

(5) $\Diamond\neg \mathsf{T}a$.

And the sentence

(6) Aristotle taught Alexander

is formalised as

(7) $\mathsf{T}a$.

Note how both "would have taught Alexander" and "taught Alexander" are expressed by the same predicate symbol "T". What distinguishes the occurrence of "T" in (5) from that in (7) is that the former, but not the latter, lies within the scope of a modal operator. In terms of Kripke's semantics for modal logic, (5) holds true at the actual world if and only if there is a possible world v such that "$\mathsf{T}a$" is not true at v. This means that the occurrence of "T" in (5) refers not to how things are, but to how things might have been — which is the function of the subjunctive mood. On the other hand, the occurrence of "T" in (7) obviously refers to how things stand in the actual world, and that is what the indicative mood is for.

What these formal considerations show is that, for a predicate, being subjunctive simply means being within the scope of some modal operator, whereas being indicative means not lying within such a scope. There is thus no need to view indicative and subjunctive versions of a predicate as syntactically distinct — the only distinction needed is that between an occurrence of a predicate lying within or without the scope of a modal operator. The reason why natural

---

of "His father is rich": "Whose father?", one would reply to "The man who would have taught Alexander was rich" by asking "The man who would have taught Alexander *if what*?"

language exhibits the redundancy of having, in addition to modal operators (with associated scopes), also syntactically distinct expressions for indicative and subjunctive predicates, is that it is not as precisely regimented as the formal idiom of modal logic: sometimes it would just be difficult to find out what the scope of a natural language modal operator is. Consider, for instance, the sentence

(8) Under certain counterfactual circumstances, some people who are unhappy would have been happy.

Were there no distinction in terms of mood between the predicates, we would have no choice but to parse the sentence according to its sequential structure, reading it as something like

(9) $\Diamond\exists x\,(\neg\mathsf{H}x \wedge \mathsf{H}x)$,

which, being logically false, is certainly not what (8) expresses. But the indicative "is unhappy" in (8) makes it clear that this predicate occurrence is not within the scope of the modal operator. And the subjunctive mood of "would have been happy" conversely indicates that this predicate occurrence does lie within the scope of the possibility operator. Hence, we must render (8) as

(10) $\exists x\,(\neg\mathsf{H}x \wedge \Diamond\mathsf{H}x)$.

The grammatical moods are thus nothing but parsing aids for the translation of ordinary modal discourse into the canonical notation of modal predicate logic.

Accordingly, the analysis of §2 misses the mark: By Russell's theory of descriptions, sentence (4) is short for

(11) Under certain counterfactual circumstances, precisely one man taught Alexander, and some man taught Alexander and would not have taught Alexander.

The grammatical moods of the predicates enable us to see that the true logical structure of (11) is

(12) Precisely one man taught Alexander, and some man taught Alexander but would, under certain counterfactual circumstances, not have taught Alexander,

or, formally,

(13) $\exists!x\,\mathsf{T}x \wedge \exists x\,(\mathsf{T}x \wedge \Diamond\neg\mathsf{T}x)$.

This is because we have two occurrences of the indicative "taught" in (11), which must therefore lie outside the scope of the modal operator, and one occurrence of the subjunctive "would have taught", which accordingly must lie within the scope of the possibility operator.

Now (13) is just longhand for

(14) $[(\iota x)(\mathsf{T}x)]\Diamond\neg\mathsf{T}(\iota x)(\mathsf{T}x)$,

that is, for the result of substituting, in (5), the description "$(\iota x)(\mathsf{T}x)$" *with wide scope* over the modal operator. But Kripke (1980, p. 13) acknowledges that his argument will not work if the description is given wide scope over the possibility operator, and points out that it is the reading with narrow scope which is relevant for his purposes. That is, Kripke is concerned with the sentence

(15) $\Diamond[(\iota x)(\mathsf{T}x)]\neg\mathsf{T}(\iota x)(\mathsf{T}x)$,

which expands to

(16) $\Diamond(\exists!x\ \mathsf{T}x \wedge \exists x\ (\mathsf{T}x \wedge \neg\mathsf{T}x))$.

The translation of this sentence into the vernacular, however, reads

(17) Under certain counterfactual circumstances, precisely one man would have taught Alexander, and some man would have taught Alexander and would not have taught Alexander,

which, given Russell's theory of descriptions, is just (3) ("under certain counterfactual circumstances, the man who would have taught Alexander would not have taught Alexander"). But, just as Kripke claims, (17), and hence (3), is indeed false. The logical analysis given here shows that no equivocation is involved at all: (15) arises from (5) through the substitution of the description "$(\iota x)(\mathsf{T}x)$" for "a" with narrow scope, just as (14) is the result of substituting *the very same description* for "a" with wide scope. "The man who taught Alexander" and "the man who would have taught Alexander" are of identical logical form — the second is simply the form which the first assumes, in surface grammar, when inserted into the scope of a modal operator. The objection put forward in §2 is therefore without any force."

This appears to be a powerful argument against the objection from grammatical mood. Nevertheless, it is itself fallacious, as I shall now attempt to show.

§4. Objection I rebutted. It is a tacit presupposition of objection I that standard modal predicate logic is an adequate tool for the logical analysis of the kind of modal discourse that we are concerned with. This presupposition is, however, not fulfilled.

Take the following piece of modal talk from Kripke's *Naming and Necessity*:

> Consider a counterfactual situation in which, let us say, fool's gold or iron pyrites was actually found in various mountains in the United States, or in areas of South Africa and the Soviet Union. Suppose that all the areas which actually contain gold now, contained pyrites instead (...) (Kripke 1980, p. 124).[6]

In short, what Kripke is saying here might be expressed thus:

(18) It might have been the case that every area which contains gold would have contained iron pyrites.

To take another example:

(19) Under certain counterfactual circumstances, everyone who has flown to the moon would not have flown to the moon.

These are perfectly innocent, intelligible, and arguably true statements. The careful distinctions of mood do not arise from mere pedantry: While (19) is probably true (imagine a course of history under which everyone who in fact became an astronaut would have become a race driver instead), its relative

(20) Under certain counterfactual circumstances, everyone who has flown to the moon would have been female

is likely to be false, since there do not seem to be possible circumstances under which, say, Neil Armstrong would have been a woman. Nevertheless, the sentence

(21) Under certain counterfactual circumstances, everyone who would have flown to the moon would have been female

is probably true, and the difference in truth value between (20) and (21) can only be due to the differences in mood. Granted, we often use additional devices to indicate whether a predicate is used *realiter*, that is, with reference to the actual world, or *irrealiter* (with respect to some counterfactual situation). For instance, (19) might be expressed by the more verbose

(22) Under certain counterfactual circumstances, everyone who has in fact flown to the moon would not have flown to the moon.

But the crucial semantic feature is still the distinction of moods: For in

(23) Under certain counterfactual circumstances, everyone who would in fact have flown to the moon would not have flown to the moon,

[6] Interestingly, the adverb "actually", on its first occurrence in the quotation, does not force evaluation of the relevant subsentence with respect to the actual world. The same is true for: "Under certain circumstances, no-one would believe in aliens, although there would actually be aliens." David Lewis has spotted an ambiguity of "actually" here (1970, section IX, 1983, section B), but, as I argue below, there is a contextual feature determining which world will be invoked by this adverb: viz., the mood of the predicate to which it is attached.

the expression "in fact" does not have the power to make the subjunctive predicate "would have flown to the moon" refer to the actual world. The addition of such modifiers may thus sometimes *facilitate* the proper interpretation of modal statements, but the orientation towards either actual or counterfactual circumstances is effected, first and foremost, by the indicative and subjunctive moods.

So let us try to formalise (19) within modal logic. The predicate "has flown to the moon" being indicative, it should, according to the analysis suggested in the preceding section, lie outside the scope of the possibility operator. If that is so, then we have no choice but to place the universal quantifier "everyone" outside the modal scope, too.[7] This leads to the following tentative paraphrase of (19):

(24) $\forall x\,(\mathsf{F}x \rightarrow \Diamond\neg\mathsf{F}x)$.

But this is not what (19) says. In terms of possible worlds, (19) requires there to be one world in which everyone who has in fact flown to the moon would not have done so — an $\exists\forall$-combination. (24), on the other hand, says that, concerning each individual who has flown to the moon separately, there is a world, possibly depending on the individual at issue, in which that individual would not have flown to the moon — a $\forall\exists$-combination.[8]

The trouble with (19) is that the predicate "has flown to the moon" is *both* indicative *and* within the scope of the modal operator. This shows immediately that something is amiss with the Kripkean defence: the subjunctive's function is not *just* to indicate that the respective predicate lies within the scope of a modal operator, but rather that it is also "modally bound" to it, i.e., that it is semantically relativised to the possible worlds conjured up by the operator. Standard modal logic, however, automatically relativises *all* predicates in the scope of a modal operator to such possible worlds, due to the stipulation that, for a given world w and formula ϕ,

$$w \models \Diamond\phi \text{ if and only if for some world } v,\ v \models \phi,$$

which is the reason for the inexpressibility of (19) in such a framework.

This problem is highly germane to our discussion, because it suggests that a formal system adequately capturing ordinary modal discourse would have to distinguish indicative and subjunctive predicates *typographically*, whereas standard modal logic only distinguishes indicative and subjunctive *occurrences* of a predicate symbol by means of the syntactical criterion of modal scope. A

[7] In the framework of Hintikka's IF-logic, there may be other choices. There, one could perhaps argue that "has flown to the moon" lies within the priority scope, but not within the binding scope, of the modal operator. See, for instance, Hintikka (1997).

[8] Indeed, no sentence of quantified S5 has the same truth conditions as (19). This was conjectured by Hazen (1976) and later proved by Hodes (1984c). For a simple direct proof, see Wehmeier (2003).

suitably revised modal logic, such as the system S5* of §7, will therefore admit *indicative* descriptions having *narrow* scope with respect to modal operators. In any case, it seems clear that the Kripkean explanation of the indicative-subjunctive distinction is not successful as it stands.

§5. Objection II: Indicative predicates shifting extensions. "Granted," the Kripkean might say, "in the formulation you have given it, the modal argument fails to deliver a contradiction. But we can easily change the formulation and produce a valid argument. Take

(25) Possibly, Aristotle did not teach Alexander,

or

(26) It is not necessary that Aristotle taught Alexander.

In contradistinction to what you have claimed about the roles of the grammatical moods, we here have indicative predicates shifting their extensions at other possible worlds. If we now perform the substitution you suggest, we obtain

(27) Possibly, the man who taught Alexander did not teach Alexander,

and

(28) It is not necessary that the man who taught Alexander taught Alexander.

If the modal operators "possibly" and "it is necessary that" make the indicative predicates shift their extensions in (25) and (26), then (27) and (28) are both false when the description is given narrow scope with respect to the modal operators. And there the argument is, again."

§6. Objection II rebutted. Before discussing the objection itself, let me point out how far-reaching our observation concerning the logical significance of the indicative-subjunctive distinction is. It is, for example, highly germane to the analysis of counterfactual conditionals. Thus, when we say

(29) If everyone who is rich had been poor, then someone who is poor would have been rich,

we again cannot explain the semantics of the indicative predicates by claiming that they are not within the scope of the conditional. Perhaps it is even clearer in the case of counterfactuals than in the case of the modalities that it is the interplay between the conditional operator and the subjunctive that effects extension-shifting to other possible worlds.

Consider now the following counterfactual:

(30) If not Oswald, but someone else had shot Kennedy, then Ruby would have killed the man who would have shot Kennedy, not the man who (in fact) shot Kennedy.

Since, as (29) shows, indicative predicates can occur within the scope of the conditional without shifting their extensions, we may give the indicative description narrow scope, without affecting the truth value of (30). It is then also clear that "the man who shot Kennedy" may be replaced *salva veritate* by the name "Oswald".

Counterfactuals, however, seem to be the basis for our intuitions concerning metaphysical possibility and necessity. Thus, we infer

(31) Under certain circumstances Nixon would have gotten Carswell through (Kripke 1971, p. 175)

from

(32) If Nixon had bribed a certain Senator, he would have gotten Carswell through (Kripke 1971, p. 176).

Thus, we naturally arrive at a modal statement by abstracting from, or perhaps quantifying away, the protasis of a counterfactual conditional. This close connection between counterfactuals and modal statements is severed by formulations using "possibly", "it is necessary that" and their cognates, with indicative predicates intended to shift extensions. Thus, even if objection II could be sustained, it would pertain to a quite singular phenomenon of limited evidential value.

But in fact, objection II can hardly be sustained. The main reason is that "possible" and "necessary", in ordinary language (and even within philosophical jargon), have so many meanings that the semantics assumed by modal logicians for sentences like (25) and (26) must be regarded as stipulated (that is, stipulated to fit standard possible worlds semantics) rather than modelled upon natural language.

To elaborate a bit: Clearly, the most natural reading for (25) is

(33) For all we know, it may well be that Aristotle did not teach Alexander,

taking the possibility to be epistemic in nature. In the case of necessity, the situation is even worse: In addition to Kripke's list of philosophical meanings of "necessary" ("a priori", "analytic", "certain"), it can mean "obligatory" ("it is necessary that I pick up my daughter from school"), or "urgently desirable" ("it is necessary that you be here in time").

Were the meanings of "possibly" and "necessarily" sufficiently clear, there would be no need to resort to subjunctive formulations in order to explain their meanings. But, for instance, Forbes (1985, p. 2), deems it necessary to explain what he means by "it is possible that P":

> As a rough elucidatory guide, "it is possible that *P*" in the broadly logical sense means that there are ways things might have gone, no matter how improbable they may be, as a result of which it would have come about that *P*.

Incidentally, I should like to point out that, in this natural language example, it is quite clearly inadmissible to substitute one and the same sentence for both occurrences of the predicate letter "*P*": On its first occurrence, it must be replaced with an indicative clause, but on the second occurrence, with a subjunctive one. For clearly, the correct formulation is

(34) There are ways things might have gone, no matter how improbable they may be, as a result of which it would have come about that Jones *would have been* rich,

not

(35) There are ways things might have gone, no matter how improbable they may be, as a result of which it would have come about that Jones *is* rich.

The artificiality of formulations of type (25) and (26) emerges even more clearly when the expressiveness problem discussed in Section 4 is taken into account. For how should one express

(36) Under certain circumstances, everyone who has flown to the moon would not have flown to the moon,

when only "possibly", "necessarily", and indicative predicates are available? It has been suggested that the adverb "actually" could help; but how natural (and comprehensible) are the sentences

(37) Possibly, everyone who has actually flown to the moon has not flown to the moon

and

(38) Necessarily, some of the things that are actually red are shiny?

It seems to me that we do not use such sentences. Instead, we use phrases like (36) or

(39) No matter how things might have gone, some of the things that *are* red *would have been* shiny.

I maintain, therefore, that objection II is based on contrived philosophers' jargon whose semantics has been stipulated to fit standard possible worlds semantics, whereas our initial formulation is well attested in ordinary language.

§7. A modal logic of indicativity and subjunctivity. The considerations of Section 4 suggest that an adequate logical analysis of the modal argument requires a revised system of modal predicate logic. In particular, it must be possible to express (19) in such a system. Given that this sentence is of $\exists\forall$-form, a translation into logical notation should start with "$\Diamond\forall x$". What comes next? Something similar to "$\mathsf{F}x \rightarrow \neg\mathsf{F}x$", just that, as was argued in Section 4, the two occurrences of the predicate "F" must be typographically distinguished so as to mark one as indicative, and the other as subjunctive.

An obvious constraint on any language of modal predicate logic is that its non-modal part should simply be the language of ordinary predicate logic. Therefore, indicative predicates are to be expressed by the ordinary predicate symbols of non-modal predicate logic (which, after all, formalises ordinary, indicative discourse), and it is the subjunctive for which we need to introduce a new notation. This constraint rules out the standard solution to the expressiveness problem, viz. the introduction of an "actually" operator "$\boldsymbol{A}$".[9] For consider the following example:

(40) Someone has flown to the moon, but under certain counterfactual circumstances, everyone who has flown to the moon would not have flown to the moon.

With an actuality operator, it would have to be formalised as[10]

(41) $\exists x\ \mathsf{F}x \wedge \Diamond\forall x\ \big(\boldsymbol{A}(\mathsf{F}x) \rightarrow \neg\mathsf{F}x\big)$,

but this is clearly not a faithful representation of the ordinary language sentence (40): In (40), there are two occurrences of the indicative predicate "has flown to the moon", and on both of these occurrences, the predicate has exactly the same semantic function, viz., to refer to how things stand in the real world. There is one occurrence of "would have flown to the moon", which is syntactically distinguished by being subjunctive, and semantically distinguished by referring not to the real, but to some counterfactual situation. In (41), however, the first occurrence of "has flown to the moon" in (40) is modelled by "$\mathsf{F}x$", and so is the occurrence of "would have flown to the moon", whereas the second occurrence of "has flown to the moon" corresponds to "$\boldsymbol{A}(\mathsf{F}x)$". This does not appear to be a transparent logical analysis — why should the two occurrences of the predicate "has flown to the moon", which function

[9] Logics with actuality operators have been studied rather extensively in the literature. See, for instance, Segerberg (1973), Hazen (1976), Crossley and Humberstone (1977), Peacocke (1978), Davies and Humberstone (1980), Hodes (1984c, 1984a, 1984b), and Forbes (1985, 1989).

[10] The sentence $\exists x\ \boldsymbol{A}(\mathsf{F}x) \wedge \Diamond\forall x\ (\boldsymbol{A}(\mathsf{F}x) \rightarrow \neg\mathsf{F}x)$, while having the same truth conditions as (41), is ruled out by the constraint that the non-modal first conjunct of (40) should be expressed in modal logic by its formalisation in ordinary predicate logic. Thanks to Tom van der Beek for indicating the need for clarification here.

in precisely the same way semantically, be modelled in *two* typographically distinct ways, as "$\mathsf{F}x$" and "$\mathcal{A}(\mathsf{F}x)$"? And why is the subjunctive predicate "would have flown to the moon", referring to some counterfactual world, represented in exactly the same way as the indicative predicate "has flown to the moon" (as it occurs first in (40)), which refers to the actual world?[11]

The idea suggests itself to introduce a typographical marker for the subjunctive. Let us symbolise the subjunctive version of the predicate "F" by "F*". We may then formalise (40) as

$$(42) \quad \exists x\, \mathsf{F}x \wedge \Diamond\forall x\, \left(\mathsf{F}x \rightarrow \neg\mathsf{F}^*x\right),$$

and this represents adequately the syntactic relations between the various occurrences of predicates in (40). Also, the intuition that the modal operators act on or bind those predicates in its scope which are subjunctive, but leave the others untouched[12], is brought out well by (42).

A system of this kind has in fact been introduced by Humberstone (1982)[13]. He employs a sentential subjunctivity operator "S", writing "$\mathsf{S}(\mathsf{F}x)$" instead of our "F^*x". I should like to propose a different solution here, essentially for the following reasons: Sentential operators can usually be iterated, but Humberstone's semantics cannot accommodate iterations of "S" (and consequently, he does not count expressions with iterated subjunctivity operators as well-formed). From the point of view of natural language, this is of course no loss, since there is no such thing as iterated subjunctivity; still, one wonders whether, under such conditions, a sentential operator is the adequate formalisation. Similarly, his semantics cannot handle modal operators within the scope of "S". Again, the exclusion of such expressions from the class of formulae is not objectionable as such, but the necessity of this measure suggests that subjunctivity does not work like a sentential operator. I therefore propose to stick with the notation introduced above: distinguish subjunctive

---

[11]Of course, the Kripkean already faces trouble when the expressiveness problem is solved by adding an "actually" operator, as pointed out e.g., by Stanley (1997, section V). But for the descriptivist there remains the curious presence of two very similar designating expressions, "$(\iota x)(\mathsf{T}x)$" and "$(\iota x)(\mathcal{A}(\mathsf{T}x))$", both ordinarily referring to the man who taught Alexander, only one of which (and in fact the more complicated one) providing the name "Aristotle" with the right meaning (see also the discussion in Soames (1998)). Such a duplication of terms is avoided by the logic S5* introduced below.

[12]The alternative, having modal operators act on *all* predicates within their scope indiscriminately, and introducing an operator-inhibiting device such as an actuality operator to undo this default binding, seems to be less intuitive. See also §10 below.

[13]Humberstone's insightful paper is unfortunately mutilated by some grave misprints. A corrected version can be found in chapter 1 of Humberstone (2000), together with a number of "updates and afterthoughts". These should be consulted for more on the relation between Humberstone's original system and our S5*, in particular with regard to the importance of introducing a subjunctive quantifier.

predicates[14] from their indicative relatives by flagging them with an asterisk "*". But now for the formal details.[15]

Fix, for the remainder of this paper, a set P of predicate symbols (each of fixed finite arity) and a set C of individual constants. The corresponding language L(P, C), or L for short, of S5* has as primitive symbols, besides the elements of P and C, the equality symbol =, some complete set of propositional connectives ($\neg$ and $\wedge$, say), infinitely many individual variables $x_1, x_2, \ldots$, the existential quantifier $\exists$, the subjunctive existential quantifier $\exists^*$ (read: "there would have been"), the possibility operator $\Diamond$, and, for each element P of P, the subjunctive P^* of P.[16] A term is an individual variable or an individual constant from C. The formulae of L are defined inductively by the following clauses:

- If P is an n-ary predicate symbol from P and $t_1, \ldots, t_n$ are terms, then $Pt_1 \ldots t_n$ and $P^*t_1 \ldots t_n$ are formulae.
- If s and t are terms, then $(s = t)$ is a formula.
- If F and G are formulae, then so are $\neg F$ and $(F \wedge G)$.
- If F is a formula and x is an individual variable, then $\exists xF$ and $\exists^* xF$ are formulae.
- If F is a formula, then so is $\Diamond F$.

An occurrence of a subjunctive predicate P^* or a subjunctive quantifier $\exists^*$ in some formula is said to be free if it is not within the scope of any diamond. An L-formula F is subjunctively closed (s-closed for short) if it contains no free occurrences of subjunctive predicates (including subjunctive quantifiers). F is closed with respect to individual variables (i-closed) if no individual variable occurs free in F (in the usual sense of freedom). An L-sentence is an L-formula which is both s-closed and i-closed.

A Kripke structure for S5* is just an ordinary S5 model[17] (with possibly varying domains). More precisely, a Kripke structure K is a tuple

$$\left(K, @, (D_w)_{w\in K}, (P_w)_{w\in K}^{P\in \mathsf{P}}, (c_\mathsf{K})_{c\in \mathsf{C}}\right),$$

consisting of a set K of possible worlds, a distinguished element @ of K (called the "actual" world), a K-indexed family $(D_w)_{w\in K}$ of sets, at least one of which is non-empty (so D, the union of all D_w, will be non-empty), a doubly indexed

[14] As will be noticed in the formal definition below, the quantifiers are here, following Frege, also regarded as (second-order) predicates.

[15] Logical symbols will often be used autonymously in the more technical sections of this paper, and concatenation is indicated by juxtaposition. No Quinean quasi-quotation is used. This should not cause any confusion.

[16] Since the equality symbol will always be interpreted by the true identity relation, there is no need to introduce the subjunctive equality predicate =*.

[17] S5 appears to be the logic of choice for the kind of modality Kripke investigates in his (1980). There is no reason why the present approach should not be extended to other systems of modal logic.

family $(P_w)^{P \in \mathsf{P}}_{w \in K}$ of relations, where for $P \in \mathsf{P}$ n-ary and arbitrary $w \in K$, P_w is a subset of D^n, and a C-indexed family $(c_{\mathsf{K}})_{c \in \mathsf{C}}$ of elements of D.[18]

A variable assignment for K is a mapping σ from the set of individual variables into D. Given an individual variable x, we also write $\mathsf{K}(x, \sigma)$ for $\sigma(x)$. For individual constants c from C, we also write $\mathsf{K}(c, \sigma)$ for c_{K}. We now define, by recursion, a relation $\mathsf{K} \models_w F[\sigma]$, holding between Kripke structures K, worlds w of K (this w shall be called the "subjunctive world"), variable assignments σ for K and L-formulae F, as follows:

- $\mathsf{K} \models_w Pt_1 \dots t_n[\sigma]$ iff $(\mathsf{K}(t_1, \sigma), \dots, \mathsf{K}(t_n, \sigma)) \in P_@$
- $\mathsf{K} \models_w P^*t_1 \dots t_n[\sigma]$ iff $(\mathsf{K}(t_1, \sigma), \dots, \mathsf{K}(t_n, \sigma)) \in P_w$
- $\mathsf{K} \models_w (s = t)[\sigma]$ iff $\mathsf{K}(s, \sigma) = \mathsf{K}(t, \sigma)$
- $\mathsf{K} \models_w \neg F[\sigma]$ iff not $\mathsf{K} \models_w F[\sigma]$
- $\mathsf{K} \models_w (F \wedge G)[\sigma]$ iff $\mathsf{K} \models_w F[\sigma]$ and $\mathsf{K} \models_w G[\sigma]$
- $\mathsf{K} \models_w \exists x\, F[\sigma]$ iff for some $e \in D_@$, $\mathsf{K} \models_w F[\sigma\{x := e\}]$[19]
- $\mathsf{K} \models_w \exists^* x\, F[\sigma]$ iff for some $e \in D_w$, $\mathsf{K} \models_w F[\sigma\{x := e\}]$
- $\mathsf{K} \models_w \Diamond F[\sigma]$ iff for some $u \in K$, $\mathsf{K} \models_u F[\sigma]$

The evaluation of an i-closed formula F is obviously independent of the choice of a variable assignment; we therefore define $\mathsf{K} \models_w F$, for i-closed F, to mean $\mathsf{K} \models_w F[\sigma]$ for every assignment σ. Likewise, the evaluation of an s-closed formula F is independent of the choice of the subjunctive world, hence, for s-closed F, we let $\mathsf{K} \models F[\sigma]$ mean: for all $w \in K$, $\mathsf{K} \models_w F[\sigma]$. For sentences F, we define $\mathsf{K} \models F$ to mean $\mathsf{K} \models_w F[\sigma]$ for all worlds $w \in K$ and all assignments σ. As in ordinary predicate logic, the meaningful statements of S5* are given by the sentences — just like "is greater than 0", or perhaps "it is greater than 0" is not meaningful in itself, neither is "Aristotle would have been fond of dogs" (in both cases, we need metalinguistic devices to bestow a truth value on the expressions: a variable assignment in the first, and a subjunctive world in the second case). And while it seems quite natural, in the case of an actuality operator, to assume the actual world as its default referent, no such intuition would support a stipulation according to which the actual world should serve, by default, as *subjunctive* world.

The way this semantics works shall be illustrated by formalising sentence (19). It can be expressed as

(43) $\Diamond \forall x\, (Fx \rightarrow \neg \mathsf{F}^* x)$.

Let us see what it means for (43) to be valid in a Kripke structure K. We have:

$$\mathsf{K} \models \Diamond \forall x\, (Fx \rightarrow \neg \mathsf{F}^* x)$$

[18] It could be argued that P_w should always be a subset of $(D_w)^n$. I do not wish to take a stand on this issue here and simply follow what seems to be common practice.

[19] $\sigma\{x := e\}$ is the function which is like σ except that it maps the variable x to e.

iff for all $w \in K$, all assignments σ,

$$\mathsf{K} \models_w \Diamond \forall x\ (Fx \to \neg \mathsf{F}^* x)[\sigma]$$

iff for all $w \in K$ and all σ there is some $u \in K$ such that

$$\mathsf{K} \models_u \forall x\ (Fx \to \neg \mathsf{F}^* x)[\sigma]$$

iff for all σ there is some u such that for all $e \in D_@$,

$$\mathsf{K} \models_u (Fx \to \neg \mathsf{F}^* x)\big[\sigma\{x := e\}\big]$$

iff for all σ there is some u such that for all $e \in D_@$,

$$\text{not } K \models_u Fx\big[\sigma\{x := e\}\big] \text{ or not } \mathsf{K} \models_u \mathsf{F}^* x\big[\sigma\{x := e\}\big]$$

iff for all σ there is some u such that for all $e \in D_@$,

$$e \notin F_@ \quad \text{or} \quad e \notin F_u$$

iff there is some $u \in K$ such that for every e in $D_@ \cap F_@$, e is not in F_u.

This is exactly what is required of a formalisation of (19). Let us now examine what form the modal argument takes in the setting of S5*.

§8. The modal argument in S5***.** In S5*, sentence (1) ("under certain counterfactual circumstances, Aristotle would not have taught Alexander") is formalised as

(44) $\Diamond \neg \mathsf{T}^* a$.

Now we are to substitute the description "$(\iota x)(\mathsf{T}x)$" for the name "a". Kripke insists on giving the description narrow Russellian scope with respect to the diamond:

(45) $\Diamond\big[(\iota x)(\mathsf{T}x)\big]\neg \mathsf{T}^*(\iota x)(\mathsf{T}x)$.

According to Russell's theory of descriptions, this may be expanded to

(46) $\Diamond\big(\exists! x\ \mathsf{T}x \wedge \exists x\ (\mathsf{T}x \wedge \neg \mathsf{T}^* x)\big)$.

Let us see what it means for (46) to hold in a structure K:

$$\mathsf{K} \models \Diamond\big(\exists! x\ \mathsf{T}x \wedge \exists x\ (\mathsf{T}x \wedge \neg \mathsf{T}^* x)\big)$$

iff for all w and σ,

$$\mathsf{K} \models_w \Diamond\big(\exists! x\ \mathsf{T}x \wedge \exists x\ (\mathsf{T}x \wedge \neg \mathsf{T}^* x)\big)[\sigma]$$

iff for all w and σ there is some u such that

$$\mathsf{K} \models_u \big(\exists! x\ \mathsf{T}x \wedge \exists x\ (\mathsf{T}x \wedge \neg \mathsf{T}^* x)\big)[\sigma]$$

iff there is some u such that $\text{card}(\mathsf{T}_@ \cap D_@) = 1$ and for some $e \in D_@$, $e \in \mathsf{T}_@$ and $e \notin \mathsf{T}_u$ iff $\text{card}(\mathsf{T}_@ \cap D_@) = 1$ and for some $e \in D_@$: $e \in \mathsf{T}_@$ and there is some u with $e \notin \mathsf{T}_u$ iff for all w and σ,

$$\mathsf{K} \models_w \exists! x\ \mathsf{T}x[\sigma] \text{ and for some } e \in D_@,\ \mathsf{K} \models_w \mathsf{T}x \wedge \Diamond \neg \mathsf{T}^* x\big[\sigma\{x := e\}\big]$$

iff for all w and σ,

$$\mathsf{K} \models_w \exists! x\ \mathsf{T}x \wedge \exists x\ (\mathsf{T}x \wedge \Diamond\neg\mathsf{T}^*x)[\sigma]$$

iff

$$\mathsf{K} \models \exists! x\ \mathsf{T}x \wedge \exists x\ (\mathsf{T}x \wedge \Diamond\neg\mathsf{T}^*x).$$

This last sentence is nothing but the result of substituting "$(\imath x)(\mathsf{T}x)$" for "a" in (44) *with wide scope* over the modal operator:

(47) $[(\imath x)(\mathsf{T}x)]\Diamond\neg\mathsf{T}^*(\imath x)(\mathsf{T}x).$

Now if, as is the case when @ is our world, "$a = (\imath x)(\mathsf{T}x)$" and (44) hold, then (47), and hence (46) and (45), are true: the substitution leading from (44) to (45) is *salva veritate* (which is what was to be expected from the informal objection to the modal argument in §2), whether the description is given wide or narrow scope over the modal operator.

Sentence (45) corresponds to our natural language formulation (4) ("Under certain counterfactual circumstances, the man who *taught* Alexander would not have taught Alexander") of §2. What about (3) ("Under certain counterfactual circumstances, the man who *would have taught* Alexander would not have taught Alexander")? It can be formalised as

(48) $\Diamond[(\imath^* x)(\mathsf{T}^*x)]\neg\mathsf{T}^*(\imath^* x)(\mathsf{T}^*x),$

which I shall use as shorthand for

(49) $\Diamond(\exists^* x\ \forall^* y\ (\mathsf{T}^*y \equiv x = y) \wedge \exists^* x\ (\mathsf{T}^*x \wedge \neg\mathsf{T}^*x)).$

This, just like (3), is evidently false. Does this fact provide a way out for the Kripkean, in that it at least proves "$(\imath^* x)(\mathsf{T}^*x)$" to be a non-rigid designator non-synonymous with "a"? No: "$(\imath^* x)(\mathsf{T}^*x)$" is not a designator, since "$a = (\imath^* x)(\mathsf{T}^*x)$", not being s-closed, has no determinate truth conditions, unless a subjunctive world has been chosen in advance. Informally speaking, this manifests itself in the incoherence of the question: "Who is the man who would have taught Alexander?", when no possible situation has been determined with respect to which the subjunctive could be interpreted. By way of analogy, consider the variable "x" and the formula "$a = x$". It, too, has no determinate truth conditions, unless a variable assignment has been specified in advance, and accordingly, we do not call the free variable (as such, i.e., in the absence of a variable assignment) a "designator".

The analysis of the modal argument in terms of S5* therefore seems to vindicate the informal objection outlined in §2 — indicative descriptions are inter-substitutable with coreferential names.

§9. The disappearance of non-extensionality and non-rigidity. The irrelevance of Russellian scope with respect to the possibility operator that was observed in the preceding section is of course just an instance of a more general phenomenon:

LEMMA 1. *Let F be an s-closed* S5*-formula, and A any* S5*-formula. Then for any structure* K *we have:*

(a) $\mathsf{K} \models \Box[(\iota x)F]A_x[(\iota x)F] \equiv [(\iota x)F]\Box A_x[(\iota x)F]$ *and*
(b) $\mathsf{K} \models \Diamond[(\iota x)F]A_x[(\iota x)F] \equiv [(\iota x)F]\Diamond A_x[(\iota x)F]$.

PROOF. We consider only the case of the box. Let K be a structure, $w \in K$ and σ any assignment for K. Then all of the following are equivalent:

- $\mathsf{K} \models_w \Box\exists x\,(\forall y\,(F_x[y] \equiv x = y) \wedge A)[\sigma]$
- $(\forall u \in K)\,(\exists e \in D_@)\,(\mathsf{K} \models_u \forall y\,(F_x[y] \equiv x = y))[\sigma\{x := e\}]$ and $\mathsf{K} \models_u A[\sigma\{x := e\}])$
- $(\forall u \in K)\,(\exists e \in D_@)\,(\{f \in D_@ : \mathsf{K} \models_u F[\sigma\{x := f\}]\} = \{e\}$ and $\mathsf{K} \models_u A[\sigma\{x := e\}])$
- $(\forall u \in K)\,(\exists e \in D_@)\,(\{f \in D_@ : \mathsf{K} \models_w F[\sigma\{x := f\}]\} = \{e\}$ and $\mathsf{K} \models_u A[\sigma\{x := e\}])$ (because $\mathsf{K} \models_u F[\sigma\{x := f\}]$, F being s-closed, does not depend on u)
- $(\exists e \in D_@)\,[(\{f \in D_@ : \mathsf{K} \models_w F[\sigma\{x := f\}]\} = \{e\})$ and $(\forall u \in K)\,\mathsf{K} \models_u A[\sigma\{x := e\}]]$
- $\mathsf{K} \models_w \exists x\,(\forall y\,(F_x[y] \equiv x = y) \wedge \Box A)[\sigma]$ ⊣

Three closely related corollaries may be drawn from the lemma.

First, it is well known for non-modal predicate logic that, under the hypothesis that $\exists!xF$, the scope of the description $(\iota x)F$ in any given formula A is semantically irrelevant. By the lemma, it follows that, given $\exists!xF$ for s-closed F, it does not matter how the description $(\iota x)F$ is scoped within S5* formulae $A_x[(\iota x)F]$. The second corollary is that S5* is extensional, as far as s-closed terms and formulae are concerned. We have, for example, that the schema

$$[(\iota x)F]\big(y = (\iota x)F\big) \longrightarrow G_x[y] \equiv G_x\big[(\iota x)F\big]$$

is valid in all S5* models, where F is s-closed and the scope for $(\iota x)F$ in $G_x[(\iota x)F]$ is arbitrary. This follows by giving the description $(\iota x)F$ in $G_x[(\iota x)F]$ maximal scope, which, by the first corollary, does not affect the truth value of $G_x[(\iota x)F]$.

The third corollary, finally, is that there is no non-rigid designation in S5*. Descriptions that are not s-closed do not qualify as *designators* at all (since their semantic values depend on the subjunctive world chosen), and s-closed descriptions designate *rigidly*: It is clearly S5*-valid, for s-closed F, that

$$\big[(\iota x)F\big]\big(y = (\iota x)F\big) \longrightarrow \big[(\iota x)F\big]\Box y = (\iota x)F;$$

by the first corollary we have that

$$\big[(\iota x)F\big]\Box y = (\iota x)F$$

is equivalent to

$$\Box[(\iota x)F]y = (\iota x)F,$$

and hence

$$[(\iota x)F](y = (\iota x)F) \longrightarrow \Box[(\iota x)F]y = (\iota x)F.$$

§10. S5* **and ordinary predicate logic.** There is an ongoing debate concerning the relation of the modal idiom to languages with explicit quantification over possible worlds — see e.g., Lewis (1968), Hazen (1976), Forbes (1985, 1989), Cresswell (1990), Melia (1992), Forbes (1992), Chihara (1998) —, so it may be instructive to see how S5* relates to such a language. Let me first introduce some notation.

$\mathsf{L}_2(\mathsf{P}, \mathsf{C})$, or L_2 for short, is the two-sorted language of ordinary first-order logic given by

- two sorts of first-order variables: x, y, z, possibly indexed, of individual type, and $\alpha_0, \alpha_1, \alpha_2, \ldots,$ of world type;
- corresponding to each n-ary predicate symbol $P \in \mathsf{P}$, an $n+1$-ary predicate symbol P°, whose first argument place is of world type and whose remaining argument places are of individual type;
- the individual constants from C;
- the binary equality predicate $=$, both of whose argument places are of individual type; and
- a fresh binary predicate symbol I, whose first argument place is of individual type and whose second place is of world type.

Now we may associate, with each Kripke structure K as defined in §7, a classical structure $\mathfrak{M}_\mathsf{K}$ for L_2 by letting the world variables range over the set K of worlds of K, the individual variables over the union D of the D_w of K, interpreting every $c \in \mathsf{C}$ by the element c_K of D, stipulating that eIw hold iff e is an element of D_w, and letting P° hold of $(w, e_1, \ldots, e_n)$ iff P_w holds of $(e_1, \ldots, e_n)$ in K. $\mathfrak{M}_\mathsf{K}$ will then satisfy the sentence $(\forall x)\ (\exists \alpha_1)\ (xI\alpha_1)$; and in fact every classical structure for L_2 which validates this sentence can be viewed as a Kripke structure in the obvious way.

We can now translate the L-formulae F of S5* into L_2-formulae $(F)_2$ as follows:

- $(Pt_1 \ldots t_n)_2 :\equiv P^\circ \alpha_0 t_1 \ldots t_n;$
- $(P^* t_1 \ldots t_n)_2 :\equiv P^\circ \alpha_1 t_1 \ldots t_n;$
- $(s = t)_2 :\equiv (s = t);$
- $(\neg F)_2 :\equiv \neg (F)_2;$
- $(F \wedge G)_2 :\equiv ((F)_2 \wedge (G)_2);$
- $(\exists x F)_2 :\equiv \exists x\ (xI\alpha_0 \wedge (F)_2);$

- $(\exists^* x F)_2 :\equiv \exists x\,(xI\alpha_1 \wedge (F)_2)$;
- $(\Diamond F)_2 :\equiv \exists \alpha_1\,(F)_2$.

The image of the translation $(\ )_2$ is the fragment of L_2 consisting of the formulae which contain as world variables at most α_0 and α_1 (the former occurring free only) and within which every individual quantifier $\exists x$ occurs only bounded, that is, in the form $\exists x\,(xI\alpha_0 \wedge \cdots)$ or $\exists x\,(xI\alpha_1 \wedge \cdots)$. The translation of an s-closed formula of S5* is an L_2-formula containing no world variable other than α_0 free.

Calling $(\ )_2$ a translation is justified by the following fact:

Given a Kripke structure K, a world w of K, a variable assignment σ for K and a formula F of S5*, $\mathsf{K} \models_w F[\sigma]$ holds if and only if $\mathfrak{M}_\mathsf{K} \models (F)_2[\sigma \cup \tau]$, where τ is any assignment of worlds to world variables mapping α_0 to @ and α_1 to w.

S5* is thus a mere notational variant of a certain fragment of L_2.[20] The semantics are isomorphic: the subjunctive world can be seen as the variable assignment for α_1, which is all we need to evaluate formulae in the image of $(\ )_2$ (α_0, functioning rather like a world constant, always being assigned @ as its value).

The crucial difference between ordinary S5 and S5* is that, technically speaking, the modal operators of S5, when interpreted as world quantifiers, cannot *selectively* quantify into just *some* of the argument positions, since these are all marked by the same world variable. This is somewhat disguised by the usual, roundabout way of translating S5 into L_2 (of first defining the relativisation of an S5 formula to an arbitrary world variable, and then defining the translation of a formula as its relativisation to the actual world, as e.g., in Forbes 1989, p. 10); but one can give a completely straightforward translation $(\ )^\circ$ of standard S5 (in the language given by P and C as above) into L_2 as follows:

- $(Pt_1, \ldots t_n)^\circ := P^\circ \alpha_0 t_1, \ldots t_n$
- $(s = t)^\circ := (s = t)$
- $(\neg F)^\circ := \neg F^\circ$
- $(F \wedge G)^\circ := (F^\circ \wedge G^\circ)$
- $(\exists x F)^\circ := \exists x\,(xI\alpha_0 \wedge F^\circ)$
- $(\Diamond F)^\circ := \exists \alpha_0\, F^\circ$.

We clearly have that a Kripke structure K (defined as before) models (in the sense of S5) an S5-formula F under the assignment σ if and only if the associated classical structure $\mathfrak{M}_\mathsf{K}$ satisfies F° under the assignment $\sigma \cup \tau$, where τ is any assignment of worlds to world variables mapping α_0 to @.

---

[20] In fact, S5* can be extended to a logic $S5^\omega$ intertranslatable with the *full* language L_2, in a way somewhat analogous to the extension of "actually" logics by numerically indexed actuality and modal operators (see Peacocke (1978) and Forbes (1989)).

Given this translation, it is easy to see that the equivocation involved in the modal argument corresponds to an overloading of the world variable α_0. First of all, α_0 plays the role of a name for the actual world (because it is standardly assigned @ as its value). But second, it is also used as a bound variable over possible worlds. This is bound to lead to fallacy; let us examine the modal argument, as it presents itself in S5 and L_2, respectively:

The substitution, in the context of S5, of "$(\iota x)(\mathsf{T}x)$" for "a" (narrowly scoped over the modal operator) in

(50) $\Diamond\neg\mathsf{T}a$

translates into L_2 as the substitution of

(51) $(\iota x)(xI\alpha_0 \wedge \mathsf{T}^\circ\alpha_0 x)$,

again with narrow scope over the existential quantifier, for "a" in

(52) $\exists\alpha_0\neg\mathsf{T}^\circ\alpha_0 a$.

In (51), α_0 occurs free only, that is, exclusively as a name of the actual world. In (52), however, the same variable becomes bound, thereby losing the individuality it enjoyed in (51). And it is entirely obvious that the substitution of "$(\iota x)(xI\alpha_0 \wedge \mathsf{T}^\circ\alpha_0 x)$" for "$a$" in (52) (with narrow scope) is not legitimate, for elementary reasons — the variable α_0, free in "$(\iota x)(xI\alpha_0 \wedge \mathsf{T}^\circ\alpha_0 x)$", having to become bound in the result of such a substitution.

Non-rigid designation, then, arises from considering the variable α_0 in the description $(\iota x)(xI\alpha_0 \wedge \mathsf{T}^\circ\alpha_0 x)$ once as being assigned the standard value @, and once as being a bound variable. It exploits the apparent connection between α_0 as a free and as a bound variable; but the connection really arises only through the bad strategy of using *one* variable to serve all purposes. There simply is no semantic link between α_0 as free and as bound variable; bound variables are, as it were, devoid of all individuality.

It may be instructive to revisit the "actually" operator $\boldsymbol{A}$ at this point. Let S5$\boldsymbol{A}$ be S5 extended by the actuality operator $\boldsymbol{A}$ and the actuality quantifier $\exists^a$, as in Hazen (1990). Then S5$\boldsymbol{A}$ can also be translated into L_2; but to bring out the peculiarity of actuality, I should like to present a translation into a slightly different language first.

Let L_3 be the language L_2 augmented with a new sentential operator sub. The language $L_{\boldsymbol{A}}$ is the fragment of L_3 inductively characterised by the clauses:

- If P is an n-ary relation symbol from P and $t_1, \ldots, t_n$ are terms, then $P^\circ\alpha_0 t_1 \ldots t_n$ is in $L_{\boldsymbol{A}}$.
- If s and t are terms (of individual type), then $(s = t)$ is in $L_{\boldsymbol{A}}$.
- If F and G are in $L_{\boldsymbol{A}}$, then so are $\neg F$ and $(F \wedge G)$.
- If F is in $L_{\boldsymbol{A}}$ and x is an individual variable, then $\exists x\ (xI\alpha_0 \wedge F)$ and $\exists x\ (\mathrm{sub}(xI\alpha_0) \wedge F)$ are in $L_{\boldsymbol{A}}$.

- If F is in L_A, then so is $\exists\alpha_0 F$.
- If F is in L_A, then so is $\mathrm{sub}(F)$.

The semantics of sub is given by the stipulation

$$\mathfrak{M} \models \mathrm{sub}(F)[\sigma \cup \tau] \quad \text{iff } \mathfrak{M} \models F\left[\sigma \cup \tau\{\alpha_0 := @\}\right],$$

where σ is an assignment for the individual variables and τ an assignment for the world variables.

Clearly, L_A is just a notational variant of the language of S5A. Given an S5A-formula F, its translation $(F)_A$ into L_A is defined recursively by the clauses

- $(Pt_1 \ldots t_n)_A := P^\circ\alpha_0 t_1 \ldots t_n$
- $(s = t)_A := (s = t)$
- $(\neg F)_A := \neg(F)_A$ and $(F \wedge G)_A := ((F)_A \wedge (G)_A)$
- $(\exists x F)_A := \exists x\, (xI\alpha_0 \wedge (F)_A)$
- $(\exists^a x F)_A := \exists x\, (\mathrm{sub}(xI\alpha_0) \wedge (F)_A)$
- $(\Diamond F)_A := \exists\alpha_0\, (F)_A$
- $(A(F))_A := \mathrm{sub}(F)_A$.

Let us look at the translation of, say,

(53) $\Diamond\forall^a x\, \left(A(Fx) \rightarrow \neg Fx\right)$

into L_A, which is

(54) $\exists\alpha_0\, \forall x\, \left(\mathrm{sub}(xI\alpha_0) \rightarrow (\mathrm{sub}(F^\circ\alpha_0 x) \rightarrow \neg F^\circ\alpha_0 x)\right)$.

As is obvious from the semantics of sub, this operator shields the variable α_0 off from quantification: Although, within the subformula occurrence of, e.g., $\mathrm{sub}(F^\circ\alpha_0 x)$, the variable occurs within the scope of the existential quantifier $\exists\alpha_0$, it is not bound by it. Now in the canonical notation of L_2, such a shielding operator is not available, and we must therefore take recourse to replacing the variable α_0 by a fresh one, α_1, say, which is not bound by any quantifier; still, we must assign this fresh variable *the same value* as α_0, since it is to simulate a *free* occurrence of α_0. In other words: α_1 is nothing but the free variable α_0 disguised from the quantifiers.

More precisely, to extend the translation $(\ )^\circ$ (of S5 into L_2) to S5A, add to the clauses for S5 given above the clauses

- $(A(F))^\circ := F^\circ[\alpha_0/\alpha_1]$,
- $(\exists^a x F)^\circ := \exists x\, (xI\alpha_1 \wedge (F)^\circ)$

where, for any L_2-formula G, $G[\alpha_0/\alpha_1]$ is the result of replacing every free occurrence of α_0 in G by α_1. Then the following holds:

A Kripke structure K models an S5A-formula F under the assignment σ, if and only if the associated classical structure $\mathfrak{M}_\mathsf{K}$ satisfies $(F)^\circ$ under the assignment $\sigma \cup \tau$, where τ is any assignment of worlds to world variables mapping both α_0 and α_1 to @.

The introduction of a second world variable is clearly an improvement over the case of simple S5. But the roles of free and bound variables are still not kept strictly apart: On its *free* occurrences, the variable α_0 is identified with the invariably free variable α_1 (referring to the actual world) through the constraint on the variable assignment τ. And this must be so; for consider the S5*A*-formula

$$(55) \quad \diamond\left[\forall^a x\, (A(Fx) \rightarrow \neg Fx)\right] \wedge Fc$$

and its translation into L_2,

$$(56) \quad \exists\alpha_0 \left[\forall x \left(xI\alpha_1 \rightarrow \left(F^\circ\alpha_1 x \rightarrow \neg F^\circ\alpha_0 x\right)\right)\right] \wedge F^\circ\alpha_0 c.$$

Since the first and third occurrences of "F" in (55) are to be evaluated with respect to the actual world, *both* α_0 and α_1 must be assigned @ to make (56) an adequate translation of (55). This strange handling of world variables reflects the misgivings expressed in §7 concerning the representation of the indicative predicate "has flown to the moon" by *two* constructions in (41): "$A(Fx)$" when occurring within the scope of a modal operator, and "Fx" otherwise. To sum up, it seems preferable to use S5* instead of S5*A* not only from an informal, but also from a technical point of view.[21]

§**11. Conclusion.** What do these results mean for the philosophy of language? There seem to be at least three issues that deserve discussion. First, if the modal argument fails, what is the status of the description theory of proper names? Second, why is it so widely believed that metaphysically modal contexts should be non-extensional? And finally, are there further philosophical arguments that become dubious when due attention is paid to the indicative-subjunctive distinction?

The first point is rather simple. The conclusion to be drawn from the failure of the modal argument is just that the description theory is back in the race again. There are, to be sure, two other popular arguments against it — the semantic and epistemic arguments —, but the first does not seem to be very strong, and the second can easily be evaded by moving from the crude form of the description theory considered here (every name being backed up by one fixed definite description) to the so-called cluster theory, postulating that,

[21] It should be noted that S5*A* is interpretable in S5*. A given formula F of S5*A* is translated into S5* as follows: First delete all occurrences of "*A*" which are not within the scope of any modal operator. Say that a predicate letter or quantifier is directly within the scope of an "actually" operator if it is not separated from it by a modal operator. Attach an asterisk to every predicate letter and quantifier which lies in the scope of a modal operator, but not directly within the scope of any of the remaining "actually" operators. Finally, erase all remaining occurrences of "*A*". This translation is clearly not one-one: Both "Ta" and "A(Ta)" are translated into the single S5*-formula "Ta".

associated with every name, there is some more or less vague *set* of descriptions giving the sense of the name.[22]

So let me turn to the second point, concerning the non-intensionality of modal contexts. It might seem that our results flatly contradict the analytic tradition. Thus Quine, in whose writings the topic figures prominently, has argued as follows (see Quine 1953, p. 143): While, for instance, "9 is necessarily greater than 7" is true, replacement of "9" by the coreferential expression "the number of planets" yields the falsehood "the number of planets is necessarily greater than 7".

Now, if it is metaphysical necessity that is at stake here, we can easily show that there is an equivocation between an indicative and a subjunctive description involved: From

(57) No matter how things might have gone, 9 would have been greater than seven

and

(58) The x which numbers the planets is 9,

all that follows is the true assertion

(59) No matter how things might have gone, the x which numbers the planets would have been greater than seven,

and not the false statement

(60) No matter how things might have gone, the x which would have numbered the planets would have been greater than 7.

But, as a matter of fact, it is not metaphysical modality that Quine is aiming at, and he is very explicit about the kind of modality involved:

> The general idea of strict modalities is based on the putative notion of analyticity as follows: a statement of the form "Necessarily . . . " is true if and only if the component statement which "necessarily" governs is analytic, and a statement of the form "Possibly . . . " is false if and only if the negation of the component statement which "possibly" governs is analytic. (Quine 1953, p. 143)

Quine is thus concerned with analyticity. But analyticity and metaphysical necessity are distinct notions: The former is truth in virtue of meaning, whereas the latter is concerned with ways the *world* might have been. Clearly, if a sentence is true solely in virtue of the meanings of its constituents, there is no guarantee that the replacement of some constituent by a coreferential one should preserve this property of analyticity. Metaphysical possibility

[22] That this move suffices to evade the epistemic argument has been argued convincingly by Dummett (1973, pp. 135–137).

and necessity, however, are not in any obvious way concerned with meanings, and so there is little reason to believe that such contexts should be non-extensional. Quine's argument is, therefore, apparently correct, but does not reveal anything about contexts of metaphysical possibility and necessity. The widespread belief that the latter should not be extensional appears to be grounded in a confusion of analyticity (and perhaps other notions) and metaphysical necessity.

Concerning other arguments relying on modal claims, I should like to discuss two philosophical theses whose credibility appears to be weakened when due attention is paid to the indicative-subjunctive distinction. The first is Kaplan's "compulsive talker" argument against Reichenbach's token-reflexive analysis of the indexical "I". Kaplan writes:

> But is it true, for example, that
>
> (A) "I" means the same as "the person who utters this token"?
>
> It is certainly true that
>
> I am the person who utters this token.
>
> But if (A) correctly asserted a synonymy, then it would be true that
>
> (B) If no one were to utter this token, I would not exist.
>
> Beliefs such as (B) could make one a compulsive talker. (Kaplan 1989, pp. 519–520; I have altered Kaplan's number labels to letter labels).

What, exactly, is going on here? Presumably, the argument is meant roughly like this: I am the person who utters this token. Hence, if I exist, then the person who utters this token exists. By Russell's theory of descriptions, if I exist, then there is someone who utters this token. So, if no-one utters this token, then I do not exist. Now it is supposed to follow from the assumption of synonymy that if no-one were to utter this token, then I would not exist, i.e., (B). This consequence would clearly be unacceptable.

But is it really (A) that is responsible for this problem? Consider the following variant of Kaplan's argument: "The person who utters this token" means the same as "the person who utters this token". Hence, if the person who utters this token exists, then the person who utters this token exists. By Russell's theory of descriptions, if the person who utters this token exists, then there is someone who utters this token. So if no-one utters this token, then the person who utters this token does not exist. But now what? It simply does not seem to follow that, if no-one *were* to utter this token, then the person who *utters* this token would not exist. At best, it follows that, if no-one *were* to utter this token, then the person who *would utter* this token would not exist. But if "I" means "the person who *utters* this token", then it does *not* mean "the person who *would utter* this token", and so the last inference in Kaplan's original argument seems to be unwarranted.

The second philosophical thesis that, under the analysis proposed here, appears to be problematic, is the claim that there are necessities knowable only *a posteriori*. Take the example of Hesperus and Phosphorus. If Hesperus is Phosphorus, then they could not possibly have been distinct planets: One planet cannot be two. And the identity can certainly only be known *a posteriori* — without recourse to experience, it would have been impossible to find out that Hesperus is Phosphorus.

Let me digress for a second. "Hesperus is Phosphorus" is, by common consent, a singular statement. It is not a general truth, like "For every prime number, there is a greater prime number". Yet, "Hesperus is Phosphorus" is trivially logically equivalent to the sentence

(61) "Everything is such that Hesperus is Phosphorus",

where a redundant quantifier has been prefixed. This, of course, makes neither the identity statement itself nor the statement prefixed with the vacuous quantifier a general truth. Adding the quantifier is a cheap trick, because it does not bind any variable in its scope.

Now consider the statement

(62) No matter how things might have gone, Hesperus would always have been Phosphorus.

Here, the necessity operator is vacuous: there are no predicates which it could bind — except, perhaps, the equality predicate, but there seems to be a clear intuition that this is a rigid predicate, having true identity as its extension at every possible world (an intuition explicitly endorsed by Kripke). But apart from "would have been (identical to)", there are no predicates in the embedded sentence, *a fortiori* no subjunctive predicates. Just as we do not call (61) a universal statement just because it has an initial (vacuous) universal quantifier, we should not call (62) a statement of necessity.

This vacuity of the necessity operator in (62) is somewhat disguised in S5, where the box is treated as a purely sentential operator. This makes genuine necessity statements take the same form as vacuous ones like (62). To take an example, consider

(63) No matter how things might have gone, Hesperus would have been the planet which would have appeared brightest in the evening sky.

This is a genuine, if false, necessity statement: The predicate "would have appeared brightest in the evening sky" is bound to the modal operator. In S5*, this is symbolised by

(64) $\Box h = (\iota^* x)(\mathsf{F}^* x)$.[23]

In ordinary S5, however, this is rendered simply as

(65) $\Box h = (\iota x)(Fx)$,

which assimilates it to the S5 sentence

(66) $\Box h = p$,

only that, instead of the name p, (65) contains the description $(\iota x)(Fx)$. The fact that the box actually binds the predicate "F" in (65), but not the name "p" in (66), is concealed in this notation.[24]

It is not exactly clear to me what we are to conclude from this observation concerning the (vacuous) necessity of identity. In any case, the argument for the existence of *a posteriori* necessities does seem to be in need of supplementation, if it can be sustained at all.

Acknowledgments. Concerning the intuitive objection to the modal argument set forth in §2, my main intellectual debt is to Ulrich Pardey. From his ordinary language investigations (1994, pp. 141–48, especially 142–45), I learned about the significance of the grammatical moods for logical analysis and about the systematic neglect of this distinction in mainstream analytic philosophy of language. It was he who encouraged me to take the intuitive objection seriously, and who helped me get clear about it in numerous conversations and discussions. Helge Rückert, Matthias Schirn, Göran Sundholm and Albert Visser read earlier versions of this paper and provided many helpful comments. Special thanks to Lloyd Humberstone and to Richard Holton for constructive criticism, discussion, and encouragement. Finally, I wish to thank the organizers of the Munich *Intensionality* conference, specifically Reinhard Kahle, for putting together such an inspiring event; thanks also to an anonymous referee for this paper who suggested valuable improvements here and there.

REFERENCES

Charles S. Chihara [1998], ***The worlds of possibility — modal realism and the semantics of modal logic***, Clarendon Press, Oxford.

Max J. Cresswell [1990], ***Entities and indices***, Kluwer, Dordrecht.

John N. Crossley and Lloyd Humberstone [1977], *The logic of "actually"*, ***Reports on Mathematical Logic***, vol. 8, pp. 11–29.

[23] Here and in what follows, I omit the descriptions' scope indicator. The descriptions are always to be thought of as having narrow scope with respect to the modal operator.

[24] Note that the sentence "$\Box h = p$", taken as an expression of S5 and of S5*, respectively, translates into the language L_2 as "$\forall\alpha_0 h = p$" and "$\forall\alpha_1 h = p$", respectively, in each case revealing the presence of an empty quantifier.

MARTIN DAVIES AND LLOYD HUMBERSTONE [1980], *Two notions of necessity*, ***Philosophical Studies***, vol. 38, pp. 1–30.

MICHAEL DUMMETT [1973], ***Frege — philosophy of language***, Duckworth, London.

GRAEME FORBES [1985], ***The mataphysics of modality***, Oxford University Press, Oxford.

GRAEME FORBES [1989], ***Languages of possibility***, Basil Blackwell, Oxford.

GRAEME FORBES [1992], *Melia on modalism*, ***Philosophical Studies***, vol. 68, pp. 57–63.

HENRY W. FOWLER [1983], ***A dictionary of modern English usage***, 2nd ed., Oxford University Press, Oxford, revised by Sir Ernest Gowers.

ALLEN HAZEN [1976], *Expressive completeness in modal language*, ***Journal of Philosophical Logic***, vol. 5, pp. 25–46.

ALLEN HAZEN [1990], *Actuality and quantification*, ***Notre Dame Journal of Formal Logic***, vol. 31, pp. 498–508.

JAAKKO HINTIKKA [1997], *No scope for scope?*, ***Linguistics and Philosophy***, vol. 20, pp. 515–44.

HAROLD T. HODES [1984a], *Axioms for actuality*, ***Journal of Philosophical Logic***, vol. 13, pp. 27–34.

HAROLD T. HODES [1984b], *On modal logics which enrich first-order* S5, ***Journal of Philosophical Logic***, vol. 13, pp. 423–54.

HAROLD T. HODES [1984c], *Some theorems on the expressive limitations of modal languages*, ***Journal of Philosophical Logic***, vol. 13, pp. 13–26.

LLOYD HUMBERSTONE [1982], *Scope and subjunctivity*, ***Philosophia***, vol. 12, pp. 99–126.

LLOYD HUMBERSTONE [2000], *Propositional attitudes: Some logical issues*, Typescript.

DAVID KAPLAN [1989], *Demonstratives*, ***Themes from Kaplan*** (Joseph Almog, John Perry, and Howard Wettstein, editors), Oxford University Press, New York and Oxford, pp. 481–563.

SAUL A. KRIPKE [1971], *Identity and necessity*, ***Identity and individuation*** (Milton K. Munitz, editor), New York University Press, New York, Reprinted in A. W. Moore (ed.), ***Meaning and Reference***, Oxford University Press, New York and Oxford, 1993, pp. 162–91. Page references in the text are to the latter edition, pp. 135–64.

SAUL A. KRIPKE [1980], ***Naming and necessity***, Harvard University Press, Cambridge, Mass., originally published (without the preface) in D. Davidson and G. Harman (eds.), ***Semantics of Natural Language***, North Holland, Dordrecht, 1972, pp. 253–355.

DAVID LEWIS [1968], *Counterpart theory and quantified modal logic*, ***Journal of Philosophy***, vol. 65, pp. 113–26, Reprinted in his ***Philosophical Papers, Volume I***, Oxford University Press, Oxford, 1983, pp. 26–39.

DAVID LEWIS [1970], *Anselm and actuality*, ***Noûs***, vol. 4, pp. 175–88, Reprinted in his Philosophical Papers, Volume I, Oxford University Press, Oxford, 1983, pp. 10–20.

DAVID LEWIS [1983], *Postscripts to "Anselm and actuality"*, ***Philosophical papers***, vol. I, Oxford University Press, Oxford, pp. 21–25.

JOSEPH MELIA [1992], *Against modalism*, ***Philosophical Studies***, vol. 68, pp. 35–56.

STEPHEN NEALE [1990], ***Descriptions***, The MIT Press, Cambridge, Mass..

ULRICH PARDEY [1994], ***Identität Existenz und Reflexivität***, Beltz Athenäum, Weinheim.

CHRISTOPHER PEACOCKE [1978], *Necessity and truth theories*, ***Journal of Philosophical Logic***, vol. 7, pp. 473–500.

WILLARD V. QUINE [1953], *Reference and modality*, ***From a logical point of view***, Harvard University Press, Cambridge, Mass., pp. 139–59.

KRISTER SEGERBERG [1973], *Two-dimensional modal logic*, ***Journal of Philosophical Logic***, vol. 2, pp. 77–96.

ARTHUR F. SMULLYAN [1948], *Modality and description*, ***The Journal of Symbolic Logic***, vol. 13, pp. 31–37.

SCOTT SOAMES [1998], *The modal argument: Wide scope and rigidified descriptions*, ***Noûs***, vol. 32, pp. 1–22.

Jason Stanley [1997], *Names and rigid designation*, ***A companion to the philosophy of language*** (Bob Hale and Crispin Wright, editors), Blackwell, Oxford, pp. 555–85.

Kai F. Wehmeier [2003], *World travelling and mood swings*, ***Foundations of the formal sciences II: Applications of mathematical logic in philosophy and linguistics*** (Benedikt Löwe, Wolfgang Malzkorn, and Thoralf Räsch, editors), Kluwer (Trends in Logic), Dordrecht, pp. 257–260.

DEPARTMENT OF LOGIC AND PHILOSOPHY OF SCIENCE
SCHOOL OF SOCIAL SCIENCES
3151 SOCIAL SCIENCE PLAZA A
UNIVERSITY OF CALIFORNIA, IRVINE
IRVINE, CA 92697–5100, USA
E-mail: wehmeier@uci.edu

COERCION VS. INDETERMINACY IN OPAQUE VERBS

THOMAS EDE ZIMMERMANN

§1. **Introduction.** This paper is about the semantic analysis of *opaque* verbs such as **seek** and **owe**, which allow for unspecific readings of their indefinite objects.[1] One may be looking for a good car without there being any car that one is looking for; or, one may be looking for a good car in that a specific car exists that one is looking for. It thus appears that there are two interpretations of these verbs — a specific and an unspecific one — and one may wonder how they are related. The present paper is a contribution to this question.

§2. History.

2.1. Paris. The time of the holy inquisition. Opaque verbs differ in their semantic behaviour from ordinary verbs. This phenomenon was already known to the medieval logician Buridanus:

> I posit the case that for a good service you performed for me, I promised you a good horse. [...] And since I owe you this, until I have paid that concerning the payment of which I have obligated myself [...], you could rightly take action against me to bring about payment to you of a horse, which you could not do if I did not owe you. [...] But the opposite is argued in a difficult way. [Buridanus (1966 [1350]: 137)]

The following modern version of the opposite argument is less verbose than the original:

> Let us then have our horse-coper arguing again. "If I owe you a horse, then I owe you something. And if I owe you something, then there is something I owe you. And this can only be a thoroughbred of mine: you aren't going to say that in virtue of what I said there's something else I owe you. Very well, then: by your claim, there's one of my thoroughbreds I owe you. Please tell me which one it is." [Geach (1965: 430)]

[1]This should be taken as the definition of *opacity* that is used in this paper; hence my use of the term differs from Quine's, who uses it in the sense of *intensional*. It should be noted that, although indefinites are criterial for opacity (in my sense of the term), they are not the only noun phrases that behave exceptionally when serving as objects to opaque verbs. However, for the purpose of the present investigation, it suffices to consider (singular) indefinite objects.

Intensionality
Edited by Reinhard Kahle
Lecture Notes in Logic, 22

The two arguments are based on two different ways of reading the sentence under debate (1) — an obvious, unspecific interpretation and a somewhat remote[2], specific one.

(1) **I owe you a horse.**

2.2. Harvard. The McCarthy era. While Buridanus had no solution to offer — indeed, he discussed the puzzle under the heading *Insolubilia* — Quine, making the same observation with verbs like **seek**, came up with an ingenious explanation:[3]

(2) **Ernest is seeking a lion.**

According to Quine, the puzzle has two sources. The first is a general structural ambiguity found in propositional attitude reports and related constructions where indefinites, which express existential quantification, may or may not logically outscope the attitude verb under which they are embedded. Thus, e.g., (3a) is ambiguous with the readings (3b) and (3c):

(3a) **Tom believes that someone denounced Catiline.**

(3b) **Tom believes that someone is such that he denounced Catiline.**

(3c) **Someone is such that Tom believes that he denounced Catiline.**

This ambiguity conspires with the lexical meaning of the opaque verb **seek** (or **owe**, as the case may be) which superficially behaves like a transitive verb but must be interpreted as abbreviating a propositional attitude, as confirmed by the paraphrase **try to find** (or **must give**). Hence paraphrasing (2) as (4a) makes the sentence susceptible to the variable scope effect:

(4a) **Ernest is trying for it to be the case that Ernest finds a lion.**

(4b) **Ernest is trying for it to be the case that a lion is such that Ernest finds it.**

(4c) **A lion is such that Ernest is trying for it to be the case that Ernest finds it.**

The two ingredients in Quine's analysis of opacity, then, are:

Q1 Lexical decomposition

Sentences with opaque verbs abbreviate propositional attitude reports.

Q2 Scopal variability

As always in attitude reports, the indefinite [object] may take different scopes.

---

[2]The remoteness of the specific reading seems to be a lexical idiosyncrasy of the verb **owe**. With other opaque verbs, including **seek**, specific readings come quite naturally. I will return to this phenomenon in Section 7.

[3]Cf. Quine (1960: 151–156); the key idea already appears in Quine (1956), written around 1952. To be fair, Buridanus did make a tentative attempt to solve the puzzle; Geach (1965) offers a modern reconstruction and some criticism.

Apart from analyzing the two interpretations of an opaque verb as a case of structural ambiguity, Quine's analysis also explains why sentences with opaque predicates differ from those with ordinary, *transparent* predicates in two further respects:

- The object position is *not quantificational.*
- The object position is *intensional.*

The first point concerns the observation that an indefinite in the object position of an opaque verb cannot be read as being existentially quantified: taken unspecifically, (1) does not imply that there is a horse, as little as (2) implies that lions exist. The explanation is quite simple: although the indefinite does express existential quantification, its being under the scope of an attitude predicate generally prevents it from projecting its existential impact to the whole sentence.

The second point highlights the fact that, on the unspecific interpretation, substitution of co-extensional indefinites does not always preserve truth values: Ernest may be looking for a twenty-year old lion without being looking for a twenty-year old circus lion, even though all lions of that age may happen to be circus lions. Again, the explanation is straightforward: the scope of an attitude predicate is an intensional position.

2.3. California. Summer of love. Montague, using techniques of higher-order modal logic, turned Quine's account of opacity into a surfacecompositional analysis[4], the starting point of which is a possible worlds adaptation of Quine's paraphrase formulated in intensional type theory; under the assumption that **try-for-it-to-be-the-case-that** denotes a binary relation between individuals and propositions, **seek a unicorn** receives the following logical analysis:[5]

---

[4]Cf. Montague (1969, 1973).

[5]The notation deviates from Montague's original account(s) in some obvious ways:

- Following the tradition of Cresswell (1973) (and, indeed, modal logic), logical translations are *proposition-based*, i.e. index (= world-time point) dependence is reserved to truth values; connectives and quantifiers operate on propositions and are interpreted accordingly. The resulting formulae can do with fewer index variables and are thus more readable. Also, as already noted by Montague (1970a: 218f.), a proposition-based interpretation avoids the complications of two separate layers of extension and intension; cf. Kaplan (1975) for related discussion.
- The logical types assigned to translations of lexical items are as low-ordered as possible; his elegant unified type assignments (i.e. one type per syntactic category), frequently forced Montague to resort to unnecessarily high types — a strategy which Partee (1997: 75) aptly called *generalizing to the worst case.*

Non-obvious notational conventions are as follows: t is the type of *propositions*, not truth-values; bold-face letters are (mnemonic) non-logical constants; simultaneous application to a sequence of arguments stands proxy for successive application in the reverse order ("Currying"). A fuller specification of the logical notation can be found in the appendix.

(5) $\lambda x \boldsymbol{T}(x, (\exists y)[\boldsymbol{U}(y) \wedge \boldsymbol{F}(x, y)])$.

Surface compositionality, i.e. a word-by-word analysis, is then achieved by applying the Fregean strategy of *meaning subtraction*[6], obtaining the meaning of **seek** by separating the quantifier expressed by **a unicorn** from the property denoted by **seek a unicorn**. Such a separation can be carried out using a series of lambda-abstractions:

$$
\begin{array}{ll}
 & (5) \\
\equiv & \lambda x \boldsymbol{T}(x, \underline{[\lambda Q(\exists y)[\boldsymbol{U}(y) \wedge Q(y)]]}(\lambda y \boldsymbol{F}(x, y))) \\
\equiv & \underline{\underline{[\lambda \mathfrak{Q} \lambda x \boldsymbol{T}(x, \mathfrak{Q}(\lambda y \boldsymbol{F}(x, y)))]}}(\underline{\lambda Q(\exists y)[\boldsymbol{U}(y) \wedge Q(y)]})
\end{array}
$$

The first step isolates the (underlined) meaning of the indefinite **a unicorn** as contributing to the (varying) implicit attitude object denoted by **find a unicorn**; the second step separates that meaning from the rest, which in turn may serve as an analysis of **seek**. Given the types in (5), it turns out that **seek** denotes a relation between an individual and a quantifier, i.e. its meaning is of type $((et)t)(et)$, viz.:

(6) $\lambda \mathfrak{Q} \lambda x \boldsymbol{T}(x, (\mathfrak{Q} y) \boldsymbol{F}(x, y))$

where "$(\mathfrak{Q} y)\varphi$" abbreviates "$\mathfrak{Q}(\lambda y \varphi)$". Since the resulting type is independent of the paraphrase, Montague's analysis is more general than Quine's:

> As far as "seeks" and "owes" are concerned, circumlocution involving infinitives is possible. It is not, however, in the case of all English verbs sharing the logical peculiarities of "seeks" and "owes" [...]
>
> Montague (1969: 177)

This claim has been debated[7]. In any case, Montague's analysis of opacity reduces unspecificity to scope; whenever opacity occurs, the logical type of

---

[6]If the meaning $|\alpha|$ of an expression α is not known (to the semanticist), it can be constructed by considering expressions β that combine with α into expressions γ_β, i.e. (roughly): $\gamma_\beta = \alpha + \beta$ — provided that (a) the meanings of all β and γ_β are known and (b) they behave compositionally, i.e. that γ_β and $\gamma_{\beta'}$ have the same meaning whenever β and β' do: $|\alpha|$ is a function assigning to any $|\beta|$ the value $|\gamma_\beta|$. The strategy may be called *meaning subtraction* because, intuitively, the meanings of α and β add up to the meanings $|\gamma_\beta|$, just like α and β add up to γ_β, and hence $|\alpha|$ is obtained by taking off (or abstracting) $|\beta|$ from $|\gamma_\beta|$. Arguably, the strategy is a reconstruction of Frege's (1884: X) infamous *context principle*, and one which is quite consistent with — in fact dependent on — the principle of compositionality, contradicting the impression given in Janssen (1997: 420f.) and elsewewhere.

[7]More precisely, what has been debated — to wit, by Larson *et al.* (1999) — is the contention that there are any unspecificity-inducing verbs (i.e. opaque verbs in the above sense) that cannot be reduced to propositional attitudes. On the other hand, in the above quotation Montague apparently related opacity to failure of existential import, not unspecificity, citing purported counter-examples like **worship** — for which he has also been (rightly) criticized by, among others, Kripke (as reported in Bennett (1974: 82ff.)).

the verb is higher than that of the indefinite object.[8] The main ingredients of Montague's analysis are:

M1 Lexical type assignment

The (relevant) argument position of an opaque verb is defined for the meanings of indefinites (existential quantifiers), i.e. the verb expresses an intensional third-order relation.

M2 Structural ambiguity

For the unspecific reading, an opaque verb takes the indefinite as its argument; for the specific reading the indefinite is quantified into that position by a general scoping mechanism.

The scoping mechanism generalizes Quine's variability in that it also covers other cases of scope ambiguity. According to it, (7a) has two distinct syntactic structures, or logical forms, viz. (7b) and (7c):

(7a) **Jones seeks a unicorn.**

(7b)

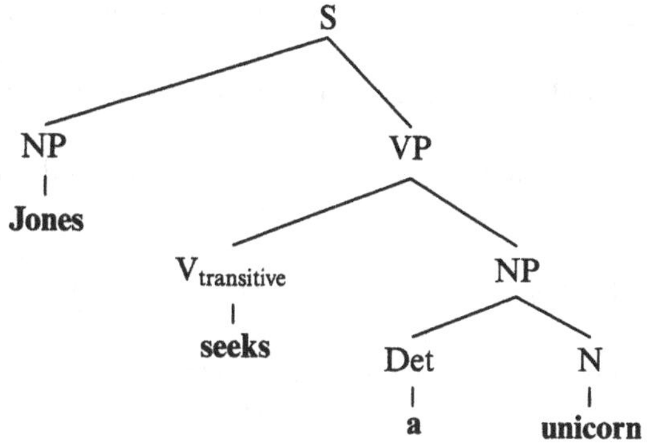

(7c)

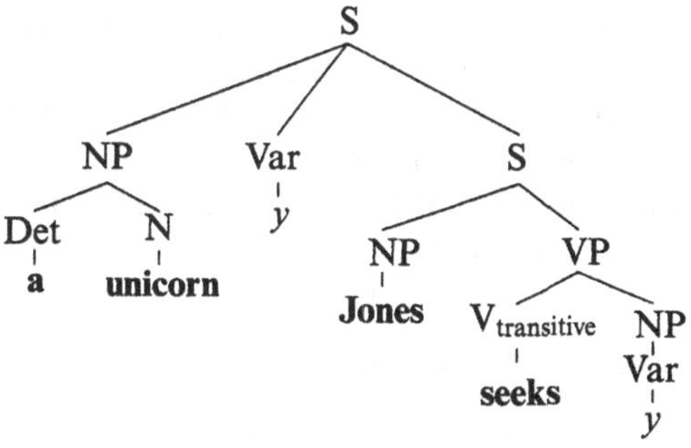

The first structure just reverses the above abstraction process by applying the

[8]... though not necessarily the other way round: due to the spurious types obtained by Montague's strategy of *generalizing to the worst case* (cf. fn. 5), any transparent verb can be re-categorized so as to take scope over a quantified object.

meaning of the verb to the quantifier expressed by the object and combining the result with the subject, thus arriving at:

(8) $\boldsymbol{T}(j, (\exists y)[\boldsymbol{U}(y) \wedge \boldsymbol{F}(j, y)])$.

The construction (7c) is interpreted as in (9c), i.e. by applying the quantifier (9a) expressed by the noun phrase (**a unicorn**) to the property obtained by abstracting the (syntactic) variable or *trace* (y) from the open formula (9b) corresponding to the scope (**Jones seeks** y):

(9a) $\lambda Q(\exists y)[\boldsymbol{U}(y) \wedge Q(y)]$

(9b) $\boldsymbol{T}(j, \boldsymbol{F}(j, y))$

(9c) $[\lambda Q(\exists y)[\boldsymbol{U}(y) \wedge Q(y)]](\lambda y \boldsymbol{T}(j, \boldsymbol{F}(j, y)))$

$\equiv$ $(\exists y)[\boldsymbol{U}(y) \wedge \boldsymbol{T}(j, \boldsymbol{F}(j, y))]$

As it stands, the interpretation (9) of the specific reading of (7a) requires an individual variable to combine with the opaque verb, which is at odds with its being an operation on quantifiers. This seemingly technical complication may be seen as reflecting a deeper conceptual distinction to which I will come in Section 4.

The technical complication can be mastered by standard techniques. Like any individual term, the bound variable may be re-interpreted as a quantifier, thanks to the categorial operation known as *Montague Lifting*, yielding $[\lambda P\, P(y)]$; alternatively, the verb itself may be re-interpreted as transparent, by the operation of *Argument Lowering*, resulting in $[\lambda y \lambda x \boldsymbol{S}(x, \lambda P\, P(y))]$, whenever "$\boldsymbol{S}$" translates the opaque verb. Both strategies lead to the desired result (9b).[9] In Section 4 we will see a reason for preferring (a variant of) the latter over the former.

M1 is general enough to cover cases of multiple opacity, as witnessed by a usage of **owe**, which according to some speakers may be read unspecifically with regard to both its direct and its indirect object. This usage is captured by either of the following analyses that only differ from each other in the relative scopes they assign to the objects — not a substantial difference given the scoping mechanism and the fact that, as far as indefinites are concerned, their relative scopes do not matter anyhow:[10]

(10) $\lambda \mathfrak{D} \lambda \mathfrak{Q} \lambda x \boldsymbol{O}(x, (\mathfrak{D} z)(\mathfrak{Q} y)\boldsymbol{G}(x, y, z))$

(10′) $\lambda \mathfrak{D} \lambda \mathfrak{Q} \lambda x \boldsymbol{O}(x, (\mathfrak{Q} y)(\mathfrak{D} z)\boldsymbol{G}(x, y, z))$

---

[9] See Hendriks (1993) for the full story.

[10] They may, once information structure (topic and focus) is taken into account, which I am ignoring here. — Notation update: "$\boldsymbol{O}$" stands for the propositional attitude of being obliged to see to it that a given proposition becomes true, and "$\boldsymbol{G}$" expresses the relation of giving as holding between the donor, the object transferred, and the recipient.

As a case in point, I may promise to treat a student (i.e. some student or other) to a beer if I do not manage to put up my class-notes on the web before the end of the vacation. If I then fail to deliver the notes, according to the analysis (10), I thereby owe a student a beer — even though I do not owe a beer to any student in particular. It appears that many speakers do not accept this judgement, reserving unspecificity for the direct object position. Their usage of **owe** is captured by the following alternative analysis:

(10″) $\lambda z \lambda \mathfrak{Q} \lambda x \boldsymbol{O}(x, (\mathfrak{Q} y)\boldsymbol{G}(x, y, z))$.

The formula shows that, in the Montagovian account, unspecificity is a matter of types and scope and not of paraphrases in terms of attitudes; both objects relate to argument positions of the embedded predicate $\boldsymbol{G}$ but only one of them, the direct object, corresponds to a higher-order variable, expressing unspecificity. It is not clear how the Quinean paraphrase strategy would handle such cases. In the Montagovian approach, however, unspecificity and propositional paraphrase are completely independent of each other, in that either may occur without the other. One direction was already indicated in the above quotation. The reverse, i.e. that decomposability does not necessarily involve unspecificity, can be shown by reference to the following, quite plausible lexical decomposition of the verb **kill** (as **C**ause to **D**ie):

(11) $\lambda y \lambda x \boldsymbol{C}(x, \boldsymbol{D}(y))$.

Despite this reduction to a propositional attitude (in the usual, broad sense), the analysis of **kill** does not induce unspecificity; a quantified object (like an indefinite) will always have scope over the verb.

Although Quine's and Montague's explanations of Buridan's puzzle proceed along the same lines in that they both assume a scope ambiguity, the details are somewhat different. In particular, in Montague's account, the above-mentioned side-effects of unspecificity — non-quantificationality and intensionality — cannot be explained in terms of underlying propositional attitudes. Rather, they are reflected in the logical type of the argument corresponding to the unspecific object.[11] Montague's account of opacity, which has been widely accepted in the semantic community[12], still has a few problems, of which we now turn to two.

---

[11]This is only true for Montague's original, truth-value based types; cf. Zimmermann (2001: 516–20) for details.

[12]... widely, though not universally:

- Larson *et al.* (1999) argue for a *clausal* analysis of opacity, i.e. one which relies on syntactic reducibility to propositional attitudes; Forbes (2000) argues for what may be dubbed a *de re* analysis of opacity, i.e. one according to which the (extensional) quantifier denoted by the indefinite is related to the attitude subject via a suitable psychological relation (a way of thinking). The problems addressed in the present paper do not arise under these approaches, of which I am, however, skeptical for independent reasons that I cannot go into here.

§3. Higher-order opacity.

3.1. Something inferred. The following inferences appear to be valid — even if the premise in (13) is understood unspecifically:[13]

(12) **Geach is selling a textbook.**

∴ **Geach is selling something.**

(13) **Geach is looking for a textbook.**

∴ **Geach is looking for something.**

In anybody's account, (12) is a matter of Logical Form, unrestricted existential quantification (expressed by **something**) being more general than its restricted form (**a textbook**). On the Montagovian approach (as well as on Quine's), this cannot be true of (13): in the unspecific reading, the object position of the opaque verb in the premise is not quantified over. It thus appears as if somehow the lexical meaning of **look for** must come into play. This conjecture is corroborated by the following plausible inference:

(14) **Geach is looking for a textbook.**

∴ **Geach is looking for a book.**

Here the quantifier in the conclusion (expressed by **a book**) is not unrestricted; but it is less restrictive than the one in the premise. This observation suggests that the relation $\boldsymbol{L}$ expressed by **look for** has a certain *monotonicity* property: **seek** may be monotonically increasing in its unspecific argument, presumably due to the underlying attitude $\boldsymbol{T}$ being closed under implication; in other words, the closure assumption (MON_T) implies the monotonicity (MON_S) of **seek**:[14]

$$(\text{MON}_T) \quad \Box(\forall x)(\forall p)(\forall q)[p \Rightarrow q \rightarrow [\boldsymbol{T}(x,p) \rightarrow \boldsymbol{T}(x,q)]]$$

$$(\text{MON}_S) \quad \Box(\forall x)(\forall P)(\forall Q)[P \sqsubseteq Q \rightarrow [\boldsymbol{S}(x,\exists_P) \rightarrow \boldsymbol{S}(x,\exists_Q)]]$$

– Zimmermann (1993) argues for a *property* analysis of opacity, i.e. one according to which indefinites express properties (as in dynamic semantics along the lines of Kamp (1981) and Heim (1982)) and opacity does not occur with (other) quantificational objects. The problems addressed in the present paper arise equally under that approach — basically because the property analysis and Montague's account, if restricted to indefinites, are intertranslatable, using type-shifting techniques à la Partee (1987).

[13]Though Geach (1965) does not address the inference pattern as such, he makes ample use of it — as in his account of the horse-monger (see above), who starts his reasoning with an instance of that pattern.

[14]"$\Rightarrow$" and "$\sqsubseteq$" respectively denote strict implication and sub-propertyhood, both of which can be defined in terms of material implication, necessity, and universal quantification. "$\exists_P$" is short for the quantifier expressed by an indefinite restricted by P, i.e. $\lambda Q(\exists x)[P(x) \wedge Q(x)]$.

Hence if L holds between an individual x and a (restricted existential) quantifier $\mathfrak{Q}$, then it also holds for any (restricted existential) quantifier $\mathfrak{D}$ such that $\mathfrak{Q} \subseteq \mathfrak{D}$. Since the object in the premise of (14) expresses the property of properties that apply to at least one textbook and since any such property applies to at least one book, the quantifier denoted by the object in the conclusion is indeed a super-property of the one denoted by **a textbook**. Hence monotonicity would explain the inference — and also that in (13). However, as will be argued below, there is reason to doubt that **seek** is monotonic.

In any case, monotonicity does not always help. Consider:

(15) **Nicholas wants a free trip on the Concorde.**

∴ **Nicholas wants a trip on the Concorde.**

This inference does not seem valid: as we all know, trips on the Concorde are excessively expensive, and hence the conclusion is likely to be false even though the premise is certainly true.[15] Hence **want**, though no doubt opaque, does not display the same kind of monotonicity behaviour as **look for**. On the other hand, the following inference goes through nevertheless:

(16) **Nicholas wants a free trip on the Concorde.**

∴ **Nicholas wants something.**

The natural explanation is to construe (16) as an instance of Existential Generalization as applied to the unspecific reading of the premise. In other words, (16) ought to be read in analogy with (17) rather than with (12):

(17) **Geach is selling *Methods of Logic*.**

∴ **Geach is selling something.**

The explanation gains force from the observation that the conclusion of (16) can be elucidated by a locution like ... **namely, a free trip on the Concorde**, just as the conclusion of (17) can be expanded by ... **namely, *Methods of Logic***. If this reasoning is correct, then the object of (18) [= the conclusion of (16)] must have a reading as a quantifier over the meanings of unspecific objects of opaque verbs; in other words, **something** must be able to quantify over quantifiers.

(18) **Nicholas wants something.**

The phenomenon of (existential) quantification over quantifiers is restricted to sentences like (18), where the quantifier is "pronominal". As a case in

[15] Asher (1987), to whom the example is due, and Heim (1992) provide explanations in terms of desire as a propositional attitude. This kind of explanation would be available under any account of opacity (including Montague's) as long as it is *consistent with* lexical decomposition.

point, the conclusion (19) of the above inference (15) does not allow for a higher-order interpretation — at least not, if the inference is indeed faulty:

(19) **Nicholas wants a trip on the Concorde.**

3.2. Something relativized. Further evidence for the existence of higher-order readings of sentences with opaque verbs comes from examples involving relative clauses:

(20) **Geach is looking for something Quine is looking for.**

Semantic folklore has it that the relative clause expresses a property first obtained by abstracting from the missing object (represented by a trace variable) and then combined with the noun by intersection.[16] As always in relative clauses, the trace stands in for an individual term so that the verb must first be argumentlowered[17] before combining with it. Then abstraction can take place at the relative clause level, resulting in:

(21) $\lambda y \boldsymbol{T}(\boldsymbol{q}, \boldsymbol{F}(\boldsymbol{q}, y))$.

Since the relative clause is restrictive (as indicated by the absence of an overt relative pronoun), its interpretation requires a certain amount of re-bracketing; due to compositionality reasons, the relative clause must be attached to the abstract noun **-thing** rather than the entire quantifier **something**.[18] As this head noun is void of content, intersection with it has no effect and (21) also serves as the argument to the binary quantifier denoted by **some**. The resulting (standard) interpretation of the object in (20) is:

(22) $\lambda Q(\exists y)[\boldsymbol{T}(\boldsymbol{q}, \boldsymbol{F}(\boldsymbol{q}, y)) \wedge Q(y)]$.

Given Montague's scoping mechanism *M2*, we now obtain two readings of (20), depending on whether the object takes scope over the opaque verb (22′) or not (22″):

(22′) $(\exists y)[\boldsymbol{T}(\boldsymbol{q}, \boldsymbol{F}(\boldsymbol{q}, y)) \wedge \boldsymbol{T}(\boldsymbol{g}, \boldsymbol{F}(\boldsymbol{g}, y))]$

(22″) $\boldsymbol{T}(\boldsymbol{g}, (\exists y)[\boldsymbol{T}(\boldsymbol{q}, \boldsymbol{F}(\boldsymbol{q}, y)) \wedge \boldsymbol{F}(\boldsymbol{g}, y)])$

On the first, wide-scope reading, (20) says that Quine and Geach are looking for the same (specific) object; on the second, unlikely reading, it says that Geach is after anything specifically sought by Quine.

---

[16] This interpretation of relative clauses was already proposed by Quine (1960: 110ff.) and later adopted by Montague (1970b).

[17] Alternatively, the trace may be Montague-lifted, of course.

[18] This complication is well known. See Heim & Kratzer (1998: 82f.) for discussion and references.

TABLE 1. Types and denotations of **some-thing**.

		Lower Order	Higher Order
something	*type*	$(et)t[=q]$	$(qt)t$
	denotation	$\lambda Q(\exists x)Q(x)$	$\lambda\Pi(\exists\Omega)\Pi(\Omega)$
some	*type*	$(et)((et)t)$	$(qt)((qt)t)$
	denotation	$\lambda P\lambda Q(\exists x)[P(x)\wedge Q(x)]$	$\lambda\Sigma\lambda\Pi(\exists\Omega)[\Sigma(\Omega)\wedge\Pi(\Omega)]$
-thing	*type*	et	qt
	denotation	$\lambda y(y=y)$	$\lambda\Omega(\Omega=\Omega)$

3.3. Something higher-order. Neither of the two above readings of (20) covers a situation in which both Geach and Quine are looking for a textbook on medieval logic without either of them looking for any particular book. It is obvious how to formalize this reading using higher-order quantification:

(23) $(\exists\Omega)[T(q,(\Omega y)F(q,y)) \wedge T(g,(\Omega y)F(g,y))]$.

On the Montagovian account of opacity, such higher-order readings are readily available given the following assumption:[19]

HOQ Higher-order quantification

The word **something** *is ambiguous, being both an ordinary quantifier over individuals and a higher-order quantifier over ordinary quantifiers.*

More precisely, *HOQ* should be read as assigning to **something** an (ordinary) quantifier of type $q = (\underline{e}t)t$ as well as a higher-order reading of type $(\underline{\underline{q}}t)t$, where the domain of individuals is replaced by the domain of (ordinary unary) quantifiers. Adapting the standard semantics of relative clauses, let us divide the higher-order variant of **something** into determiner and noun, both of which will have to be typeadapted accordingly. The following table specifies all types and denotations according to *HOQ*:

Together, *HOQ* and the quantification mechanism *M2* thus predict two sources of ambiguity: the ***L*** vs. ***H*** readings of **something** on the one hand, and the *Narrow scope* vs. *Wide scope* interpretations of the object on the other. One may expect these parameters to vary freely in sentences like:

(24) **Geach is looking for something.**

However, one of the four combinations, viz. ***NH***, leads to a *type clash*: the verb, being of type $q(et)$, cannot cope with an argument of type $(qt)t$. (24) thus ends up with three readings:

[19]Cf. Moltmann (1997: 20) for a more general proposal and Zimmermann (1993: 171ff.) for a similar one in a different framework (cf. fn. 12), where "higher-order" means *expressing a property of properties*. For simplicity, I assume that the two uses of **something** constitute separate readings; this will not affect the present discussion.

TABLE 2. Readings of (24).

Order→ / *↓Scope*	*Low*	*High*
Narrow	$\boldsymbol{T}(\boldsymbol{g}, (\exists y)\boldsymbol{F}(\boldsymbol{g}, y))$	–
Wide	$(\exists y)\boldsymbol{T}(\boldsymbol{g}, \boldsymbol{F}(\boldsymbol{g}, y))$	$(\exists \mho)\boldsymbol{T}(\boldsymbol{g}, (\mho y)\boldsymbol{F}(\boldsymbol{g}, y))$

The ***L*** readings are just as in the above treatments (8) and (9c) of (7a) [**Jones seeks a unicorn**], only with the relativization to unicorns lifted. It may be noted in passing that, as it stands, the ***H*** reading is not entirely adequate. Rather, higher-order existential quantification should somehow be restricted because otherwise the quantifier could be instantiated by **nothing**, thus allowing for unwelcome inferences like:

(25) **I owe you nothing.**

∴ **I owe you something.**

Presumably, the restriction would have to be on (ordinary) existential quantifiers;[20] I leave the matter open, because it will not be of any importance in the following.

In order to obtain (23) using the lexical hypothesis *HOQ*, one only needs to generalize the above relative clause treatment to higher-order traces. Applying the same re-bracketing as before, the relative clause in (20) would then receive the following interpretation:

(26) $\lambda\mho\, \boldsymbol{T}(\boldsymbol{q}, (\mho y)\boldsymbol{F}(\boldsymbol{q}, y))$.

(26) can be obtained by directly combining the trace (the variable "$\mho$" of type q) with the opaque verb — it is just of the right type. Again, the head noun **-thing** is semantically trivial, so that the object of (20) receives the following higher-order interpretation:

(26′) $\lambda\Pi(\exists\mho)[\boldsymbol{T}(\boldsymbol{q}, (\mho y)\boldsymbol{F}(\boldsymbol{q}, y)) \wedge \Pi(\mho)]$.

Of course, this unary (higher-order) quantifier is due to the ***H*** interpretation of the determiner **some**. Finally, quantifying (26′) into the matrix **Geach is looking for $\mho$** leads to the desired result (23). To summarize, *HOQ* assigns to (20) the readings given in Table 3

Table 3 lists precisely those readings of (20) that can be obtained by letting the two parameters — type of [**some**]**thing**, scope of the object — vary as much as possible. Hence they are the readings one would initially expect, given the

[20]This is what would be expected under the property analysis (cf. fn. 12), according to which opacity only arises in connection with existential quantifiers (or their property counterparts). Note that the narrow-scope reading the Montagovian account of opacity assigns to the premise of (25) is at best marginal.

TABLE 3. Readings of (20).

	L	H
N	$T(g, (\exists y)[T(q, F(q, y) \wedge F(g, y))])$	–
W	$(\exists y)[T(q, F(q, y))\wedge T(g, F(g, y))]$	$(\exists\Omega)[T(q, (\Omega y)F(q, y))\wedge T(g, (\Omega y)F(g, y))]$

above setup. However, closer inspection of the interpretation mechanisms reveals that there are more complex ways of combining them. In particular, one may, as it were, activate the *W*ide scope parameter twice over by scoping the *H*igher-order reading of **something** over the opaque verb and at the same time assigning the variable bound by the quantifier wide scope. The result would be:

(27) $(\exists\Omega)[T(q, (\Omega y)F(q, y)) \wedge (\Omega y)T(g, F(q, y))]$.

(27) is true of a situation in which Geach happens to be looking for his favourite pencil, whereas Quine is just after some instrument or other to jot down a note. And more combinations along these lines are conceivable. It may well be that none of them constitutes a genuine reading of (20), so that the parameters underlying Table 3 are indeed correct.

Let us finally note that, given the (WH) reading (25) of (20), **seek** is unlikely to be monotonic in the sense indicated further above. For if it were, any two sentences of the forms **Geach is looking for *a*[*n*]** N and **Quine is looking for *a*[*n*]** N' would jointly imply (20). This fact may be illustrated by a specific example. Suppose **seek** were monotonic and (28) and (29) were true for their respective narrow and wide scope readings (28′) and (29′):

(28) **Geach is looking for a pen.**

(28′) $T(g, (\exists y)[P(y) \wedge F(g, y)])$

$\equiv$ $S(g, \lambda Q(\exists y)[P(y) \wedge Q(y)])$

(29) **Quine is looking for a book.**

(29′) $(\exists y)[B(y) \wedge T(q, F(q, y))]$

$\equiv$ $(\exists y)[B(y) \wedge S(q, \lambda P\, P(y))]$

Now, clearly, the following inclusions hold, where b is some specific book witnessing (29′):

(30) $[\lambda Q(\exists y)[P(y) \wedge Q(y)]] \subseteq [\lambda Q(\exists x)Q(x)]$

(30′) $[\lambda P\, P(b)] \subseteq [\lambda Q(\exists x)Q(x)]$

Hence, by monotonicity, (28′) and (29′) respectively imply:

(28″) $\boldsymbol{S}(\boldsymbol{g}, \lambda Q(\exists x)Q(x))$

$\equiv$ $\boldsymbol{T}(\boldsymbol{g}, (\underline{\underline{\exists}} y)\boldsymbol{F}(\boldsymbol{g}, y))$

(29″) $\boldsymbol{S}(\boldsymbol{q}, \lambda Q(\exists x)Q(x))$

$\equiv$ $\boldsymbol{T}(\boldsymbol{q}, (\underline{\underline{\exists}} y)\boldsymbol{F}(\boldsymbol{q}, y))$

But then the denotation of $\underline{\underline{\exists}}$ of **something** satisfies the quantificational matrix of (25), thus verifying (20).

One may summon pragmatics to protect monotonicity from this absurdity. Maybe if (28) and (29) are true, then so is (20), but it would nevertheless be misleading to utter the sentence. It is by no means obvious that any such reasoning bears scrutiny. In particular, in fleshing it out one would somehow have to draw a line between illicit generalizations such as (28′) and (29′) and perfectly natural monotonicity inferences like (13) and (14) above. And though this line must be drawn with or without monotonicity, in a pragmatic account one would have to rely on rational principles of effective communication, whereas without monotonicity the line could be drawn conventionally and/or conceptually. In the absence of any particular proposal, the alleged monotonicity of **seek** should therefore be taken with caution.

§4. Specificity.

4.1. Eliminating essential propositions. The ***WL*** reading of (24) above is true if, and only if, the referent of the subject (Geach) bears a certain psychological attitude (of trying) to a proposition of a certain form (that Geach find a specific individual), viz.:

(31) $\boldsymbol{F}(\boldsymbol{g}, y)$.

Propositions of this form are *essential*, covering precisely the worlds in which a given individual (y) has a given property (being found by Geach); in other words, essential propositions are those of the form $P(x)$, for fixed (non-trivial) properties P and individuals x.[21] However, it has been argued — correctly, I believe — that essential propositions are well beyond the cognitive grasp of ordinary human beings.[22] Take Geach. Only five minutes ago, he

[21] Essential propositions are as close as possible worlds semantics can get to Kaplan's (1989) *Russellian* or *singular* propositions (which is why they are sometimes referred to as such). Note that the ***WL*** reading in Table 2 does not require the essential proposition to be unique: there may be more than one specific individual sought by Geach. Nor is the form of the essential proposition(s) unique: in the case at hand, it may also be described as the set of worlds in which Geach ($= x$) has the property P of seeking the particular object y.

[22] Lewis (1981). The argument depends on the — initially plausible, but hotly debated — assumption that attitudes and intentions can be characterized in purely qualitative terms, which I will simply take for granted.

held his copy y of Buridan's *Sophismata* in his hand, but now it looks like it disappeared. So he starts looking for y. Does he thus bear the attitude of trying towards the set $f(y)$ of indices at which he (Geach) finds y? No — at least not if that attitude is reconstructed in the following way:

(T) **try** expresses a relation holding between an individual x and a proposition p at an index i if and only if, at i, x performs an action the goal of which is to bring about a situation (index) of which p holds.

However, in order for Geach to direct his action(s) to $f(y)$, it appears that he would have to be able to distinguish the indices of which $f(y)$ holds from those of which $f(y)$ does not hold — which he cannot because, from his perspective, a possible world that differs from ours only in that some other copy z of the *Sophismata* took y's place would be indistinguishable from reality. Hence it is not y that Geach is after but *his copy of* Sophismata. In general, if some person x appears to be reported as standing in the relation of trying (or any other psychological attitude) to an essential proposition like $f(y)$, the report should be construed as being about x standing in a relation to some proposition $f(c)$, where c is an individual concept suitably connecting x and y. More specifically, and following Kaplanian lines[23], the ***WL*** reading of (24) reports Geach (i) to have an internal *vivid name* for (or a *de re access* to) y and (ii) to try to find the individual with that name (or thus accessed). Formalization of (i) can then proceed via a ternary predicate ***VN*** relating a concept (i.e. access) N (of type *et*) and two individuals (Geach and his copy of *Sophismata*, in reverse order)[24], whereas (ii) is an ordinary *de dicto* attitude ascription involving the same concept N. The following formula thus captures the truth-conditions of the ***WL*** reading of (24) more adequately than the formalization in Table 2:

(32) $(\exists y)(\exists N)[\boldsymbol{VN}(N, y, \boldsymbol{g}) \wedge \boldsymbol{T}(\boldsymbol{g}, (\exists z)[N(z) \wedge \boldsymbol{F}(\boldsymbol{g}, z)])]$.

Similarly, (33) is a more adequate formalization of the *de re* reading of (7a) than was Montague's (9c):

(7a) **Jones seeks a unicorn.**

(9c) $(\exists y)[\boldsymbol{U}(y) \wedge \boldsymbol{T}(\boldsymbol{j}, \boldsymbol{F}(\boldsymbol{j}, y))]$

(33) $(\exists y)(\exists N)[\boldsymbol{U}(y) \wedge \boldsymbol{VN}(N, y, \boldsymbol{j}) \wedge \boldsymbol{T}(\boldsymbol{j}, (\exists z)[N(z) \wedge \boldsymbol{F}(\boldsymbol{j}, z)])]$

---

[23]Kaplan (1969). A more sophisticated version, augmenting propositions and accesses by subjective perspectives, has been developed by Lewis (1979). To avoid unnecessary distractions, I want to steer clear of these complications. For the same reason I have been assuming that the infinitives in the paraphrases stand for propositions rather than properties.

[24]The precise nature of ***VN*** shall not concern us here. Suffice it to say that ***VN*** is likely to be context-dependent (hence the italics) and that $\boldsymbol{VN}(N, y, x)$ and $\boldsymbol{VN}(N, z, x)$ together imply $N(y)$ and $y = z$; in other words, N must be a partial individual concept of y. More substantial conditions can be gleaned from the literature from Kaplan (1969) onward; see especially Aloni (2001: ch. 2) for a thorough discussion and a specific proposal.

The question is how to arrive at (32) given the syntactic input (7c):**

(7c)

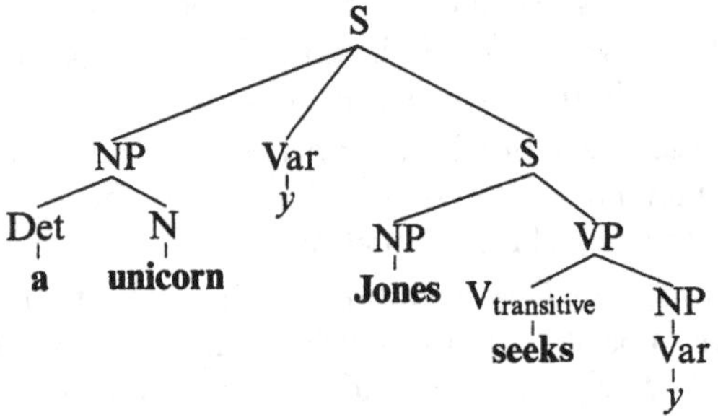

4.2. Specificity by coercion. As already noted in Section 2.3, the combination of the trace variable y in (7c) with the opaque verb leads to a type mismatch — the former being of type e while the latter expects a quantifier of type $(et)t$. In the simplistic Montagovian interpretation of the *de re* construction there are two equivalent ways out of the embarrassment:

ML Montague Lift

The Montague-lifted version of an individual x of type e is the following quantifier of type q:

$$\lambda P\, P(x)$$

AL Argument Lowering

The argument-lowered version of a relation $\mathfrak{R}$ *of type* $q(et)$ *is the following relation of type* $e(et)$*:*

$$\lambda y \lambda x \mathfrak{R}(x, \lambda P\, P(y))$$

When applied to the decomposable verb **seek**, these two (alternative) type shifting strategies yield:

(34a)

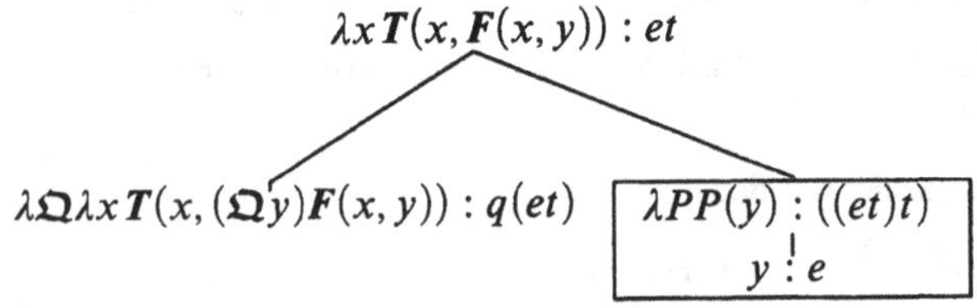

Montague Lifting

(34b)

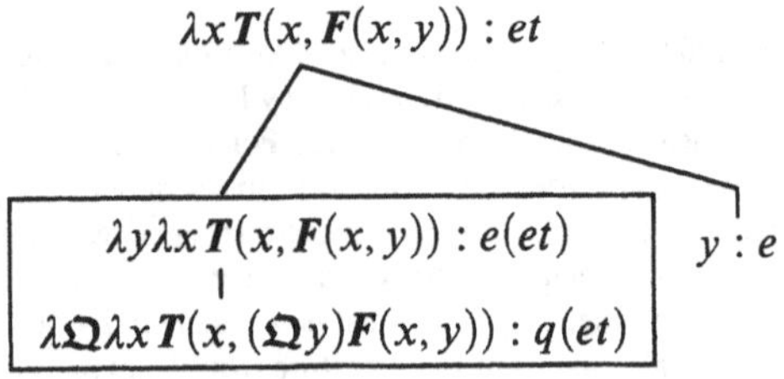

Argument Lowering

Either strategy is meant to apply when a more straightforward combination of meanings (like application, composition, or intersection) is unavailable. In other words, (34a) and (34b) are semantic *coercion mechanisms*[25] that see to it that one of the initially unfitting expressions is re-interpreted in a way that allows for a straightforward combination with the other one. In the case at hand, the specific construal of opaque verbs, the two relevant expression are (a) the opaque verb as obtained by analyzing the unspecific use and (b) the variable bound by its quantified object. The two strategies only differ with regard to which of (a) and (b) must give way. But they agree both on their starting point and on the result.

In the more sophisticated Kaplanian *de re* approach, the type mismatch is the same. However, the strategy of type-shifting the argument of the opaque verb is unavailable. Not only does ***ML*** yield the wrong result — this is what this whole section has been all about; it is even beyond repair. The reason is that, on the one hand, the neo-Kaplanian *de re* construal requires the opaque verb to combine with an individual to form a property, viz. the following:

(35) $\lambda x(\exists N)[\boldsymbol{VN}(N,y,x) \wedge \boldsymbol{T}(x,(\exists z)[N(z) \wedge \boldsymbol{F}(x,z)])]$.

On the other hand, there is no compositional way of re-interpreting the variable y as a quantifier y^+ such that the result (36) of applying S[eek] to y^+ is logically equivalent to (35).[26]

(36) $\lambda x \boldsymbol{T}(x, y^+(\lambda y \boldsymbol{F}(x,y)))$

$\equiv$ $\boldsymbol{S}(y^+)$

To see this, it suffices to consider indices at which $\boldsymbol{T}$ denotes the universal relation holding between all individuals and propositions, while the extension of $\boldsymbol{VN}$ is empty. Then, whichever quantifier y^+ may be, (36) and (35) will denote the universal and the empty property (of individuals) respectively.

Even though ***ML*** does not carry over, compositionality is not lost on the improved *de re* construal (33). For it turns out that the strategy employed in (34b) can be adapted. This is immediately clear from the fact that the desired result (35) is equivalent to:

(37) $\lambda x(\exists N)[\boldsymbol{VN}(N,y,x) \wedge \boldsymbol{S}(x, \lambda P(\exists z)[N(z) \wedge P(z)])]$.

---

[25] I am using the term *coercion* in a more general way than Pustejovsky (1993), whose use of the term would cover Montague Lifting, but not Argument Lowering. See Blutner (2002) for a recent survey of coercion and related mechanisms. The approach taken there — *underspecification* to be resolved by contextual determination of hidden parameters — does not seem to be applicable in cases of type variance as considered in the present paper.

[26] One may, however, reverse functor-argument structure by re-interpreting y as an operation of type $(q(et))(et)$ that yields (35) when applied to $\boldsymbol{S}$, viz.:

$$\lambda \mathfrak{R} \lambda y \lambda x(\exists N)[\boldsymbol{VN}(N,y,x) \wedge \mathfrak{R}(x, \lambda P(\exists z)[N(z) \wedge P(z)])]$$

Note how this operation, to which we will return in Section 6.2, mimicks *De Re* Lowering.

Hence one may refine (34b) to obtain the perfectly compositional:

(38) $$\lambda x(\exists N)[\boldsymbol{VN}(N,y,x)\wedge \boldsymbol{S}(x,\lambda P(\exists z)[N(z)\wedge P(z)])]:et$$

$$\lambda y\lambda x(\exists N)[\boldsymbol{VN}(N,y,x)\wedge \boldsymbol{S}(x,\lambda P(\exists z)[N(z)\wedge P(z)])]:e(et) \qquad y:e$$

$$\boldsymbol{S}:((et)t)(et)$$

De Re Lowering

More precisely, and more generally, the operation of *De Re Lowering* is a type shift turning an attitude to unspecific objects into a binary relation between individuals:

DRL **De Re *Lowering***

The de-re-lowered version of a relation $\mathfrak{R}$ *of type* $q(et)$ *is the following relation of type* $e(et)$*:*

$$\lambda y\lambda x(\exists N)[\boldsymbol{VN}(N,y,x)\wedge \mathfrak{R}(x,\lambda P(\exists z)[N(z)\wedge P(z)])]$$

This comparatively simple (*sic!*) compositional derivation of the specific reading may come as a surprise given the complications of quantifying into overt attitude contexts.[27]

In its unspecific use, then, the verb **seek** has its *notional* sense, expressing a relation between a person and a concept, represented by the indefinite; in the specific reading the verb appears in a purely *objectual* sense, expressing a relation between the attitude subject and an individual quantified over by the indefinite.[28] The choice between the two uses is one of grammatical environment — or so it seems; if the verb directly combines with a quantificational object, its notional sense will be activated; but when combined with an individual term (like a variable bound by the object), then the verb contributes its objectual sense. In the case of our example **seek**, (39a)[= (6) above] represents the notional sense, whereas (39b) is the neo-Kaplanian account of its objectual sense:

(39a) $\lambda\mathfrak{Q}\lambda x\,\boldsymbol{T}(x,(\mathfrak{Q}y)\boldsymbol{F}(x,y))$

(39b) $\lambda y\lambda x(\exists N)[\boldsymbol{VN}(N,y,x)\wedge \boldsymbol{T}(x,(\exists z)[N(z)\wedge \boldsymbol{F}(x,z)])]$

According to the coercion view, which is the standard view, the objectual sense (39b) can and must be derived from the notional sense. But the coercion view is not the only one.

---

[27]Cf. Cresswell & von Stechow (1982: 525ff.). Quantifying into opaque verbs is simpler in that the position quantified over is fixed by the lexical meaning of the verb, whereas in overt clausal embeddings there is not even an upper limit to the *number* of quantifiers to outscope the attitude. However, a more general *de re* mechanism would be needed for *unspecific* opaque objects containing names or descriptions. We will have to return to these matters in Section 6.2.

[28]The terms *notional* and *objectual* are Quine's (1956) and Forbes's (2000), respectively.

4.3. Specificity by indeterminacy. *De Re* Lowering is considerably more complex than either of the traditional strategies exemplified in (34). And although it, too, is a type coercion mechanism, it is one of a peculiar kind. For, unlike Montague Lifting and Argument Lowering, *De Re* Lowering is not a *logical* operation in that it introduces non-logical material in the form of the relation $\boldsymbol{VN}$. This fact may be taken as evidence that $\boldsymbol{DRL}$ is of a lexical rather than a syntactic nature. The interpretation of syntactic constructions and processes is usually taken to be a matter of logical combination, substantial content being confined to the lexicon. Although this separation is rarely made explicit, it does underlie most work in logically oriented natural language semantics.

The facts that the objectual sense of an opaque verb is quite remote from its notional sense and that it cannot be obtained by logical combination may indicate that type coercion is not involved at all. After all, one of the premises of the coercion account was that a type clash occurs between (a) the opaque verb *as obtained by analyzing the unspecific use* and (b) the variable bound by its quantified object. In other words, the coercion approach takes it for granted that the natural function of an opaque verb is as a host to a quantifier, where it gives rise to the unspecific reading, and that the specific usage must be derived from that. This may be, and in fact has been, disputed.[29] Of course, a mere lexical ambiguity seems unlikely, because the same variation between a notional and an objectual sense appears to occur in many opaque verbs.[30] Instead, the two uses of the predicate could be two facets of one single overall meaning of **seek**. The idea is this: among the things eligible for being sought are both ordinary individuals (of type e) and unspecific objects (quantifiers of type q), and what it means to be sought by someone depends on the type of these things: (!) an *individual* is sought by trying to find whatever fits the vivid name the subject has of that individual; a *quantifier* is sought by trying to make the preoperty of being found by the subject acceptable to (i.e. an element of the extension of) the quantifier. This idea is fleshed out by merging the two types of objects sought into one denotation of **seek**:

$$\text{(39c)}\quad \lambda\boldsymbol{\Omega}\lambda x\,\boldsymbol{T}(x,(\boldsymbol{\Omega}y)\boldsymbol{F}(x,y)) \oplus \lambda y\lambda x(\exists N)[\boldsymbol{VN}(N,y,x) \wedge \boldsymbol{T}(x,(\exists z)[N(z)\wedge \boldsymbol{F}(x,z)])]$$

$$\equiv\quad \boldsymbol{S}\oplus \lambda y\lambda x(\exists N)[\boldsymbol{VN}(N,y,x)\wedge \boldsymbol{S}(x,\lambda P(\exists z)[N(z)\wedge P(z)])]$$

[29] During the past decade, various colleagues suggested to me that specific readings of opaque verbs may not be the results of type coercion but rather instances of semantic indeterminateness; I am not sure I can remember all sources (let alone the earliest one). The present investigation was kicked off by a discussion during the Rutgers seminar mentioned in the Acknowledgements, where the indeterminacy view was brought up and defended by Matthew Stone.

[30] . . . though not in all of them. As announced in fn. 2 we will return to this point in Section 7.

The formula merely merges the notional sense (39a) of **seek** with the neo-Kaplanian account (39b) of its objectual sense, using a *type-transcendent* version $\oplus$ of set-theoretic union. The precise nature of the formula and the language it belongs to need not concern us here, as long as it can be made sense of.[31] And that sense clearly captures the characterization (!) of the extension of **seek**.

According to (39c), then, the difference between the notional and the objectual sense of **seek** — or any other opaque verb — is not conceptual, or lexical. The meaning of the word **seek** is a certain relation between individuals and all kinds of objects. What precisely it means for an individual to stand in that relation to a given object may very much depend on the nature of that object and, in particular, on its logical type. But that does not mean that which relation the verb denotes depends on that object. The relation is always the same, but it is *indeterminate* as to the kind of objects that it connects. In (39c) this relation is defined using the same notation as in the standard coercion account. It would have been more faithful to the spirit of indeterminacy, of course, to represent the type-transcendent relation by a single letter, "Σ" say; but (39c) will largely facilitate the comparison in the sections to follow. Still, its disjunctive form should be viewed as an artefact of the representation and should not distract from the fact that, according to the indeterminacy view, only one relation holds between subjects and different kinds of objects.

Though (39c) is in line with the idea informally expressed in (!), it is not obvious that idea itself is in line with pre-theoretic intuition. In fact, if (40) and (41) happen to be true for their respective unspecific and specific readings, the sense in which Quine is related to the abstract object of an unspecific pen (represented by a quantifier) appears to differ from that in which Geach is related to his particular writing utensil:

(40) **Quine seeks a pen.**

(41) **Geach seeks a pen.**

What makes (41) true is the fact that there is this specific object that Geach is quite literally looking for. However, no such object — whether unspecific, abstract, intensional, or what-have-you — exists in Quine's case. The following paraphrase of (41) into semi-technical, Quinean jargon helps bringing out the point:

(42) **A pen is such that Geach <u>seeks</u> it.**

---

[31] Obviously, (39c) is not a formula of the type-theoretic language we have been using so far: the union of two relations of different types is type-transcendent, i.e. it cannot itself have a type in the ordinary sense. However (39c) can be expressed in a straightforward extension of that language, allowing for unions of types as functional domains. Details are given in the appendix.

It seems that the underlined sense of **seek** cannot be quite the same as the one featuring in the *unspecific* reading of (40): the relation (expressed by unspecific **seek**) in which Quine is reported to stand to whatever he is after does not seem to be the relation in which Geach stands to the particular pen he is reported to be looking for. Again, Geach is quite literally looking for this object; but Quine is not looking for any object — whether pen or quantifier — in *that literal sense*. True, he is looking for something — an unspecific pen to be sure — but not in the way in which Geach is looking for his dear old Conway Stewart. At least these are my intuitions. But then good theories need not square with pre-theoretic intuitions, as long as they are sound and correct in their predictions.

According to the indeterminate interpretation (39c) of **seek**, matters are quite different. Here Quine and Geach do stand in the same relation to an unspecific (and abstract) and a specific (and concrete) object, respectively. The difference is purely circumstantial, not linguistic, let alone conceptual. It is simply that the way Quine, or anyone, goes about looking for an unspecific object is rather different from the way in which Geach, or anyone, is looking for a specific object. In that respect, seeking is like peeling. Peeling an egg is quite different from peeling a potato. But that does not mean that there are two concepts of peeling, one for eggs and one for potatoes (and yet another one for salami). The difference only lies in the objects. And just like eggs and potatoes are radically different objects when it comes to peeling them, so are specific and unspecific objects (or individuals and quantifiers) when it comes to seeking them. In particular — so the indeterminacy view goes — there is only one concept of seeking.

Let us sum up. In the coercion view, the unspecific and the specific construal of an opaque verb are distinct relations. In the case of **seek**, the former is the relation $\boldsymbol{S}$ given in (39a); the latter, which is derived from it and to which we will refer as "$\boldsymbol{S}^*$", is specified in (39b). Intuitively, the difference shows in that standing in $\boldsymbol{S}$ to something is not the same as standing to it in $\boldsymbol{S}^*$. In the indeterminacy view, $\boldsymbol{S}$ and $\boldsymbol{S}^*$ are parts of the same relation $\boldsymbol{\Sigma}(=\boldsymbol{S} \cup \boldsymbol{S}^*)$, and the purported difference is an illusion due to the fact that the kinds of objects to which one can stand in $\boldsymbol{S}$ are different from the kinds of objects to which one can stand in $\boldsymbol{S}^*$. It seems that intuition is not too helpful when it comes to deciding between the two approaches to the lexical meaning of opaque verbs. Let us therefore look for (!) more decisive evidence.

§5. Higher-order specificity.

5.1. Higher-order findings. In the indeterminacy view, the opaque verb **seek** unequivocally expresses a relation $\boldsymbol{\Sigma}$ that may hold between a subject x and an object O independently of O's type — x and O just have to be suitably related.

Although it may be hard to come up with a complete characterization of this relation, there are some obvious conditions, satisfaction of which guarantees that Σ holds between a given x and O. And this is where, according to the indeterminacy view, types may come in. Some such (sufficient) conditions on being related by Σ can only be met by objects of suitable types. For instance, Σ holds between x and O if x is **T**rying for the property of being **F**ound by x to be in the extension of O:

- *Unspecificity Criterion*

$$\boldsymbol{T}(x, O(\lambda y \boldsymbol{F}(x, y))) \qquad [\equiv \boldsymbol{S}(x, O)]$$

If the criterion is met, Σ holds between x and O and a sentence expressing this is perceived as being unspecific; hence the name. Obviously, the Unspecificity Criterion can only be met if the object O can be applied to properties of individuals. In a misleading way, this is borne out by the formalization. Read as an expression of ordinary type logic, the formula is only well-formed if "O" denotes a quantifier. Otherwise a type clash would ensue: "$\boldsymbol{F}$" needs an individual expression as its second argument so that the argument of "O" comes out as being of type *et*; and since "$\boldsymbol{T}$" takes a proposition as its right argument, O must be of type $(et)t$. However, if we follow the lead of the indeterminacy approach, relations like the denotation $\boldsymbol{F}$ of **find** need not be homogeneous, only connecting objects of fixed types. In particular, the type of the bound variable "y" need not be determined by the predication "$\boldsymbol{F}(x, y)$". Reading in this more liberal way, we find that, in order to meet the Unspecificity Criterion, an object O must be some sort of quantifier (or operator), but not necessarily one over individuals. For instance, the ***H***igher-order reading of **something** could satisfy the Unspecificity Criterion, provided that ***F***inding that quantifier could be made any sense of. On the other hand, an individual definitely cannot satisfy the Unspecificity Criterion (in lieu of "O"), simply because, I take it, it cannot be applied as an operator to a property. From the indeterminacy point of view, this explains why individual-denoting objects cannot be construed unspecifically.

The specific construal, too, may be seen as a special case of satisfying the indeterminate relation Σ. We may formulate the corresponding criterion to be met by a subject x and an object O in Kaplanian terms:

- *Specificity Criterion*

$$(\exists N)[\boldsymbol{VN}(N, O, x) \wedge \boldsymbol{T}(x, (\exists z)[N(z) \wedge \boldsymbol{F}(x, z)])] \qquad [\equiv \boldsymbol{S}^*(x, O)]$$

Again it may first seem that the condition on O determines its type: read as a formula of ordinary type logic, the criterion demands O to be an individual; for this is the type of objects occupying the third argument position of the ***V***ivid

*N*ame relation. However, as before, one may wonder whether the condition extends to objects of other types. For this to be the case, the attitude subject x would have to have a vivid name N, i.e. some sort of qualitative description, of something that is not an individual. Since N is a description of O, the third argument of the embedded ***F***ind predicate (occupied by z) should be of the same type as O. Hence, whether such a case ever arises again largely depends on whether the achievement of ***F***inding can be directed to objects that are not individuals.

The type underspecification in the above criteria opens the possibility of one object satisfying both, possibly in relation with different subjects. In particular, it is conceivable that a subject x and an individual quantifier $\boldsymbol{\Omega}$ of type q satisfy the Unspecificity Criterion, while some (possibly distinct) individual y and the same quantifier $\boldsymbol{\Omega}$ meet the Specificity Criterion. If, moreover, $\boldsymbol{\Omega}$ is denoted by some indefinite ***a*[*n*]** N, then, according to the indeterminacy view, both X **seeks *a*[*n*]** N and Y **seeks *a*[*n*]** N would be true, where X and Y are names of x and y, respectively. Moreover, since both x and y bear $\boldsymbol{\Sigma}$ to $\boldsymbol{\Omega}$, then so would the co-predication X **and** Y **seek *a*[*n*]** N be true[32]. On the other hand, one may expect the coercion analysis to predict such co-predications as zeugmatic, the apparent common satisfaction of ... **seeks *a*[*n*]** N by x and y being a case of equivocation. It would thus seem that such cases may form a test ground for the two rivalling approaches to the lexical semantics of opacity.

If such cases exist. The following two fictitious scenarios are as close as I have come to finding them:

SCENARIO 1. Geach hears that Quine has developed a serious case of *meinongitis*, a disease that leads victims to believe in all kinds of weird objects, like gold mountains, or cardinal numbers. In fact, rumour has it that Quine spends his time on gigantic search operations in order to prove the existence of such objects. In order to find out whether there is anything to these stories, Geach snitches Quine's diary and scans it for evidence of search preparation: page for page, *Geach is looking for something Quine is looking for*.

SCENARIO 2. Geach hears that Quine has developed a serious case of *meinongitis*, a disease that makes one believe in all kinds of weird objects, like gold mountains, or cardinal numbers. In fact, rumour has it that Quine donated a large sum for the protection of the endangered species of mermaids and that he has just left for an expedition to discover the last specimens in their natural habitat. Alarmed by the bewildering news, Geach decides to search Quine's office for pertinent evidence. What he is after is some kind of record of

[32] *Co-predication* is distributive application of a given predicate to a given group; the term originates with Copestake & Briscoe (1995), I believe.

Quine's quest for mermaids: in a sense, *Geach is looking for something Quine is looking for*.

In either of the stories the final sentence is intended to describe Geach as trying to find an unspecific object of Quine's intentions.[33] Though both utterances are clearly marginal, the second one seems even harder to get, presumably due to a difference in specificity. For, unspecific though Quine's object of desire in Scenario 2 may be, it is not unspecific for Geach, who, somewhat paradoxically, may be described as looking for a specific unspecific object — or, less paradoxically: as specifically looking for an object of an unspecific quest. Not so in the other case, where Geach is not committed to finding a specific such object, as long as it is proof of Quine's unfortunate state of mind (or brain). Hence, with apparent redundancy, Geach may be described as looking for an unspecific object — or, less oddly: as unspecifically looking for an object of an unspecific quest.

Let us first take a look at Scenario 1. Its final sentence (20) has already been considered. However, the readings that we have come across so far all miss the point of the above story:

(20) **Geach is looking for something Quine is looking for.**

(22′) $(\exists y)[\boldsymbol{T}(\boldsymbol{q}, \boldsymbol{F}(\boldsymbol{q}, y)) \wedge \boldsymbol{T}(\boldsymbol{g}, \boldsymbol{F}(\boldsymbol{g}, y))]$

(22′*) $(\exists y)(\exists N)(\exists M)[\boldsymbol{VN}(N, y, \boldsymbol{q}) \wedge \boldsymbol{T}(\boldsymbol{q}, (\exists z)[N(z) \wedge \boldsymbol{F}(\boldsymbol{q}, z)])$
$\wedge\, \boldsymbol{VN}(M, y, \boldsymbol{g}) \wedge \boldsymbol{T}(\boldsymbol{g}, (\exists z)[M(z) \wedge \boldsymbol{F}(\boldsymbol{g}, z)])]$

(22″) $\boldsymbol{T}(\boldsymbol{g}, (\exists y)[\boldsymbol{T}(\boldsymbol{q}, \boldsymbol{F}(\boldsymbol{q}, y)) \wedge \boldsymbol{F}(\boldsymbol{g}, y)])$

(22″*) $\boldsymbol{T}(\boldsymbol{g}, (\exists y)(\exists N)[\boldsymbol{VN}(N, y, \boldsymbol{q}) \wedge \boldsymbol{T}(\boldsymbol{q}, (\exists z)[N(z)$
$\wedge\, \boldsymbol{F}(\boldsymbol{q}, y)]) \wedge \boldsymbol{F}(\boldsymbol{g}, y)])$

(23) $(\exists \boldsymbol{\Omega})[\boldsymbol{T}(\boldsymbol{q}, (\boldsymbol{\Omega} y)\boldsymbol{F}(\boldsymbol{q}, y)) \wedge \boldsymbol{T}(\boldsymbol{g}, (\boldsymbol{\Omega} y)\boldsymbol{F}(\boldsymbol{g}, y))]$

Given Scenario 1, (22″) is false, and a Kaplanian *de re* construal (22″*) does not save it: plainly, Geach's goal is not to find objects Quine is *specifically* looking for, but rather concerns Quine's putative *unspecific* searches. Also, the wide-scope readings (22′) (or its cumbersome Kaplanian version (22′*)) and (23) can be discarded because each implies that, in some sense, Quine is looking for something — which need not be the case according to the scenario. Rather, the relevant reading of (20) is one according to which

[33]In order to facilitate the comparison with the analyses provided in the previous sections, I have chosen an example involving the same attitude twice over. This is not essential. Something like **Geach is looking for something Quine owes him** would have done equally well: just imagine a scenario with Geach scanning through promissory notes and trying to ascertain that Quine still owes him something.

Geach is unspecifically looking for unspecific objects of Quine's intention. To see what this means, one should consider the following continuation:

SCENARIO 1 (continued). After hours of searching, he finally discovers the incredible truth in the entry for July 17, 1953, which reads: "Still no relief from mermaid deprivation — heading for Copenhagen now." So the search is over: *Geach found something Quine had been looking for.*

The object of Geach's act of finding, then, is an unspecific object of Quine's acts of seeking. Since, according to the Montagovian account, unspecific objects in this sense are (existential) quantifiers, it follows that the relevant notion of finding relates persons to quantifiers. Hence this particular, abstract sense of **find** must be of type $q(et)$, which is precisely what he had been after. Let us formalize this particular use of **find** as relating a specific and an unspecific object by the constant "$\boldsymbol{F}^{\#}$". The present tense version (43) of the last sentence of the continuation can then be treated as in (44), which involves a higher-order existential quantifier:

(43) **Geach finds something Quine is looking for.**

(44) $(\exists \mathfrak{Q})[\boldsymbol{T}(\boldsymbol{q}, (\mathfrak{Q}y)\boldsymbol{F}(\boldsymbol{q}, y)) \wedge \boldsymbol{F}^{\#}(\boldsymbol{g}, \mathfrak{Q})]$

Since Geach's goal was reached by finding an unspecific object of Quine's quest, it also follows that the relevant notion of seeking involves the abstract notion $\boldsymbol{F}^{\#}$ of finding. In other words, the missing reading of (20) can be formalized as

(45) $\boldsymbol{T}(\boldsymbol{g}, (\exists \mathfrak{Q})[\boldsymbol{T}(\boldsymbol{q}, (\mathfrak{Q}y)\boldsymbol{F}(\boldsymbol{q}, y)) \wedge \boldsymbol{F}^{\#}(\boldsymbol{g}, \mathfrak{Q})])$.

A similar consideration shows that the intended reading (45′) of the final sentence of Scenario 2 also involves higher-order finding $\boldsymbol{F}^{\#}$:

(45′) $(\exists \mathfrak{Q})[\boldsymbol{T}(\boldsymbol{q}, (\mathfrak{Q}y)\boldsymbol{F}(\boldsymbol{q}, y)) \wedge \boldsymbol{T}(\boldsymbol{g}, \boldsymbol{F}^{\#}(\boldsymbol{g}, \mathfrak{Q}))]$.

Marginal though it may be, (45′) is precisely the kind of reading relevant for a co-predication test: it meets the Unspecificity Criterion, ascribing to Geach a specific attitude to an object that at the same time satisfies the Unspecificity Criterion with respect to Quine. However, if (45′) is a reading of (20), then, *a fortiori* so is its unspecific variant (45), which is less strained to begin with. It will turn out, though, that the compositional derivation of (45) suffices to rule out the applicability of any co-predication test along the lines indicated. Hence we will ignore (45′) for most of the present section, only briefly returning to its marginal character towards the end. Let us now turn to the question of how to obtain (45) as an analysis of (20).

5.2. Seeking high, seeking low: ambiguity. The motivation behind the present investigation into these marginal examples is to analyze them in terms of indeterminacy and then apply a co-predication test. Before we can do this, however, we had better made sure that the higher-order findings are in fact cases of indeterminacy at all; otherwise (45) and (45′) would not instantiate the Specificity and Unspecificity Criteria in the first place. In other words, even if the specificity/ unspecificity contrast is accounted for in terms of indeterminacy, the one between $\boldsymbol{F}$ and $\boldsymbol{F}^{\#}$ may be of a different kind.[34] In fact, the simplest way of deriving (45) is to stipulate an ambiguity in the transparent verb **find** — $\boldsymbol{F}$ vs. $\boldsymbol{F}^{\#}$ — and have it carry over to the opaque verb **seek**, which then also gets two readings, according to whether the object to be found is an individual or a quantifier. In the first case, we have the familiar relation between individuals and quantifiers; in the second case, the latter quantify over the objects of **find** in the $\boldsymbol{F}^{\#}$-sense, i.e. they are higher-order quantifiers:

$(\boldsymbol{S}_{\Downarrow})$ $\lambda\boldsymbol{\Omega}\lambda x\,\boldsymbol{T}(x, (\boldsymbol{\Omega}y)\boldsymbol{F}(x, y))$ [= (6)]

$(\boldsymbol{S}_{\Uparrow})$ $\lambda\Pi\lambda x\,\boldsymbol{T}(x, (\Pi\boldsymbol{\Omega})\boldsymbol{F}^{\#}(x, \boldsymbol{\Omega}))$

The idea that the $\Uparrow/\Downarrow$-ambiguity of **seek** reflects an underlying ambiguity in **find** may be independently motivated by the above abstract construal (44) of (43). However, if **find** were ambiguous between $\boldsymbol{F}$ and $\boldsymbol{F}^{\#}$, i.e. (judging from the types) between a transparent and an opaque reading, one would expect the following sentence to be ambiguous between a transparent, a specific, and an unspecific reading[35]:

(49) **Geach finds a mermaid.**

(49t) $(\exists y)[\boldsymbol{M}(y) \wedge \boldsymbol{F}(\boldsymbol{g}, y)]$

(49s) $(\exists y)(\exists N)[\boldsymbol{VN}(N, y, \boldsymbol{g}) \wedge \boldsymbol{F}^{\#}(\boldsymbol{g}, \lambda P(\exists z)[N(z) \wedge P(z)])]$

(49u) $\boldsymbol{F}^{\#}(\boldsymbol{g}, \lambda P(\exists y)[\boldsymbol{M}(y) \wedge P(y)])$

And not only that. Given that (43) is true of the above continuation of Scenario 1, (49) would have to be true too; after all, its unspecific reading (49u) instantiates the existential quantification (44). This I take to be absurd: there is no ambiguity in (49), the sentence being unequivocally false for the scenario in question. Hence, whatever the source of the purported ambiguity in **seek** may be, it cannot be an analogous ambiguity of the verb

[34] I am deeply indebted to Nicholas Asher for pointing this out to me after a presentation of a previous version to this paper.

[35] (49t) and (49s) may turn out to be equivalent, however, given a detailed truth-conditional specification of $\boldsymbol{F}^{\#}$.

find. In particular, we still require an explanation for the observation that sentences like (43) produce the same kind of higher-order effects as do their opaque counterparts like (20). Given the close connection between the two, it seems more likely that there is a single account for both. I thus conclude that the ambiguity thesis is not viable. It was not very plausible to begin with anyway: $\boldsymbol{F}^{\#}$ construals also occur in other languages, and there are analogous phenomena with other opaque verbs. So there ought to be a systematic explanation.

5.3. Seeking high-or-low: indeterminacy. If ambiguity does not explain the higher-order ($\Uparrow$) uses of **seek**, maybe indeterminacy can. I can think of two ways the meaning of **find** may look if it is indeterminate as to the order of the specific objects:

$$(\boldsymbol{F}_{\oplus}) \quad [\lambda y \lambda x \boldsymbol{F}(x, y)] \oplus [\lambda \Omega \lambda x \boldsymbol{F}^{\#}(x, \Omega)] \qquad [\equiv \boldsymbol{F} \oplus \boldsymbol{F}^{\#}]$$

$$(\boldsymbol{F}_{\vee}) \quad \lambda \Omega \lambda x[(\Omega y)\boldsymbol{F}(x, y) \vee \boldsymbol{F}^{\#}(x, \Omega)]$$

($\boldsymbol{F}_{\oplus}$) is the most straightforward way of implementing indeterminacy, making use of the type-transcendent union already encountered in Section 4.3: indeterminacy between two senses of different types comes out as the union of the two senses. ($\boldsymbol{F}_{\vee}$) emulates the same union within the higher of the two original types, after casting the transparent relation $\boldsymbol{F}$ in the type $q(et)$ of opaque verbs.[36] As it turns out, neither of the two is of any use here. In fact, both are plagued by the same problems as the ambiguity analysis of **find**, predicting that sentences like (49) would be true for the continuation of Scenario 1. Moreover, neither of them can be used in a lexical decomposition of **seek**, which would come out as one of the following:

$$(\boldsymbol{S}_{\oplus}) \quad \lambda \Omega \lambda x \underline{\boldsymbol{T}(x, \boldsymbol{F}^{\#}(x, \Omega))} \qquad [\equiv \lambda \Omega \lambda x \boldsymbol{T}(x, \boldsymbol{F}_{\oplus}(x, \Omega))]$$

$$(\boldsymbol{S}_{\vee}) \quad \lambda \Omega \lambda x[\boldsymbol{T}(x, [(\Omega y)\boldsymbol{F}(x, y) \vee \boldsymbol{F}^{\#}(x, \Omega)])] \qquad [\equiv \lambda \Omega \lambda x \boldsymbol{T}(x, \boldsymbol{F}_{\vee}(x, \Omega))]$$

The underlined part of ($\boldsymbol{S}_{\oplus}$) results from applying ($\boldsymbol{F}_{\oplus}$) to Ω and x, drawing on a generalized version of β-conversion: the domain of ($\boldsymbol{F}_{\oplus}$) covers both individuals and quantifiers, and the values assigned to either will be the value assigned by the corresponding component, which is $[\lambda \Omega \lambda x \boldsymbol{F}^{\#}(x, \Omega)]$ in the case of quantifiers.[37] But, surely, ($\boldsymbol{S}_{\oplus}$) is as inadequate a reconstruction of indeterminate **seek** as can be; in fact, it boils down to the reading ($\boldsymbol{S}_{\Uparrow}$) according to the ambiguity analysis.

---

[36] This is done by a well-known type shift mapping R of type $e(et)$ to $\lambda \Omega \lambda x(\Omega y)R(x, y)$ — the prime example of the Montagovian strategy of generalizing to the worst case mentioned in fn. 5 above. — It may be noted that the two unions are obtained by two distinct operations, viz. type-transcending merge vs. type-shifted disjunction. This is no coincidence: given the framework in the appendix, neither operation can replace the other one in these formulae.

[37] The relevant laws of the logic of type-transcendent union are listed in the appendix.

($S_\vee$) does not fare much better, for it moves the indeterminacy from the expression to the attitude subject, which is clearly inadequate. In Scenario 1, Geach is *in some sense* looking for something Quine was looking for, but he would not have been content had he found a mermaid; similarly, Quine's Meinongian search would not have been over had he come across his own diary entry containing a representation of what he was looking for.[38] Briefly, someone looking for an unspecific object is *either* trying to find (a) a corresponding specific object *or* trying to find (b) a representation of the unspecific object — and *not* trying to find something that is either (a) or (b).

Putting the indeterminacy in the expression is not hard. It suffices to give wide scope to the type-transcendent union. The following formalization therefore ought to be a better candidate for $\Uparrow/\Downarrow$-indeterminate **seek** than either of ($S_\oplus$) and ($S_\vee$):

$$(\boldsymbol{S}_{\Downarrow\oplus\Uparrow})\quad ([\lambda\mathbf{\Omega}\lambda x\,\boldsymbol{T}(x,(\mathbf{\Omega}y)\boldsymbol{F}(x,y))]\oplus[\lambda\Pi\lambda x\,\boldsymbol{T}(x,(\Pi\mathbf{\Omega})\boldsymbol{F}^{\#}(x,\mathbf{\Omega}))])\qquad [=\boldsymbol{S}_{\Downarrow}\oplus\boldsymbol{S}_{\Uparrow}]$$

One may suspect that ($S_{\Downarrow\oplus\Uparrow}$) runs into a co-predication problem. In Scenario 1, Geach is looking for something Quine is looking for *in some sense*. And, we may assume, Quine is indeed looking for something. So, redundantly speaking, *in some sense* Quine is looking for something Quine is looking for. The point is that the senses are not the same. Hence though both (50) and (51) are true for Scenario 1, (52) is not:

(50) **Geach is looking for something Quine is looking for.**

(51) **Quine is looking for something Quine is looking for.**

(52) **Both Geach and Quine are looking for something Quine is looking for.**

As the reader is invited to verify, the previous analysis ($S_\vee$) of **seek** predicts all three sentences to be true for that scenario. However, the data come out correctly for ($S_{\Downarrow\oplus\Uparrow}$), even though ($S_{\Downarrow\oplus\Uparrow}$) happens to be the only sense of **seek**. The reason is that the relevant readings of (50) and (51) differ in logical form. The only readings according to which (50) and (51) are true in the scenario are:

$$(50')\quad \boldsymbol{T}(\boldsymbol{g},(\exists\mathbf{\Omega})[\boldsymbol{T}(\boldsymbol{q},(\mathbf{\Omega}y)\boldsymbol{F}(\boldsymbol{q},y))\wedge\boldsymbol{F}^{\#}(\boldsymbol{g},\mathbf{\Omega})])\qquad[=(45)]$$

$$(51')\quad (\exists\mathbf{\Omega})[\boldsymbol{T}(\boldsymbol{q},(\mathbf{\Omega}y)\boldsymbol{F}(\boldsymbol{q},y))\wedge\boldsymbol{T}(\boldsymbol{q},(\mathbf{\Omega}y)\boldsymbol{F}(\boldsymbol{q},y))]\qquad[\equiv(\exists\mathbf{\Omega})\boldsymbol{T}(\boldsymbol{q},(\mathbf{\Omega}y)\boldsymbol{F}(\boldsymbol{q},y))]$$

To derive these readings with the analysis ($S_{\Downarrow\oplus\Uparrow}$) of **seek**, one should first

---

[38] This reasoning is not entirely waterproof: if **try** is closed under implication, then anyone trying to Find an unspecific mermaid would *a fortiori* try to Find-or-Find# the quantifier expressed by **a mermaid**. But wait for the next paragraph.

notice that they share the interpretation of the object, viz.:[39]

(53)

- $\lambda\Pi(\exists\mathfrak{Q})[\boldsymbol{S}_{\Downarrow\oplus\Uparrow}(\boldsymbol{q},\mathfrak{Q}) \wedge \Pi(\mathfrak{Q})] : (qt)t$ **something Quine is looking for**
  - $\lambda\Sigma\lambda\Pi(\exists\mathfrak{Q})[\Sigma(\mathfrak{Q}) \wedge \Pi(\mathfrak{Q})] : (qt)((qt)t)$ **some**
  - $\lambda\mathfrak{Q}\ \boldsymbol{S}_{\Downarrow\oplus\Uparrow}(\boldsymbol{q},\mathfrak{Q}) : qt$ **-thing Quine is looking for**
    - $\lambda\mathfrak{Q}\ \mathfrak{Q} = \mathfrak{Q} : qt$ **-thing**
    - $\lambda\mathfrak{Q}\ \boldsymbol{S}_{\Downarrow\oplus\Uparrow}(\boldsymbol{q},\mathfrak{Q}) : qt$ **[that] Quine is looking for**
      - $\mathfrak{Q} : q$
      - $\boldsymbol{S}_{\Downarrow\oplus\Uparrow}(\boldsymbol{q},\mathfrak{Q}) : t$ **Quine is looking for $\mathfrak{Q}$**
        - $\boldsymbol{q} : e$ **Quine**
        - $\lambda x\ \boldsymbol{S}_{\Downarrow\oplus\Uparrow}(x,\mathfrak{Q}) : et$ **is looking for $\mathfrak{Q}$**
          - $\boldsymbol{S}_{\Downarrow\oplus\Uparrow} : \{q,(qt)t\}(et)$ **is looking for**
          - $\mathfrak{Q} : q$

Again, the argument of the verb (the trace $\mathfrak{Q}$) resolves the indeterminacy in **seek** so that the object comes out as:

(53′) $\lambda\Pi(\exists\mathfrak{Q})[\boldsymbol{T}(\boldsymbol{q},(\mathfrak{Q}y)\boldsymbol{F}(\boldsymbol{q},y)) \wedge \Pi(Q)]$.

But in order to obtain (50′), the object as interpreted in (53′) must take narrow scope, whereas it has wide scope in (51′):

(50″)

- $\boldsymbol{S}_{\Downarrow\oplus\Uparrow}(\boldsymbol{g},\lambda\Pi(\exists\mathfrak{Q})[\boldsymbol{T}(\boldsymbol{q},(\mathfrak{Q}y)\boldsymbol{F}(\boldsymbol{q},y)) \wedge \Pi(Q)]) : t$ $[\equiv (50')]$ **Geach is looking for something Quine is looking for**
  - $\boldsymbol{g} : e$ **Geach**
  - $\lambda x\ \boldsymbol{S}_{\Downarrow\oplus\Uparrow}(x,\lambda\Pi(\exists\mathfrak{Q})[\boldsymbol{T}(\boldsymbol{q},(\mathfrak{Q}y)\boldsymbol{F}(\boldsymbol{q},y)) \wedge \Pi(Q)]) : et$ **is looking for something Quine is looking for**
    - $\boldsymbol{S}_{\Downarrow\oplus\Uparrow} : \{q,(qt)t\}(et)$ **is looking for**
    - $\lambda\Pi(\exists\mathfrak{Q})[\boldsymbol{T}(\boldsymbol{q},(\mathfrak{Q}y)\boldsymbol{F}(\boldsymbol{q},y)) \wedge \Pi(Q)] : (qt)t$ **something Quine is looking for**

---

[39] The tree merges syntactic and semantic nformation in an obvious way. Notation: curly brackets around types indicate indeterminacy between them; see the appendix for precise definitions.

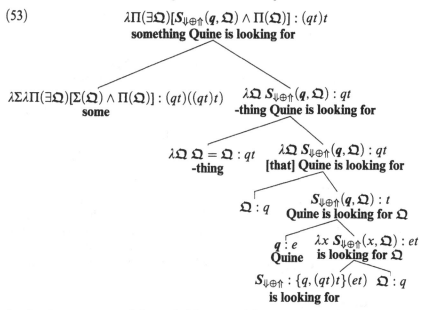

(51″) $(\exists \mathfrak{Q})[T(q, (\mathfrak{Q}y)F(q, y)) \wedge S_{\Downarrow\oplus\Uparrow}(q, \mathfrak{Q})] : t$ $[\equiv (51')]$

$\lambda\Pi(\exists\mathfrak{Q})[T(q, (\mathfrak{Q}y)F(q, y)) \wedge \Pi(Q)] : (qt)t_\mathfrak{Q}$
something Quine is looking for

$S_{\Downarrow\oplus\Uparrow}(q, \mathfrak{Q}) : t$
Quine is looking for $\mathfrak{Q}$

$q : e$ $\lambda x\, S_{\Downarrow\oplus\Uparrow}(x, \mathfrak{Q}) : et$
Quine   **is looking for $\mathfrak{Q}$**

$S_{\Downarrow\oplus\Uparrow} : \{q, (qt)t\}(et)$ $\mathfrak{Q}$
is looking for

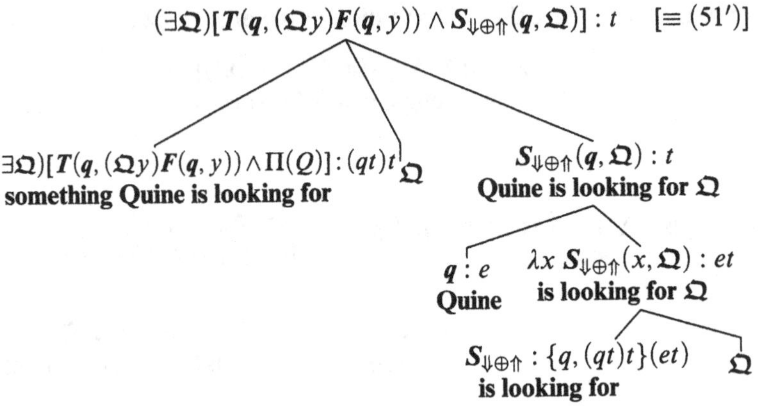

According to the indeterminacy approach to $\Downarrow$ vs. $\Uparrow$, (50″) and (51″) are the only true readings of (50) and (51). Hence it is obvious that (52) must be false, as desired. So, of the three above $\Uparrow/\Downarrow$-indeterminacy analyses of **seek**, $(S_{\Downarrow\oplus\Uparrow})$ is clearly the best. Nevertheless I am sceptical about it. The reason is that, still in the same scenario, the following three sentences should come out as *true* (in at least one sense):

(54)   **Geach is looking for something.**

(55)   **Quine is looking for something.**

(56)   **Both Geach and Quine are looking for something.**

According to $(S_{\Downarrow\oplus\Uparrow})$, (54) and (55) are implied by (50) and (51). More precisely, either has a reading that is implied by the respective readings (50′) and (51′), to wit:

(54′) $(\exists\Pi)T(g, (\Pi\mathfrak{Q})F^\#(x, \mathfrak{Q}))$ $[\equiv (\exists\Pi)S_{\Downarrow\oplus\Uparrow}(g, \Pi)]$

(55′) $(\exists\mathfrak{Q})T(q, (\mathfrak{Q}y)F(x, y))$ $[\equiv (\exists\mathfrak{Q})S_{\Downarrow\oplus\Uparrow}(q, \mathfrak{Q})]$

However, these two readings together do not imply (56). Now, (56) may still happen to be true for Scenario 1, given the analysis $(S_{\Downarrow\oplus\Uparrow})$, due to certain *lexical* properties of **seek**, like monotonicity or some connection between F and $F^\#$. But I think (56) ought to be *logically* implied by (54) and (55), which according to $(S_{\Downarrow\oplus\Uparrow})$ it is not. I take this to be enough motivation to try a different approach to the $\Uparrow/\Downarrow$-shift.

5.4. Seeking high as seeking low: coercion. Regarding Scenario 1 and its continuation, one cannot help thinking that Geach did not *literally* find an unspecific object, let alone a quantifier. Rather, what he found was a *trace* of Quine's search activities, in form of a diary entry containing a reference to a certain unspecific object. In other words, the object found by Geach *represented* a quantifier $\mathfrak{Q}$ satisfying: **Quine is looking for $\mathfrak{Q}$**. If we write "R" for the relation of representation as it holds between an unspecific object and anything that denotes that unspecific object (the italicization again indicating that this is likely to be a context-dependent variable rather than an ordinary non-logical constant), we may reduce the above *ad hoc* relation $F^\#$ to the ordinary, transparent Find relation in the following way:

(57) $F^\# = \lambda\Omega\lambda x(\exists z)[R(z,\Omega) \wedge F(x,z)].$

Using (57), the relevant reading (44) of (43) now reads:

(58) $(\exists z)(\exists\Omega)[T(q,(\Omega y)F(q,y)) \wedge R(z,\Omega) \wedge F(g,z)].$

The improvement in this formalization is that, once $F^\#$ is definable in terms of F, it may be derived from it systematically. In fact, there is quite a natural way of obtaining (58) as a reading of (43) without positing ambiguity or indeterminacy. The reason is that, given the intended reading (26′) of the object **something Quine is looking for**, (58) must derive from feeding a higher-order quantifier to the transparent verb **find** of type $e(et)$:

(26′) $\lambda\Pi(\exists\Omega)[T(q,(\Omega y)F(q,y)) \wedge \Pi(\Omega)].$

It is indeed natural to solve this apparent type conflict by coercion. This may be done by either typeshifting (i) the verb **find** from $e(et)$ to $q(et)$, or (ii) the object (26′) from $(qt)t$ to $(et)t[= q]$. In either case the latter would have to quantify over the object position of the former. For reasons that will become clear in a moment, we will only consider option (ii):[40]

(59) $(\exists z)(\exists\Omega)[R(z,\Omega) \wedge T(q,(\Omega y)F(q,y)) \wedge F(g,z)] : t$
        Geach finds something Quine is looking for

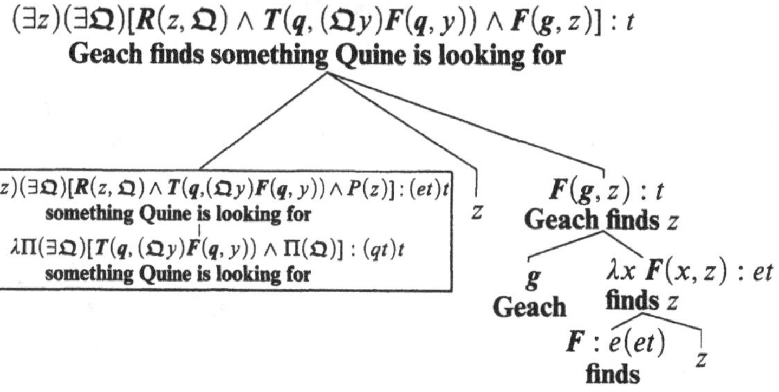

To obtain the highlighted shift in (59), the following (somewhat complicated[41]) operation can be used, which may come in handy in coercion contexts in general:

---

[40]The type shift on the verb would have mapped F onto: $\lambda\Omega\lambda x(\exists z)[R(z,\Omega) \wedge F(x,z)]$. — To keep the semantic apparatus to a minimum, quantification is expressed by a Montagovian scope mechanism. More local interpretation devices could have been used; cf. Heim & Kratzer (1998: 178ff.) for a survey.

[41]Part of the complications are an artefact of the framework. The operation comes out a lot more natural within the property approach to opacity.

QL Quantifier lowering

The lowered version of a higher-order quantifier $\mathfrak{D}$ of type $(qt)t$ is the following ordinary quantifier of type q:

$$[\lambda Q(\mathfrak{D}\mathfrak{Q})(\exists y)[\boldsymbol{R}(y, \mathfrak{Q}) \wedge Q(y)]]$$

I leave it to the reader to check that this coercion analysis solves all the problems encountered in connection with the ambiguity and indeterminacy analyses. In particular, it should be noted that ***QL*** coercion on the one hand avoids the co-predication problem of (52), because it requires the object of (50) to undergo ***QL*** if the sentence is to be true of Scenario 1, whereas (51) is only true of a literal (and redundant) construal; on the other hand, (54)–(56) come out as straightforward logical consequences of (50)–(52), involving higher-order quantification. Hence ***QL*** coercion fares a lot better than its rivals. Moreover, the fact that the $\Uparrow$-readings are rather rare and remote may be seen as further evidence for an account in terms of coercion. I will therefore assume that the $\Uparrow$-readings are indeed produced by ***QL***.

It must not be left unmentioned that the precise nature of the coercion mechanism underlying the $\Uparrow$-readings is not entirely clear. The reason is that **seek**, being opaque, does not call for coercion when combined with a higher-order quantifier; there just is no type conflict. In other words, in view of the independently motivated ***W***ide Scope ***H***igher order reading of (20) copied from Table 3 above, the application (60) of ***QL*** appears to be unmotivated:[42]

(20) **Geach is looking for something Quine is looking for.**

(***WH***) $(\exists\mathfrak{Q})[\boldsymbol{T}(\boldsymbol{q}, (\mathfrak{Q}y)\boldsymbol{F}(\boldsymbol{q}, y)) \wedge \boldsymbol{T}(\boldsymbol{g}, (\mathfrak{Q}y)\boldsymbol{F}(\boldsymbol{g}, y))]$

(60) $\boldsymbol{T}(\boldsymbol{g}, (\exists z)(\exists\mathfrak{Q})[\boldsymbol{R}(z, \mathfrak{Q}) \wedge \boldsymbol{T}(\boldsymbol{q}, (\mathfrak{Q}y)\boldsymbol{F}(\boldsymbol{q}, y)) \wedge \boldsymbol{F}(\boldsymbol{g}, z)]) : t$

Geach is looking for something Quine is looking for

$\boldsymbol{g}$
Geach

$\lambda x\ \boldsymbol{T}(x, (\exists z)(\exists\mathfrak{Q})[\boldsymbol{R}(z, \mathfrak{Q}) \wedge \boldsymbol{T}(\boldsymbol{q}, (\mathfrak{Q}y)\boldsymbol{F}(\boldsymbol{q}, y)) \wedge \boldsymbol{F}(x, z)]) : et$
is looking for something Quine is looking for

$\lambda\mathfrak{Q}\lambda x\ \boldsymbol{T}(x, (\mathfrak{Q}y)\boldsymbol{F}(x, y)) : q(et)$
is looking for

$\lambda P(\exists z)(\exists\mathfrak{Q})[\boldsymbol{R}(z, \mathfrak{Q}) \wedge \boldsymbol{T}(\boldsymbol{q}, (\mathfrak{Q}y)\boldsymbol{F}(\boldsymbol{q}, y)) \wedge P(z)] : (et)t$
something Quine is looking for

$\lambda\Pi(\exists\mathfrak{Q})[\boldsymbol{T}(\boldsymbol{q}, (\mathfrak{Q}y)\boldsymbol{F}(\boldsymbol{q}, y)) \wedge \Pi(\mathfrak{Q})] : (qt)t$
something Quine is looking for

---

[42] The problem vanishes under the clausal approach to opacity, where the syntactic structure underlying **seek** contains an occurrence of **find**, which may then trigger ***QL***. The reading (***WH***) of (20) could still be accounted for by a different scoping. — Incidentally, type shifting the verb (as in fn. 40) would have led to the wrong reading in (60), as the reader may check.

Despite this unclarity, I take the relation between $\Downarrow$- and $\Uparrow$-construals of **seek** and **find** to be governed by coercion.

The coercion analysis of higher-order finding also goes some way towards explaining the oddness of (45′) as a reading of (20).

(20) **Geach is looking for something Quine is looking for.**

(45′) $(\exists\mathbf{\Omega})[\boldsymbol{T}(\boldsymbol{q},(\mathbf{\Omega}y)\boldsymbol{F}(\boldsymbol{q},y)) \wedge \boldsymbol{T}(\boldsymbol{g},\boldsymbol{F}^{\#}(\boldsymbol{g},\mathbf{\Omega}))]$

As is apparent from its *de dicto* counterpart (60), deriving (45′) would involve quantifying in on top of the type coercion under scrutiny. In particular, given a coercion view of specificity, (45′) would thus necessitate double coercion on a single constituent, which may have to be ruled out in general.[43]

The scenarios discussed in this section had been intended to show that it is possible to stand in the specific **seek** relation to an unspecific object. However, according to the above coercion analysis of the $\Downarrow/\Uparrow$-contrast this is not the case: subjects only stand in the specific **seek** relation to representations of quantifiers, not to the quantifiers themselves. The point is best illustrated by the failures of co-predication mentioned earlier. According to the relevant readings of (50) and (51), Geach and Quine can be said to each stand in a seeking relation to the same unspecific object O (denoted by higher-order **something Quine is looking for**), but the relations are not the same: the relation Quine bears to O is denoted by **seek**; but Geach stands to O in the relation of *seeking-a-representation-of*. Hence co-predication fails precisely because Geach does not stand in any **seek** relation to the unspecific object O.

The upshot of our long discussion of these admittedly marginal examples is entirely negative. Appearances to the contrary, specific quests for unspecific objects do not constitute any evidence against the view that **seek** is indeterminate between a specific and an unspecific construal.[44]

§6. Co-predication and coordination.

6.1. Co-predication: semantics vs. pragmatics. In the previous section we used co-predication data as a test for deciding between coercion (or ambiguity) and indeterminacy to account for purported specific higher-order readings. In this section we will apply similar tests to specific vs. unspecific readings. The idea is simple enough: since the coercion approach supposes a structural ambiguity where there is none according to the indeterminacy interpretation, the two should disagree on co-predications of subjects that differ in the sense

[43] Cf. de Swart (2000).

[44] I am indebted to Nicholas Asher for making me see that the decision between coercion and indeterminacy with respect to higher-order specificity has no bearing on the decision between coercion and indeterminacy with respect to the objectual and the notional sense. This insight gave the present paper its direction.

in which they satisfy a predicate in point. However, matters are not that simple. In fact, all theories of opacity agree on the most straightforward co-predications. Imagine Geach and Quine both needing a pen to write a postcard home, but whereas Geach knows he has his own pen somewhere in his briefcase, Quine hopes that there are pens lying around in the post-office, which they are entering together. Quine immediately starts searching the place, while Geach opens his briefcase. Hence it would seem that both (61) and (62) are true:

(61) **Quine is looking for a pen.**

(62) **Geach is looking for a pen.**

However, although (63) does not seem to be outright *false*, applying it to the situation at hand is clearly somewhat strained:

(63) **Both Quine and Geach are looking for a pen.**

On the other hand, nothing is wrong with the general statement:

(64) **Both Quine and Geach are [each] looking for something.**

Let us now see how the various approaches to opacity deal with these data. It will be instructive to compare the *i*ndeterminacy analysis (65) of **seek** with the *c*oercion analysis based on *De Re* Lowering as well as the *s*tandard Montagovian approach, both of which take (66) to be the lexical meaning of **seek**:

(65) $\lambda\mathfrak{Q}\lambda x\boldsymbol{T}(x,(\mathfrak{Q}y)\boldsymbol{F}(x,y)) \oplus \lambda y\lambda x(\exists N)[\boldsymbol{VN}(N,y,x)$
$\wedge\, \boldsymbol{T}(x,(\exists z)[N(z)\wedge \boldsymbol{F}(x,z)])]$ [≡ (39c)]

(66) $\lambda\mathfrak{Q}\lambda x\boldsymbol{T}(x,(\mathfrak{Q}y)\boldsymbol{F}(x,y))$ [= (39a) = (6)]

To begin with, all three approaches agree on the true, unspecific reading of (61); and as to the true, specific reading of (62), the standard analysis with its essential propositions is the odd one out:

(61*sic*) $\boldsymbol{T}(\boldsymbol{q},(\exists y)[\boldsymbol{P}(y)\wedge \boldsymbol{F}(\boldsymbol{q},y)])$

(62*s*) $(\exists y)[\boldsymbol{P}(y)\wedge \boldsymbol{T}(\boldsymbol{g},\boldsymbol{F}(\boldsymbol{g},y))]$

(62*ic*) $(\exists y)(\exists N)[\boldsymbol{P}(y)\wedge \boldsymbol{VN}(N,y,\boldsymbol{g})$
$\wedge \boldsymbol{T}(\boldsymbol{g},(\exists z)[N(z)\wedge \boldsymbol{F}(\boldsymbol{g},z)])]$

Similarly, the readings attributed to (63) only differ *re de re*:

(63***sicn***) $\boldsymbol{T}(\boldsymbol{q},(\exists y)[\boldsymbol{P}(y)\wedge \boldsymbol{F}(\boldsymbol{q},y)])$ ***narrow scope***
$\wedge\, \boldsymbol{T}(\boldsymbol{g},(\exists y)[\boldsymbol{P}(y)\wedge \boldsymbol{F}(\boldsymbol{g},y)])$

(63***sm***) $(\exists y)(\exists z)[\boldsymbol{P}(y)\wedge \boldsymbol{P}(z)\wedge \boldsymbol{T}(\boldsymbol{q},\boldsymbol{F}(\boldsymbol{q},y))$ ***medium scope***
$\wedge\, \boldsymbol{T}(\boldsymbol{g},\boldsymbol{F}(\boldsymbol{g},z))]$

(63*icm*) $(\exists y)(\exists z)(\exists N)(\exists M)[\boldsymbol{P}(y) \wedge \boldsymbol{P}(z) \wedge \boldsymbol{VN}(N, y, \boldsymbol{q}) \wedge \boldsymbol{VN}(M, z, \boldsymbol{g}) \wedge \boldsymbol{T}(\boldsymbol{q}, (\exists y)[N(y) \wedge \boldsymbol{F}(\boldsymbol{q}, y)]) \wedge \boldsymbol{T}(\boldsymbol{g}, (\exists z)[M(z) \wedge \boldsymbol{F}(\boldsymbol{g}, z)])]$

(63*sw*) $(\exists y)[\boldsymbol{P}(y) \wedge \boldsymbol{T}(\boldsymbol{q}, \boldsymbol{F}(\boldsymbol{q}, y)) \wedge \boldsymbol{T}(\boldsymbol{g}, \boldsymbol{F}(\boldsymbol{g}, y))]$ *wide scope*

(63*icw*) $(\exists y)(\exists N)(\exists M)[\boldsymbol{P}(y) \wedge \boldsymbol{VN}(N, y, \boldsymbol{q}) \wedge \boldsymbol{VN}(M, y, \boldsymbol{g}) \wedge \boldsymbol{T}(\boldsymbol{q}, (\exists z)[N(z) \wedge \boldsymbol{F}(\boldsymbol{q}, z)]) \wedge \boldsymbol{T}(\boldsymbol{g}, (\exists z)[M(z) \wedge \boldsymbol{F}(\boldsymbol{g}, z)])]$

(63***sicn***) co-predicates unspecifically looking for a pen of Quine and Geach; according to any of the three approaches, the reading is obtained by directly combining the opaque verb with the indefinite object. The ***m*** readings attribute looking for a specific pen to both, although the pen in question may be a different one for each — as may be the particular vivid name under which it is known to them; the reading is obtained by quantifying the object into the open sentence *x* **are looking for** *y* and then quantifying the subject into it. Reversing this order of quantifiers finally leads to a ***w***ide scope reading of the object, according to which Geach and Quine are looking for the same pen, although they may describe it in different terms. The three scoping possibilities exhaust the readings predicted for (63) in any of the three accounts of opacity.

Given that the specific readings obtained by the standard approach are unlikely to be true anyway and that Quine just is not looking for any object under a specific unique description that would meet the requirements of the ***V***ivid ***N***ame condition, this only leaves the unspecific co-predication (63***sicn***) as a possible candidate for a true reading of (63). Since the unspecific reading of (62) is not *logically* entailed by its true specific reading, the only way of making it — and thus the co-predication (63***sicn***) — true, is by specific assumptions on the *lexical* meaning of **seek**. One such assumption was considered in Section 3.1 already, in connection with lexically motivated inferences: **seek** may be monotonically increasing in its unspecific argument. And although we have seen reasons to be wary of this assumption, let us briefly reconsider it. The monotonicity of **seek** may be reduced to a closure of implication in the underlying attitude $\boldsymbol{T}$; in other words, the closure assumption (67) implies the monotonicity (68) of **seek**, which in turn may have repercussions on the truth of (63***sicn***) in the situation at hand:[45]

(67) $\Box(\forall x)(\forall p)(\forall q)[p \Rightarrow q \rightarrow [\boldsymbol{T}(x, p) \rightarrow \boldsymbol{T}(x, q)]]$

(68) $\Box(\forall x)(\forall P)(\forall Q)[P \sqsubseteq Q \rightarrow [\boldsymbol{S}(x, \exists_P) \rightarrow \boldsymbol{S}(x, \exists_Q)]]$

Indeed, we may safely assume that Geach knows his missing Conway Stewart

[45] "$\Rightarrow$" and "$\sqsubseteq$" respectively denote strict implication and sub-propertyhood, both of which can be defined in terms of material implication, necessity, and universal quantification. "$\exists_P$" is short for the quantifier expressed by an indefinite restricted by P, i.e. $\lambda Q(\exists x)[P(x) \wedge Q(x)]$.

to be a pen, i.e. the vivid name ***CS*** witnessing (62*ic*) in the sense of (69), implies penhood in the sense of (70):

(69) $\boldsymbol{P}(\boldsymbol{cs}) \wedge \boldsymbol{VN}(\boldsymbol{CS}, \boldsymbol{cs}, \boldsymbol{g}) \wedge \boldsymbol{T}(\boldsymbol{g}, (\exists z)[\boldsymbol{CS}(z) \wedge \boldsymbol{F}(\boldsymbol{g}, z)])$

(70) $\Box(\forall y)[\boldsymbol{CS}(y) \rightarrow \boldsymbol{P}(y)]$

However, given the general principle (68), (69) and (70) together imply the unspecific reading of (62); thus, with (61*sic*), (63***sicn***) comes out as true. Why, then, would it still sound odd when someone commented on the postal situation as in (63)? Two hypothetical reasons spring to mind. The first builds on the fact that Geach is only looking for an *unspecific* pen insofar as he is looking for a *specific* one. Hence the unspecific reading of (62), though true, is not as informative as its specific reading (62*ic*). Invoking standard (Gricean) pragmatic reasoning, (61) would be rather misleading and thus inappropriate *if used unspecifically* — and this oddness might carry over to the co-predication (63) *of which it is part*. I can see at least two problems with this kind of reasoning: for one thing, the hearer normally cannot tell whether (61) is used unspecifically or not, and hence one premise of the Gricean reasoning is shaky; for another thing, it is not obvious why principles of informativity should carry over from one expression to others of which it is part. The other reason why (61) may be inappropriate has to do with the fact that, unlike Geach, Quine is indeed looking for any old pen, as long as it can be used for writing. Hence Geach and Quine, though both satisfying the predicate of being unspecifically seeking a pen, satisfy it in different ways, whereas the conjunction may suggest they do not. Again, I find this argument rather dubious if only for the reason that it seems to overgeneralize. For, as we have said, nothing appears to be wrong with (64) which attributes some seeking or other to both Geach and Quine and which is true even though the activities are as different as in the case of (63). The conclusion is that, in the absense of any pragmatic explanation, the inadequacy of (63) provides further evidence against the monotonicity (68) of **seek** and, indeed, the inferential closure (67) of the underlying attitude of **Trying**.

Finally, note that (64) as such does not present a problem to any analysis. Given the higher-order interpretation of **something** of Section 3.3 above, the following formalization, adequate and true, is obtained:

(71) $(\exists \boldsymbol{\Omega})[\boldsymbol{T}(\boldsymbol{q}, (\boldsymbol{\Omega} y)\boldsymbol{F}(\boldsymbol{q}, y)) \wedge \boldsymbol{T}(\boldsymbol{g}, (\boldsymbol{\Omega} y)\boldsymbol{F}(\boldsymbol{g}, y))]$.

6.2. Coordination: mixed objects. Apart from a pen, what both Geach and Quine need is a stamp. Geach had bought some the day before, so he includes them in his search of his briefcase, which may now be accurately reported by:

(72) **Geach is looking for his pen and a stamp.**

Although there is no specific stamp that Geach is looking for, the missing pen is quite specific. It is again instructive to have a look at the standard formalization of the intended, *mixed* reading of (72):

(72*s*) $T(g, [F(g, cs) \land (\exists y)[St(y) \land F(g, y)]]) : t$
Geach is looking for Geach's pen and a stamp

$g : e$
Geach

$\lambda x\ T(x, [F(x, cs) \land (\exists y)[St(y) \land F(x, y)]]) : et$
is looking for Geach's pen and a stamp

$\lambda \Omega \lambda x\ T(x, (\Omega y)F(x, y)) : q(et)$
is looking for

$\lambda P[P(cs) \land (\exists y)[St(y) \land P(y)]] : q$
Geach's pen and a stamp

$\lambda P\ P(cs) : q$
Geach's pen
$cs : e$
Geach's pen

$\lambda P(\exists y)[St(y) \land P(y)] : q$
a stamp

$\lambda Q \lambda P(\exists y)[Q(y) \land P(y)] : (et)q$
a

$St : et$
stamp

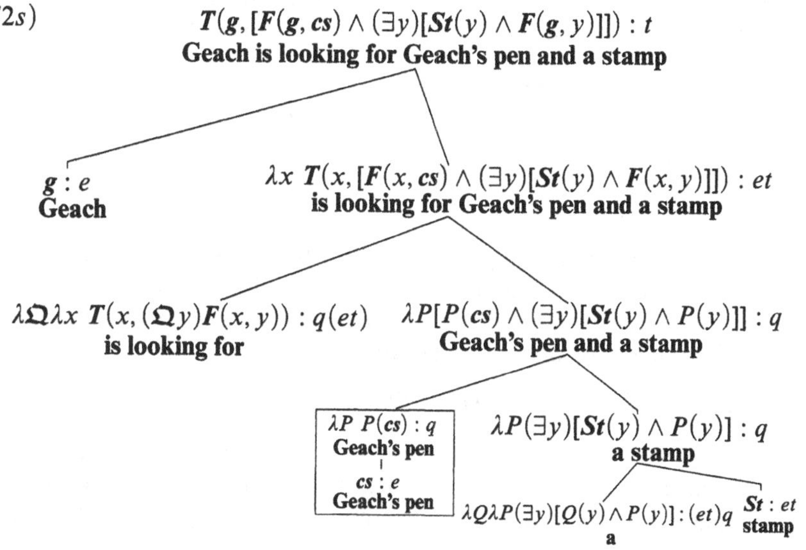

For simplicity we treat **his stamp** as a name of *cs*. The application of Montague Lifting at the place indicated is essential for this reading. In particular, it could not have been obtained by Argument Lowering, because the unspecificity of the right conjunct would have been lost. And it is this feature of the — inadequate — standard account of the reading under discussion which stands in the way of an adaptation of the coercion analysis: the latter is based on a Kaplanian *de re* interpretation which, as we have seen, does not lend itself to any version of Montague Lifting. Nor does it seem possible to find another type-shifting mechanism to this end. The reason is that, according to the coercion account, the required reading would have to invoke two distinct construals of **seek**, a specific one that involves *VN*, and the underlying unspecific one that does not.

The indeterminacy approach fares better in this respect, because in principle the same verb **seek** can take a specific and an unspecific object at the same time. However, the technicalities are somewhat delicate. To begin with, the two objects cannot be directly conjoined: **Geach's pen** is of the non-conjoinable[46] type *e*, whence it would have to be Montague-lifted first, thus becoming eligible for the unspecific sense of **seek**. In other words, conjoining the objects would lead to a variant of the standard interpretation involving essential propositions; I leave it to the readers to verify this. However, the

---

[46]I am relying on the standard type shifting terminology: a type is *conjoinable* (or *Boolean*) if it is either *t* or a pair whose right component is conjoinable. Cf. Partee (1997: 75).

desired result may be reached in an indirect way. To this end, one may first observe that the true reading of (72) boils down to the conjunction of the respective *s*pecific and *u*nspecific readings of (73) and (74):

(73) **Geach is looking for his pen.**

(73*s*) $(\exists N)[\boldsymbol{VN}(N, \boldsymbol{cs}, \boldsymbol{g}) \wedge \boldsymbol{T}(\boldsymbol{g}, (\exists z)[N(z) \wedge \boldsymbol{F}(\boldsymbol{g}, z)])]$

(74) **Geach is looking for a stamp.**

(74*u*) $\boldsymbol{T}(\boldsymbol{g}, (\exists y)[\boldsymbol{St}(y) \wedge \boldsymbol{F}(\boldsymbol{g}, y)])$

Within the indeterminacy approach it is possible to derive (72) as a conjunction of (73) and (74). The idea is that the conjoined object may express a conjunctive condition (75) on opaque predicates $\mathfrak{R}$ (dependent on the subject x), viz. that they hold of each of the conjuncts:[47]

(75) $\lambda\mathfrak{R}\lambda x[\mathfrak{R}(x, \boldsymbol{cs}) \wedge \mathfrak{R}(x, \underline{\lambda P(\exists y)[\boldsymbol{St}(y) \wedge P(y)]})]$,

where $\mathfrak{R}$ is a variable of the type $(\{e, q\}(et))$ of opaque verbs (according to the indeterminacy approach) and the underlined part is the (standard) analysis of **a stamp**. Applying (75) to the indeterminate analysis (65) of **seek** (and the result to $\boldsymbol{g}$) then yields the conjunction of (73) and (74), which is equivalent to:

(76) $(\exists N)[\boldsymbol{VN}(N, \boldsymbol{cs}, \boldsymbol{g}) \wedge \boldsymbol{T}(\boldsymbol{g}, (\exists z)(\exists y)[N(z) \wedge \boldsymbol{St}(y) \wedge \boldsymbol{F}(\boldsymbol{g}, z) \wedge \boldsymbol{F}(\boldsymbol{g}, y)])]$
$[\equiv \boldsymbol{\Sigma}(\boldsymbol{g}, \boldsymbol{cs}) \wedge \boldsymbol{\Sigma}(\boldsymbol{g}, \exists_{\boldsymbol{St}})]$,

where $\boldsymbol{\Sigma}$ is the indeterminate relation expressed by **seek**. However, the advantage of indeterminacy over coercion in the case of mixed object readings is illusory, as becomes clear when we switch from conjunction to disjunction, for which purpose we assume that Geach remembers having a couple of pencils in his briefcase that are equally suitable for his purposes. Then the following sentence is true in that Geach would be satisfied if he found ***cs*** as much as he would be content with finding any of his pencils.

(77) **Geach is looking for his pen or a pencil.**

Using the same trick of construal as in (75), (77) comes out as (78), which is a legitimate reading, but not the intended one:

(78) $\boldsymbol{\Sigma}(\boldsymbol{g}, \boldsymbol{cs}) \vee \boldsymbol{\Sigma}(\boldsymbol{g}, \exists_{\boldsymbol{St}})$
$[\equiv (\exists N)[\boldsymbol{VN}(N, \boldsymbol{cs}, \boldsymbol{g}) \wedge \boldsymbol{T}(\boldsymbol{g}, (\exists z)[N(z) \wedge \boldsymbol{F}(\boldsymbol{g}, z)])]$
$\vee\, \boldsymbol{T}(\boldsymbol{g}, (\exists y)[\boldsymbol{St}(y) \wedge \boldsymbol{F}(\boldsymbol{g}, y)])]$

[47]The formalization is inspired by the account of mixed coordinations (of indirect questions and **that**-clauses under attitude verbs like **know**) sketched in Krifka (2001: 314ff.). Apart from the type-mixing, this abstract way of interpreting conjunction is in line with the general approach taken in Hendriks (1993).

(78) may be used by a speaker who wants to express uncertainty as to what exactly Geach is looking for. In the situation at hand, this reading is false because Geach is neither looking for his pen nor for a pencil but for anything that would be either his pen (as given to him) or a pencil (of some sort). Hence there is no hope to reduce the intended reading of (77) to (78). This intended reading is readily formalized, however:

(79) $(\exists N)[\boldsymbol{VN}(N, \boldsymbol{cs}, \boldsymbol{g}) \wedge \boldsymbol{T}(\boldsymbol{g}, (\exists z)(\exists y)[N(z) \wedge \boldsymbol{F}(\boldsymbol{g}, z) \wedge \boldsymbol{St}(y) \wedge \boldsymbol{F}(\boldsymbol{g}, y)])]$.

Unlike (78), (79) does not allow for a paraphrase in terms of the indeterminate relation $\boldsymbol{\Sigma}$. How, then, does the reading (79) come about? Maybe the puzzle dissolves once the parallel between conjunction and disjunction is given up and the latter receives a non-Boolean interpretation as it has been proposed in other environments.[48] However, I think that a conceptually simpler way leads to a generalization of the *de re* mechanism to arbitrary occurrences of referential terms in objects of opaque verbs, as it would be needed in any approach to opacity: unspecifically read objects may contain proper names (e.g., in relative clauses) that must be interpreted *de re* if reference to essential propositions is to be avoided. And there seems to be no reason why this mechanism should not be applied to the case at hand. Even though I do not know the details of such a *de re* construal, I am convinced that it could somehow be provided, thus solving the puzzle of mixed objects independently of the lexical entry and type of the opaque verb. I therefore conclude that, as far as the choice between coercion-based vs. indeterminacy-induced specificity is concerned, the evidence provided by coordination and co-predication data is inconclusive. But there are other grounds for the decision.

§7. Lexical gaps.

7.1. Missing specificity.[49] According to the coercion approach, specific readings arise as a result of a type mismatch: the opaque verb, expecting a quantified object, must cope with a referential term, and it does so by adapting its lexical content. The mechanism responsible for this meaning adaptation is perfectly general, applying to any opaque verb and referential object in a canonical way. However, not all such combinations are attested. As we have already seen, some specific readings are rather remote, to say the least. In fact, the direct object of **owe** always seems to be interpreted unspecifically. Let us reconsider (the English translation of) Buridan's classical example:

(1) **I owe you a horse.**

[48] Cf. Zimmermann (2000). As far as I can see, an account of the wide-scope reading (78) in terms of non-Boolean disjunction would be more natural — and, incidentally, in the spirit of the analysis suggested in Rooth & Partee (1982).

[49] The ideas expressed in this section were developed in discussion with Joachim Sabel.

Under the coercion approach, (1) ought to have a reading according to which there is a particular horse that the speaker must give to the hearer:

(80) $(\exists y)(\exists N)[\boldsymbol{H}(y) \wedge \boldsymbol{VN}(N, y, \boldsymbol{i}) \wedge \boldsymbol{H}(y) \wedge \boldsymbol{O}(\boldsymbol{i}, (\exists z)[N(z) \wedge \boldsymbol{G}(\boldsymbol{i}, z, y)])]$.

A minor worry with this formalization is that it treats obligation on a par with psychological attitudes, which is almost certainly inadequate for independent reasons. In particular, obligations do not depend on the ways individuals mentioned in them are given to those under obligation; rather, it appears that any way of presenting the individuals would do, thus making the first conjunct in the matrix of (80) superfluous. This leads to the following, more adequate formalization, where vividity is weakened to uniqueness:

(80′) $(\exists y)(\exists N)[\boldsymbol{H}(y) \wedge (\forall u)[N(u) \leftrightarrow u = y] \wedge \boldsymbol{O}(\boldsymbol{i}, (\exists z)[N(z) \wedge \boldsymbol{G}(\boldsymbol{i}, z, y)])]$.

In the next subsection, we will see that this substitution of the ordinary *de re* construal for a more objective relation is also needed in other cases. What is of interest to us in the present context is the inadequacy of (80′), i.e. the fact that (1) appears to lack a specific reading. And this is not because such a reading could not be made sense of. In fact, the explicit paraphrase (81) of (1) is ambiguous in precisely the way expected:

(81) **I am obliged to give you a horse.**

The lack of a specific reading seems to be an idiosyncrasy of the verb **owe**, as corroborated by the oddness of sentences like the following:

(82) **I owe you this horse.**

(83) **I owe you Flavellus.**

How can this gap be explained? Why does **owe** shun specificity? (1) ought to be just as ambiguous as (81); and (82) and (83) ought to be perfectly normal. From the coercion perspective all this is quite mysterious.

At first glance, undefinedness may appear natural and helpful: why not *restrict* the meaning of **owe** such that individuals cannot occupy the object slot? The problem with this idea is that, given the coercion approach, it is vacuous: objects are quantifiers anyway, and the sense in which they may be individuals is a derived one. However, blocking all quantifiers that are derived from individuals — in the sense of (80′) or whatever may turn out to be adequate — would be too strong: any unique description of an individual corresponds to an existential quantifier, which may be expressible by an indefinite and should thus not be banned as a possible unspecific debt. And the same goes for Montague-lifted individuals, i.e. the quantifiers of the form $(\lambda P\, P(u))$, which would be obtained by a standard quantifying-in mechanism. Hence missing unspecificity remains a mystery.

Not so from the indeterminacy point of view. If no sense emerges from the combination of a verb and an individual-denoting expression, then the verb simply does not express a relation in which a subject can stand to an individual; it only takes unspecific objects as (second) arguments. Hence rather than being of the indeterminate type $e(\{e, q\}(et))$, the relation denoted by **owe** is of the simple type $e(q(et))$. Ignoring the indirect object (which I take to be specific), this is precisely the type of an opaque verb according to the coercion approach. However, with indeterminacy there is no coercion — the object must combine by functional application — and thus no specific reading will be predicted. Instead, (1) only gets one, unspecific reading, whereas (82) and (83) are instances of type clashes — just like (84) where a proposition would have to occupy the position of the direct object, which it cannot, for lexical reasons.

(84) ***I owe you that I give you a horse.**

7.2. Missing unspecificity. More lexical idiosyncrasies can be found. In particular, there is the mirror-image of missing specificity as encountered in connection with **owe**: *missing unspecificity*. Of course, there cannot be an opaque verb that *never* shows any unspecific readings; for that would meet the defining criterion of transparency. But there are verbs that do *not always* give rise to unspecificity. One of them is **resemble**, which is opaque, as the following example shows:[50]

(85) **Tom's horse resembles a unicorn.**

If Tom had a horse with a bulged forehead, (85) could well be true without there being any unicorns; the horse merely shares some features with any typical, if non-existent, representatives of the species unicorn. And, as expected, there is another, definitely false reading of (85) that says that Tom's horse bears re-semblance to a certain unicorn; this is the specific reading. Now consider:

(86) **Tom resembles a professor.**

Surprisingly, (86) is not ambiguous in the same way.[51] Certainly, there is the specific reading that would be true if Tom (my younger son) looked sufficiently like me. But the unspecific reading according to which he would have to share some properties with a typical professor — forgetfulness and short-sightedness, maybe — does not exist. Given the indeterminacy approach, this gap can be accounted for in various ways. One possibility is to force a specific

[50]The example is from Zimmermann (1993: 158), where it is used to show that opacity does not imply reducibility to propositional attitudes — thus re-establishing Montague's claim (cf. Section 2.3). In the meantime Roger Schwarzschild (p.c.) pointed out to me that " — **resembles** ..." resembles, i.e. could almost be, " — **could almost be** ...".

[51]At least, learning this from the participants in the Rutgers seminar (mentioned in the Acknowledgements) took me by surprise; I have since asked more native speakers only some of whom confirmed the judgement. In any case, I am very much indebted to my informants for this observation as well as the diagnosis in terms of personhood.

reading whenever the quantifier $\mathfrak{Q}$ denoted by an indefinite object involves personhood, in the following sense:

(87) A quantifier $\mathfrak{Q}$ of type q *involves* a property P of type (et) iff the following holds:

$$(\forall x)[(\mathfrak{Q}y)x = y \rightarrow P(x)].$$

I will use the symbol "$\gg$" to express involvement. According to (87) the quantifier denoted by **a professor** involves being a person (i.e. the property expressed by **is a person**), whereas the quantifier denoted by **a unicorn** does not. In order to force specificity for quantifiers involving personhood, one may split up the meaning of **resemble** into three subrelations:

- a relation between individuals x and y where x (specifically) resembles y;
- a relation between individuals x and quantifiers $\mathfrak{Q}$ where $\mathfrak{Q}$ involves personhood and applies to the property of (specifically) resembling x;
- a relation between individuals x and quantifiers $\mathfrak{Q}$ where $\mathfrak{Q}$ does not involve personhood and x unspecifically resembles $\mathfrak{Q}$.

The three parts can then be glued together using the type-transcendent union operation $\oplus$. More precisely, and relying on a lexical decomposition suggested by Roger Schwarzschild,[52] a plausible analysis of **resemble** looks like this:

(88) $$\lambda y \lambda x(\exists N)(\exists M)(\exists u)(\exists v)[\boldsymbol{VN}(N,x,u) \wedge \boldsymbol{VN}(M,y,v) \wedge \Diamond(\exists z)[N(z) \wedge M(z)]]$$

$$\oplus \quad \lambda\mathfrak{Q}\lambda x[\mathfrak{Q} \gg \boldsymbol{P} \wedge (\exists N)(\exists u)[\boldsymbol{VN}(N,x,u) \wedge (\mathfrak{Q}y)(\exists M)(\exists v)[\boldsymbol{VN}(M,y,u) \wedge \Diamond(\exists z)[N(z) \wedge M(z)]]]]$$

$$\oplus \quad \lambda\mathfrak{Q}\lambda x[\neg\mathfrak{Q} \gg \boldsymbol{P} \wedge (\exists N)(\exists u)[\boldsymbol{VN}(N,x,u) \wedge \Diamond(\mathfrak{Q}y)N(y)]]$$

The idea behind (88) is that specific resemblance holds between two individuals if they are *possibly identical*. Given that the latter notion trivializes when understood as modalized identity between the individuals themselves (collapsing into identity *tout court*), we need to construe it in terms of *de re* modality, thereby introducing vivid names (as used by some unspecified subject).[53] This explains the first part of the relation as well as the second one, which generalizes the first part to the worse case of a quantifier (involving

---

[52]Cf. fn. 51. The formal implementation is my own.

[53]It should be noted that the final conjuncts in the first and the third parts of (88) are, respectively, equivalent to the following conditions, which are more tedious but capture the idea of possible identity more directly:

- $\Diamond(\exists x')(\exists y')[N(x') \wedge M(y') \wedge x' = y']$
- $\Diamond(\mathfrak{Q}y)(\exists x')[N(x') \wedge x' = y]$

In this connection I may also mention that the modality expressed by "$\Diamond$" is unlikely to be metaphysical possibility, but rather some hightly restricted form of existential quantification over

personhood) in object position. The third part reflects the intuition that, in the case of existential $\mathbf{\Omega}$, unspecific resemblence is modalized *de re* predication: resembling a unicorn boils down to possibly being a unicorn. The trick is that, by the relativization in the first conjunct, this reading does not arise with quantifiers involving the property of being a person. In fact, the latter always come out as equivalent to their specific readings.

One interesting feature of the indeterminacy treatment of missing unspecificity is that it makes opacity a matter of degree, ranging from *fully opaque* verbs (like **owe**) that only admit unspecific readings, via *ordinary opaque* verbs (like **seek**) that always allow for the characteristic ambiguity, and *mixed* cases (as in **resemble**) to *transparency*, the absence of unspecific readings. It would be interesting to see how much of this range is actually made use of in natural languages, but such an investigation is well beyond the limited scope of this paper.

Despite the parallels between missing specificity and missing unspecificity, the coercion approach can in principle cope with the latter. In fact, restricted generalization to the worst case can be adaped, taking the union of the second and third parts of the relation in (88). This union is a relation of type $(q(et))$ and can be defined in more orthodox type notation, using disjunction in place of type-transcendent union:

(89) $$\lambda\mathbf{\Omega}\lambda x[[\mathbf{\Omega} \gg \boldsymbol{P} \wedge (\exists N)(\exists u)[\boldsymbol{VN}(N,x,u) \\ \wedge (\mathbf{\Omega}y)(\exists M)(\exists v)[\boldsymbol{VN}(M,y,u) \wedge \Diamond(\exists z)[N(z) \wedge M(z)]]]] \\ \vee [\neg\mathbf{\Omega} \gg \boldsymbol{P} \wedge (\exists N)(\exists u)[\boldsymbol{VN}(N,x,u) \wedge \Diamond(\mathbf{\Omega}y)N(y)]]].$$

(89) obviously yields the same readings as (88) whenever a quantified object is interpreted *in situ*. For the specific reading, one needs to employ a *de re* interpretation mechanism. The only obstacle is that resemblance is not a psychological attitude. Hence vivid names should not be relativized to the referent of the (grammatical) subject. One possibility, already indicated in the previous subsection, is to introduce an objective restriction on descriptions of *res*; another option, arguably more adequate in the case at hand, is to follow (88) and existentially generalize the attitude subject, along the following lines:

GDRL ***Generalized*** **De Re** ***Lowering***

The generalized de-re-*lowered version of a relation* $\mathfrak{R}$ *of type* $q(et)$ *is the following relation of type* $e(et)$*:*

$$\lambda y\lambda x(\exists N)(\exists u)[\boldsymbol{VN}(N,y,u) \wedge \mathfrak{R}(x,\lambda P(\exists z)[N(z) \wedge P(z)])]$$

It thus appears that, from the coercion point of view, the choice of the pertinent *De Re* Lowering operation — ***DRL*** vs. ***GDRL*** vs. some objectivized version — would have to be lexically controlled, however this may be done. In any

Logical Space; in any case it is far from being well understood (by Roger Schwarzschild, or myself, for that matter).

case, applying ***GDRL*** to the relation in (88) yields the desired result; as a somewhat tedious calculation shows, the generalized *de-re*-lowered version of (88) is equivalent to the first component of the indeterminate relation in (87).[54]

The upshot of the discussion of specificity-related lexical gaps is that, as long as the mystery of missing unspecificity cannot be resolved using coercion, the indeterminacy approach has a slight advantage.

§8. Conclusions and prospects.

8.1. Summary. Opaque verbs are verbs that admit unspecific readings of indefinite objects. The standard interpretation of opaque verbs, following ideas from Quine and Montague, takes these indefinites to denote existential quantifiers, which may be thought of as standing for unspecific objects, and the opaque verb to express a property of, or a relation to, unspecific objects — the property or relation being the *notional* sense of the verb. Opaque verbs may also be construed like ordinary transparent verbs, thereby expressing their *objectual* sense. In the standard approach to opacity, the objectual sense is derived from the notional sense by a logical transformation, a type shift, which on closer inspection turns out to be inadequate, at least in some cases. Replacing the logical transformation by a more suitable operation that takes into account the mental attitude of the subject to the specific object, naturally leads to the question of whether that extralogical operation could be part of the lexical meaning of the opaque verb, rather than being grammatically induced, as the standard approach would have it. If so, opaque verbs would lexically express one *indeterminate* sense instead of a notional sense that can be coerced into an objectual sense if need arises. On the face of it, it seems hard to decide which, if any, of these two construals of opaque verbs — the indeterminateness approach or the coercion approach — is more adequate. Three potential pieces of evidence were examined to decide the question.

The first was the phenomenon of *higher-order specificity*, seemingly involving — somewhat paradoxically — specific readings of unspecific objects. The coercion approach should have problems distinguishing the phenomenon from ordinary unspecificity, in particular in cases of co-predication. However, it turns out that higher-order specificity is itself based on an independent coercion mechanism, which allows the specificity-coercion approach to steer clear of any co-predication problems with specific higher-order readings. And the indeterminateness approach does not have any problems with higher-order specificity to being with. Hence the phenomenon does not help deciding between the two approaches.

Secondly, we looked at possibilities for conjoining specific and unspecific attitudes and objects, as they occur in *co-predications* and *mixed coordinate*

[54] The proof relies on the properties of the *V*ivid *N*ame relation mentioned in fn. 24.

objects. As to the former, it appears that both coercion and indeterminacy are too generous in admitting co-predications that are clearly felt to be inappropriate. However, it may well be that the over-generated readings can be explained away on pragmatic grounds. In any case, they would not help us decide between the theories, which agree on the pertinent data. On the other hand, mixed coordinations of specific and unspecific objects, seemed to reveal a tiny advantage of the indeterminacy approach — which, however, vanished once a fuller range of data was taken into account. For it proved that mixed disjunctions presented serious problems for both approaches — problems which could presumably be solved by adding a general *de re* mechanism. Again, the data turned out to be inconclusive.

Finally a difference in adequacy between the two approaches to opacity came up in connection with theoretically predicted, but *unattested ambiguities*. The existence of verbs (like **owe**) that do not allow for specific readings is hard to explain from the coercion point of view, whereas in the indeterminateness approach it comes out as a simple type-difference between such verbs and ordinary opaque verbs. The mirror-image phenomenon — missing specific readings in otherwise opaque verbs (like **resemble**) — can be explained in terms of undefinedness on both approaches. But the difference remains: missing specificity can only be accounted for in terms of indeterminateness.

There are bound to be more phenomena that can be used to decide the question. And it may also be that some solution to the puzzle of missing specificity in coercion terms may be found. In other words, I regard the above, largely negative results as a first step to deeper investigation.

8.2. A final speculation. The difference between the two approaches to the meaning of opaque verbs lies in the possibility of reducing one of their two senses to the other. According to the coercion approach, the objectual sense can be reduced to the notional sense; according to the indeterminacy approach, the two are merely facets of one overall meaning. The above discussion presupposed that one of these two approaches is right. However, this need not be so. Apart from the rather unattractive possibility of genuine lexical ambiguity (which clearly produces more problems than it solves), there remains the vague hope of reversing the direction of analysis: what if the objectual sense turned out to be the underlying one, with the notional derived from it by some as yet unknown mechanism? After all, this would be a reduction in the plainest sense, viz. an explanation of the (logically) complex in terms of the (logically) simple. Following this lead is definitely beyond this paper — but, I think, not too wide of the mark.

Appendix: The logic of type-transcendent union. The type-theoretic language L used in the above text is a straightforward implementation of the construction ideas mentioned in footnotes 5 and 31. Here are the most important definitions.

The *basic types* are the symbols e and t. The set of *types* T is the smallest set that contains the basic types as elements and such that for any $a_1, \ldots a_n$, $b \in T : (\{a_1, \ldots, a_n\}, b) \in T$ (for any positive integer n); in case $n = 1$, one may write "(ab)" for $(\{\{a\}, b\})$ and omit outer parentheses and commas.

For each $a \in T$, we assume there to be a (possibly empty) set of non-logical constants Con_a and an infinite set Var_a or variables. The *L-expressions* L_a of arbitrary types a are defined by the following recursion:

- $Con_a \subseteq L_a$;
- $Var_a \subseteq L_a$;
- "$\alpha(\beta)$" $\in L_a$ whenever $\alpha \in L_{Xa}$ and $\beta \in L_b$, for some $X \subseteq T$ and $b \in X$;
- "$[\lambda x \alpha]$" $\in L_a$ whenever $a = bc$, $x \in Var_b$, and $\alpha \in L_c$, for some $b, c \in T$;
- "$(\alpha \oplus \beta)$" $\in L_a$ whenever $\alpha \in L_{Xb}$, $\beta \in L_{Yb}$, and $a = (X \cup Y, b)$ for some $X, Y \subseteq T$ and $b \in T$ such that $X \cap Y = \phi$;
- "$(\neg\varphi)$"$\in L_a$ whenever $a = t$ and $\varphi \in L_a$;
- "$(\alpha \wedge \psi)$" $\in L_a$ whenever $a = t$ and $\varphi, \psi \in L_a$;
- "$(\forall x)\varphi$" $\in L_a$ whenever $a = t$, $\varphi \in L_a$, and $x \in Var_b$, for some $b \in T$;
- "$(\Box\varphi)$" $\in L_a$ whenever $a = t$ and $\varphi \in L_a$.

Other connectives and quantifiers are defined as abbreviations in the usual fashion. A pair $\mathfrak{F} = (U, W)$ is called a *frame* whenever U and W are two non-empty, disjoint sets of *urelements*; these conditions are imposed to guarantee that the domains of distinct types are disjoint. The following definitions assume a frame $\mathfrak{F} = (U, W)$. For any type $a \in T$, the domain D_a of *$\mathfrak{F}$-objects of type a* is defined by the following recursion (on the depth of types):

- $D_e = U$;
- $D_t = P(W)$;
- $D(\{a_1, \ldots, a_n\}, b) = \{f \mid f : (D_{a_1} \bigcup \ldots \bigcup D_{a_n}) \rightarrow D_b\}$;

Given a frame $\mathfrak{F} = (U, W)$, an *$\mathfrak{F}$-model* is a pair $\mathfrak{M} = (\mathfrak{F}, F)$, where $F : \bigcup_{a \in T} Con_a \rightarrow \bigcup_{a \in T} D_a$ such that $F(c) \in D_a$ whenever $c \in Con_a$. Likewise an *$\mathfrak{F}$-assignment* is a function $g : \bigcup_{a \in T} Var_a \rightarrow \bigcup_{a \in T} D_a$ such that $g(x) \in D_a$ whenever $x \in Var_a$. If g is an $\mathfrak{F}$-assignment, $x \in Var_b$, and $u \in D_b$ (for some b), then $g^{x/u} = (g \backslash \{x, g(x)\}) \bigcup \{(x, u)\}$.

Given an $\mathfrak{F}$-model $\mathfrak{M} = (\mathfrak{F}, F)$ the *denotation* $[[\alpha]]^{\mathfrak{M},g}$ *of* an L-expression a of type a (relative to an $\mathfrak{F}$-assignment g) is an $\mathfrak{F}$-object of type a defined by the following recursion (on the structure of expressions):

- $[\![c]\!]^{\mathfrak{M},g} = F(c)$, if $c \in Con_a$;
- $[\![x]\!]^{\mathfrak{M},g} = g(x)$, if $x \in Con_a$;
- $[\![\alpha(\beta)]\!]^{\mathfrak{M},g} = [\![\alpha]\!]^{\mathfrak{M},g}([\![\beta]\!]^{\mathfrak{M},g})$;
- $[\![\lambda x \alpha]\!]^{\mathfrak{M},g}(u) = [\![\alpha]\!]^{\mathfrak{M},g^{x/u}}$ whenever $x \in Var_b$ and $u \in D_b$;

- $[\![(\alpha \oplus \beta)]\!]^{\mathfrak{M},g} = [\![\alpha]\!]^{\mathfrak{M},g} \bigcup [\![\beta]\!]^{\mathfrak{M},g}$;
- $[\![\neg\varphi]\!]^{\mathfrak{M},g} = W \backslash [\![\varphi]\!]^{\mathfrak{M},g}$;
- $[\![(\varphi \wedge \psi)]\!]^{\mathfrak{M},g} = [\![\varphi]\!]^{\mathfrak{M},g} \bigcap [\![\psi]\!]^{\mathfrak{M},g}$;
- $[\![(\forall x)\varphi]\!]^{\mathfrak{M},g} = \{w \in W \mid \{u \mid w \in [\![\varphi]\!]^{\mathfrak{M},g^{x/u}}\} = D_b\}$ whenever $x \in Var_b$;
- $[\![\Box\varphi]\!]^{\mathfrak{M},g} = \{w \in W \mid [\![\varphi]\!]^{\mathfrak{M},g} = W\}$.

Using the above local interpretation as a starting point, global semantic notions (entailment, validity, logical equivalence, ...) may be defined in the usual way, by quantifying over all models. For instance, two L-expressions α and β are *logically equivalent* (and hence intersubstitutable *salva denotatione*) — notation: "$\alpha \equiv \beta$" — if $[[\alpha]]^{\mathfrak{M},g} = [[\beta]]^{\mathfrak{M},g}$ for all $\mathfrak{F}$-models $\mathfrak{M}$ and $\mathfrak{F}$-assignments g on arbitrary frames $\mathfrak{F}$. In particular, assuming the symbol "$\equiv$" to be flanked by L-expressions only, the following logical equivalences can be established by routine arguments:

- $(\alpha \oplus \beta) \equiv (\beta \oplus \alpha)$
- $((\alpha \oplus \beta) \oplus \gamma) \equiv (\alpha \oplus (\beta \oplus \gamma))$
- $([\lambda x \alpha] \oplus [\lambda y \beta])(\gamma) \equiv \alpha^{x/\gamma}$ — provided that no free variable of β occurs in α where the result $\alpha^{x/\gamma}$ of replacing free x in α by β is defined in the usual fashion.

Acknowledgments. This paper mainly develops ideas that came up in discussion in a seminar I co-taught with Roger Schwarzschild at Rutgers in the spring term 2000. I owe a great unspecific debt to Roger and all the participants. More specific thanks go to an anonymous reviewer, Nicholas Asher, Joy Grohmann (*née*, and at the time, Trombley), Joachim Sabel, Arnim von Stechow, and Matthew Stone for reasons given in the footnotes; to those native speakers who drew my attention to the data discussed in Section 7; to Jürgen Konradi for proof-reading; to Rachel Hendery for stylistic suggestions and grammatical corrections; and to the audiences on various occasions that I had to present part of the above material: the *DIP* colloquium (Amsterdam, September 2000), the *Sinn und Bedeutung 2000* (Amsterdam, December 2000), the Berlin semantics *Kreis* (May 2001), and of course the Munich workshop on intensionality.

REFERENCES

[1] M. Aloni, ***Quantification under conceptual covers***, Dissertation, University of Amsterdam, 2001.

[2] N. Asher, *A typology for attitude verbs and their anaphoric properties*, ***Linguistics and Philosophy***, vol. 10 (1987), pp. 125–198.

[3] M. Bennett, *A variation and extension of a montague fragment of English*, ***Montague grammar*** (B. Partee, editor), Academic Press, New York, 1976, pp. 119–163.

[4] R. Blutner, *Lexical semantics and pragmatics*, ***Semantics*** (F. Hamm and T. E. Zimmermann, editors), Buske, Hamburg, 2002, pp. 27–58.

[5] J. Buridanus, ***Sophisms on meaning and truth***, Translated by T. K. Scott, Appliton-Century-Crofts, New York, 1966, (= Sophismata, 1350).

[6] A. Copestake and T. Briscoe, *Semi-productive polysemy and sense extension*, ***Journal of Semantics***, vol. 12 (1995), pp. 15–67.

[7] M. J. Cresswell, ***Logics and languages***, Methuen, London, 1973.

[8] M. J. Cresswell and A. von Stechow, *De re belief generalized*, ***Linguistics and Philosophy***, vol. 5 (1982), pp. 503–535.

[9] H. de Swart, *Tense, aspect and coercion in a cross-linguistic perspectives*, ***Proceedings of the Berkeley formal grammar conference workshops*** (M. Butt and T. H. King, editors), On-line publication 2000. `cslipublications.stanford.edu/LFG/5/bfg00/bfg00-toc.html`.

[10] G. Forbes, *Objectual attitudes*, ***Linguistics and Philosophy***, vol. 23 (2000), pp. 141–183.

[11] G. Frege, ***Die Grundlagen der Arithmetik***, Koebner, Breslau, 1884, [The Foundations of Arithmetic. Translated by J. L. Austin. Oxford 1950].

[12] P. Geach, *A medieval discussion of intentionality*, ***Logic, methodology and philosophy of science*** (Y. Bar-Hillel, editor), North Holland, Amsterdam, 1965, pp. 425–433.

[13] I. Heim, ***The semantics of definite and indefinite noun phrases***, Dissertation, UMass at Amherst, 1982.

[14] ———, *Presupposition projection and the semantics of attitude verbs*, ***Journal of Semantics***, vol. 9 (1992), pp. 183–221.

[15] I. Heim and A. Kratzer, ***Semantics in generative grammar***, Oxford University Press, Oxford, 1998.

[16] H. Hendriks, ***Studied flexibility***, Dissertation, University of Amsterdam, 1993.

[17] T. M. V. Janssen, *Compositionality*, ***Handbook of logic and language*** (J. van Benthem and A. ter Meulen, editors), (With an appendix by B. H. Partee), Elsevier, Amsterdam, 1997, pp. 417–473.

[18] H. Kamp, *A theory of truth and semantic representation*, ***Formal methods in the study of language, Part 1*** (J. A. G. Groenendijk et al., editors), Mathematical Centre Tracts, Amsterdam, 1981, pp. 277–322.

[19] D. Kaplan, *Quantifying in*, ***Words and objections: Essays on the work of W. V. Quine*** (D. Davidson and J. Hintikka, editors), Reidel, Dordrecht, 1969, pp. 206–242.

[20] ———, *How to Russell a Frege-Church*, ***Journal of Philosophy***, vol. 72 (1975), pp. 716–729.

[21] ———, *Demonstratives. an essay on the semantics, logic, metaphysics and epistemology of demonstratives and other indexicals*, ***Themes from Kaplan*** (J. Almog, J. Perry, and H. Wettstein, editors), Oxford University Press, Oxford, 1989, pp. 481–563.

[22] M. Krifka, *For a structured meaning account of questions and answers*, ***Audiatur Vox Sapientiae*** (C. Féry and W. Sternefeld, editors), Akademic-Verlag, Berlin, 2001, pp. 287–319.

[23] R. Larson, M. den Dikken, and P. Ludlow, *Intensional transitive verbs and abstract clausal complementation*, Manuscript, SUNY at Stony Brook, 1999.

[24] D. Lewis, *Attitudes de dicto and de se*, ***The Philosophical Review***, vol. 88 (1979), pp. 513–543.

[25] ———, *What puzzling Pierre does not believe*, ***Australasian Journal of Philosophy***, vol. 59 (1981), pp. 283–289.

[26] F. Moltmann, *Intensional verbs and quantifiers*, ***Natural Language Semantics***, (1997), pp. 1–52.

[27] R. Montague, *On the nature of certain philosophical entities*, ***Monist***, vol. 53 (1969), pp. 159–195.

[28] ———, *English as a formal language*, ***Linguaggi nella società e nella tecnica*** (B. Visentini, editor), Edizioni di Communità, Milan, 1970a, pp. 189–223.

[29] ———, *Universal grammar*, ***Theoria***, vol. 36 (1970b), pp. 373–398.

[30] ———, *The proper treatment of quantification in ordinary English*, ***Approaches to natural language*** (J. Hintikka et al., editors), Reidel, Dordrecht, 1973, pp. 221–242.

[31] B. Partee, *Noun phrase interpretation and type shifting principles*, ***Studies in discourse representation theory and the theory of generalized quantifiers*** (J. Groenendijk et al., editors), Foris, Dordrecht, 1987, pp. 115–143.

[32] ———, *Montague grammar*, ***Handbook of logic and language*** (J. van Benthem and A. ter Meulen, editors), [With H. Hendriks], Elsevier, Amsterdam, 1997, pp. 5–91.

[33] B. Partee and M. Rooth, *Generalized conjunction and type ambiguity*, ***Meaning, use, and interpretation of language*** (R. Bäuerle et al., editors), de Grugter, Berlin, 1983, pp. 361–383.

[34] J. Pustejovsky, *Type coercion and lexical selection*, ***Semantics and the lexicon*** (J. Pustejovsky, editor), Kluwer, Dordrecht, 1993, pp. 73–94.

[35] W. V. O. Quine, *Quantifiers and propositional attitudes*, ***Journal of Philosophy***, vol. 53 (1956), pp. 177–187.

[36] ———, ***Word and object***, MIT Press, Cambridge, Mass., 1960.

[37] M. Rooth and B. Partee, *Conjunction, type ambiguity, and wide scope "or"*, ***Proceedings of the 1st west coast conference on formal linguistics*** (Stanford) (D. Flickinger et al., editors), 1982, pp. 353–362.

[38] T. E. Zimmermann, *On the proper treatment of opacity in certain verbs*, ***Natural Language Semantics***, vol. 1 (1993), pp. 149–179.

[39] ———, *Free choice disjunction and epistemic possibility*, ***Natural Language Semantics***, vol. 8 (2000), pp. 255–290.

[40] ———, *Unspecificity and intensionality*, ***Audiatur Vox Sapientiae*** (C. Féry and W. Sternefeld, editors), Akademic-Verlag, Berlin, 2001, pp. 514–533.

INSTITUT FÜR KOGNITIVE LINGUISTIK
JOHANN WOLFGANG GOETHE-UNIVERSITÄT
D-60629 FRANKFURT/M, GERMANY
E-mail: T.E.Zimmermann@lingua.uni-frankfurt.de

LECTURE NOTES IN LOGIC

General Remarks

This series is intended to serve researchers, teachers, and students in the field of symbolic logic, broadly interpreted. The aim of the series is to bring publications to the logic community with the least possible delay and to provide rapid dissemination of the latest research. Scientific quality is the overriding criterion by which submissions are evaluated.

Books in the Lecture Notes in Logic series are printed by photo-offset from master copy prepared using LaTeX and the ASL style files. For this purpose the Association for Symbolic Logic provides technical instructions to authors. Careful preparation of manuscripts will help keep production time short, reduce costs, and ensure quality of appearance of the finished book. Authors receive 50 free copies of their book. No royalty is paid on LNL volumes.

Commitment to publish may be made by letter of intent rather than by signing a formal contract, at the discretion of the ASL Publisher. The Association for Symbolic Logic secures the copyright for each volume.

The editors prefer email contact and encourage electronic submissions.

Editorial Board

David Marker, Managing Editor
Dept. of Mathematics, Statistics,
and Computer Science (M/C 249)
University of Illinois at Chicago
851 S. Morgan St.
Chicago, IL 60607-7045
marker@math.uic.edu

Vladimir Kanovei
Lab 6
Institute for Information
Transmission Problems
Bol. Karetnyj Per. 19
Moscow 127994 Russia
kanovei@mccme.ru

Steffen Lempp
Department of Mathematics
University of Wisconsin
480 Lincoln Avenue
Madison, Wisconsin 53706-1388
lempp@math.wisc.edu

Lance Fortnow
Department of Computer Science
University of Chicago
1100 East 58th Street
Chicago, Illinois 60637
fortnow@cs.uchicago.edu

Shaughan Lavine
Department of Philosophy
The University of Arizona
P.O. Box 210027
Tuscon, Arizona 85721-0027
shaughan@ns.arizona.edu

Anand Pillay
Department of Mathematics
University of Illinois
1409 West Green Street
Urbana, Illinois 61801
pillay@math.uiuc.edu

Editorial Policy

1. Submissions are invited in the following categories:

i) Research monographs
ii) Lecture and seminar notes
iii) Reports of meetings
iv) Texts which are out of print

Those considering a project which might be suitable for the series are strongly advised to contact the publisher or the series editors at an early stage.

2. Categories i) and ii). These categories will be emphasized by Lecture Notes in Logic and are normally reserved for works written by one or two authors. The goal is to report new developments quickly, informally, and in a way that will make them accessible to non-specialists. Books in these categories should include

- at least 100 pages of text;
- a table of contents and a subject index;
- an informative introduction, perhaps with some historical remarks, which should be accessible to readers unfamiliar with the topic treated;

In the evaluation of submissions, timeliness of the work is an important criterion. Texts should be well-rounded and reasonably self-contained. In most cases the work will contain results of others as well as those of the authors. In each case, the author(s) should provide sufficient motivation, examples, and applications. Ph.D. theses will be suitable for this series only when they are of exceptional interest and of high expository quality.

Proposals in these categories should be submitted (preferably in duplicate) to one of the series editors, and will be refereed. A provisional judgment on the acceptability of a project can be based on partial information about the work: a first draft, or a detailed outline describing the contents of each chapter, the estimated length, a bibliography, and one or two sample chapters. A final decision whether to accept will rest on an evaluation of the completed work.

3. Category iii). Reports of meetings will be considered for publication provided that they are of lasting interest. In exceptional cases, other multi-authored volumes may be considered in this category. One or more expert participant(s) will act as the scientific editor(s) of the volume. They select the papers which are suitable for inclusion and have them individually refereed as for a journal. Organizers should contact the Managing Editor of Lecture Notes in Logic in the early planning stages.

4. Category iv). This category provides an avenue to provide out-of-print books that are still in demand to a new generation of logicians.

5. Format. Works in English are preferred. After the manuscript is accepted in its final form, an electronic copy in LaTeX format will be appreciated and will advance considerably the publication date of the book. Authors are strongly urged to seek typesetting instructions from the Association for Symbolic Logic at an early stage of manuscript preparation.

Lecture Notes in Logic

From 1993 to 1999 this series was published under an agreement between the Association for Symbolic Logic and Springer-Verlag. Since 1999 the ASL is Publisher and A K Peters, Ltd. is Co-publisher. The ASL is committed to keeping all books in the series in print.

Current information may be found at `http://www.aslonline.org`, the ASL Web site. Editorial and submission policies and the list of Editors may also be found above.

Previously published books in the *Lecture Notes in Logic* are:

1. *Recursion theory.* J. R. Shoenfield. (1993, reprinted 2001; 84 pp.)
2. *Logic Colloquium '90; Proceedings of the Annual European Summer Meeting of the Association for Symbolic Logic, held in Helsinki, Finland, July 15–22, 1990.* Eds. J. Oikkonen and J. Väänänen. (1993, reprinted 2001; 305 pp.)
3. *Fine structure and iteration trees.* W. Mitchell and J. Steel. (1994; 130 pp.)
4. *Descriptive set theory and forcing: how to prove theorems about Borel sets the hard way.* A. W. Miller. (1995; 130 pp.)
5. *Model theory of fields.* D. Marker, M. Messmer, and A. Pillay. (1996; 154 pp.)
6. *Gödel '96; Logical foundations of mathematics, computer science and physics; Kurt Gödel's legacy. Brno, Czech Republic, August 1996, Proceedings.* Ed. P. Hajek. (1996, reprinted 2001; 322 pp.)
7. *A general algebraic semantics for sentential objects.* J. M. Font and R. Jansana. (1996; 135 pp.)
8. *The core model iterability problem.* J. Steel. (1997; 112 pp.)
9. *Bounded variable logics and counting.* M. Otto. (1997; 183 pp.)
10. *Aspects of incompleteness.* P. Lindstrom. (1997, 2nd ed. 2003; 163 pp.)
11. *Logic Colloquium '95; Proceedings of the Annual European Summer Meeting of the Association for Symbolic Logic, held in Haifa, Israel, August 9–18, 1995.* Eds. J. A. Makowsky and E. V. Ravve. (1998; 364 pp.)
12. *Logic Colloquium '96; Proceedings of the Colloquium held in San Sebastian, Spain, July 9–15, 1996.* Eds. J. M. Larrazabal, D. Lascar, and G. Mints. (1998; 268 pp.)
13. *Logic Colloquium '98; Proceedings of the Annual European Summer Meeting of the Association for Symbolic Logic, held in Prague, Czech Republic, August 9–15, 1998.* Eds. S. R. Buss, P. Hájek, and P. Pudlák. (2000; 541 pp.)
14. *Model Theory of Stochastic Processes.* S. Fajardo and H. J. Keisler. (2002; 136 pp.)
15. *Reflections on the Foundations of Mathematics; Essays in honor of Solomon Feferman.* Eds. W. Seig, R. Sommer, and C. Talcott. (2002; 444 pp.)

16. *Inexhaustibility; a non-exhaustive treatment.* T. Franzén. (2004; 255 pp.)

17. *Logic Colloquium '99; Proceedings of the Annual European Summer Meeting of the Association for Symbolic Logic, held in Utrecht, Netherlands, August 1–6, 1999.* Eds. J. van Eijck, V. van Oostrom, and A. Visser. (2004; 208 pp.)

18. *The Notre Dame Lectures.* Ed. P. Cholak. (2005, 185 pp.)

19. *Logic Colloquium 2000; Proceedings of the Annual European Summer Meeting of the Association for Symbolic Logic, held in Paris, France, July 23–31, 2000.* Eds. R. Cori, A. Razborov, S. Todorčević, and C. Wood. (2005; 408 pp.)

20. *Logic Colloquium '01; Proceedings of the Annual European Summer Meeting of the Association for Symbolic Logic, held in Vienna, Austria, August 1–6, 2001.* Eds. M. Baaz, S. Friedman, and J. Krajíček. (2005, 486 pp.)

21. *Reverse Mathematics 2001.* Ed. S. Simpson. (2005, 401 pp.)

22. *Intensionality.* Ed. R. Kahle. (2005, 265 pp.)

Printed in the United States
by Baker & Taylor Publisher Services